MATERIALS SCIENCE RESEARCH
Volume 19

ADVANCES IN MATERIALS CHARACTERIZATION II

MATERIALS SCIENCE RESEARCH

Recent volumes in the series:

A Continuation Order Plan is available for this series. A continuation order will bring delivery of each new volume immediately upon publication. Volumes are billed only upon actual shipment. For further information please contact the publisher.

Library of Congress Cataloging in Publication Data

Symposium on Advances in Materials Characterization (1984: Alfred, N.Y.)
  Advances in materials characterization II.

  (Materials science research; v. 19)
  "Proceedings of a Symposium on Advances in Materials Characterization, held July
30–August 3, 1984, in Alfred, New York"—T.p. verso.
  Includes bibliographical references and index.
  1. Materials—Testing—Congresses. 2. Materials—Surfaces—Congresses. I. Snyder,
Robert L., 1935–   . Condrate, Robert A., 1938–   . III. Johnson, P. F. IV. Title.
V. Series.
TA410.S93 1984   620.1'1   85-12300

ISBN-13: 978-1-4615-9441-3   e-ISBN-13: 978-1-4615-9439-0
DOI: 10.1007/978-1-4615-9439-0

Proceedings of a symposium on Advances in Materials Characterization,
held July 30–August 3, 1984, in Alfred, New York

© 1985 Plenum Press, New York
Softcover reprint of the hardcover 1st edition 1985
A Division of Plenum Publishing Corporation
233 Spring Street, New York, N.Y. 10013

MATERIALS SCIENCE RESEARCH • Volume 19

ADVANCES IN MATERIALS CHARACTERIZATION II

Edited by

R. L. Snyder
R. A. Condrate, Sr.
and
P. F. Johnson

New York State College of Ceramics
Alfred University
Alfred, New York

PLENUM PRESS • NEW YORK AND LONDON

PREFACE

This book represents the proceedings of the second interdisciplinary conference on materials characterization held from July
30 through August 3, 1984 at the New York State College of Ceramics
at Alfred University. The conference was the 20th in the University
Series on Ceramic Science, instituted in 1964 by Alfred University,
the University of California at Berkeley, North Carolina State
University and Notre Dame University. Volume I of the proceedings
of the first conference using this interdisciplinary approach to
materials characterization was published as "Advances in Materials
Characterization", edited by D. R. Rossington, R. A. Condrate and
R. L. Snyder, and was listed as volume 15 of the Materials Science
Research series of Plenum Press (New York, 1983).

The purpose of bringing together scientists from a wide range
of disciplines to present and discuss the latest developments in
their fields is to promote cross fertilization. The first
conference of this type and its resulting volume of proceedings
stimulated a significant dialogue between disciplines concerning the
characterization of materials, therefore indicating a need for a
continuing series of such conferences.

Characterization lies at the core of materials science.
Advancements in this science proceed rapidly after breakthroughs in
characterization techniques. This series of conferences has as
their goal the stimulation of the development of new ways of
"seeing" into the nature of materials phenomena. It is the nature
of modern science for researchers to become specialists. We become
expert at the use of a particular set of physical principles and
their applications to a limited set of problems. This usually
implies the devotion of such a large portion of our time and
energies to our own particular limited areas that we cannot follow
developments in other fields at the depth required to stimulate new
ideas in our own.

If we have learned anything in recent years it is that
characterization of a material or a phenomena is seldom accomplished

by the application of a single tool but results from a broad attack using the full arsenal of modern techniques. Thus an interdisciplinary approach to characterization has two benefits: the increasing of our awareness of the potential of other techniques and the conditions under which they are applied, and even more significantly, the application of ideas born in another discipline to our own.

The various discussions of instrumental techniques presented in both of these volumes are excellent summaries of the state-of-the-art of materials characterization at this still rather early stage of materials science. The application of the tools described here, and those yet to be developed, holds the key to the development of this infant into a mature science. It seems clear that conferences as presented in this series, and the resulting volumes of their proceedings should help catalyze such growth. However one rather myopic reviewer of volume I tried to view it as a text book!

The 32 papers included in these Proceedings cover a broad array of established and new characterization techniques. They have been broadly divided into categories determined by the type of characterization being sought. These categories are: Molecular Structure, Lattice Structure, Phase, Surface and Interfaces and Microstructure. However, as would be expected for such a broad survey there is often considerable overlap among techniques and goals of the studies.

The three day conference was preceeded by a series of hands on work shops which proved invaluable to preparing researchers, in fields outside their own, for full participation in the conference to follow. The editors would like to express their special appreciation to the following people for their efforts in preparing and conducting these workshops:

1. W. A. Lanford for "Nuclear Reaction Analysis and Rutherford Backscattering Spectrometry".
2. W. E. Votava for "Analytical Scanning Electron Microscopy".
3. R. A. Condrate and J. R. Ferraro for "Optical Spectral Techniques for Elemental, Phase and Structural Analysis".
4. R. L. Snyder for "Quantitative Phase Analysis by Automated X-ray Diffraction".
5. P. F. Johnson for "Techniques of Surface Analysis, XPS, SIMS and AES".

The editors wish to thank all speakers and contributors to the conference, and especially to those who were invited to present overview papers in their field of expertise (specifically, Dr. P. J. Bray, Dr. R. M. Barnes and Dr. B. O. Mysen).

The editors also gratefully acknowledge the conference sponsors listed below. Their financial support and commitments were invaluable constituents to the success of the conference.

U.S. Dept. of the Interior - Bureau of Mines
Brockway, Inc.
Horiba Instruments, Inc.
Instruments SA, Inc
Joel
Leybold-Heraeus Vacuum Products
Perkin-Elmer Corporation
Shott Optical Glass Co.
Siemens Corporation

The success of any conference depends not only upon the quality of papers presented and its sponsorship, but also on the many people involved in the planning and preparation stages. It is impossible to list all the people at the New York State College of Ceramics who were responsible for the efficient running of the conference, but special thanks are due to Mrs. Coral Link and Ms. Faith Orth for registration, organizational and secretarial services. Mr. William Emrick did an outstanding job of setting up most of the local arrangements, including housing, food and many details of the program for those accompanying the participants. Special thanks must go to our wives Sheila Snyder, Judy Condrate and Karen Johnson for the endless details that they kept organized, and for their planning and conducting of the program involving those accompanying conference participants. To all these people, and the many graduate students who performed the many last minute tasks that inevitably arise, the editors express their deep gratitude.

It is our sincere hope that this volume will prove to be of benefit to all people involved in materials characterization.

Alfred, New York

Robert L. Snyder
Robert A. Condrate
Paul F. Johnson

CONTENTS

STRUCTURAL AND ELEMENTAL CHARACTERIZATION

PHASE CHARACTERIZATION

MICROSTRUCTURE CHARACTERIZATION

NMR CHARACTERIZATION OF GLASSES

P. J. Bray and S. J. Gravina

Department of Physics
Brown University
Providence, RI 02912

INTRODUCTION

For over twenty five years nuclear magnetic resonance (NMR) has been used as a probe of the microstructure of glasses. More recently NMR has begun to be used as a probe of the dynamical processes in glass. This paper summarizes some of the basic and more recent applications of NMR to the understanding of glasses. After a brief review of the relevant NMR theory, several studies which demonstrate various applications of NMR to glass science will be discussed.

NMR THEORY

Nuclei with spin 1/2 or greater possess a magnetic moment given by $\mu = \mu_0 \mathbf{I}$ where $\mathbf{I}$ is the quantum mechanical spin vector and μ_0 is the Bohr nuclear magneton. (The magnitude of the spin vector $\mathbf{I}$ is given by $\sqrt{I(I+1)}$ where the scalar quantity I, the spin value, must be an integer or half-integer). If this nucleus is placed in a magnetic field $\mathbf{H}$, then the magnetic moment couples to $\mathbf{H}$ with the familiar Zeeman interaction whose energy is

$$E = -\mu \cdot \mathbf{H} = -\mu_0 \mathbf{I} \cdot \mathbf{H} = -\mu_0 H_0 m. \tag{1}$$

The magnetic quantum number m has $2I + 1$ values ranging from -I to +I in integer steps, and H_0 equals the magnitude of $\mathbf{H}$. There are, then, $2I + 1$ energy levels, and the difference in energy between adjacent levels is $\Delta E = \mu_0 H_0 = h\omega_0$ where h = Planck's constant, h, divided by 2π, $\omega_0 = 2\pi\nu_0$ and ν_0 is the resonance frequency. The Zeeman energy levels for a nucleus of spin $I = 3/2$ are displayed in Fig. 1a.

In an NMR experiment, a sample of spins is bathed in radio-frequency (RF) radiation (e.g. by placing the sample in the inductance coil of an oscillator). When

1

the frequency of radiation ν equals the resonance frequency ν_0, transitions are induced between adjacent energy levels. This leads to absorption of the RF energy of the oscillator. A plot of energy absorption versus oscillator frequency in the vicinity of resonance is shown in Fig. 1b.

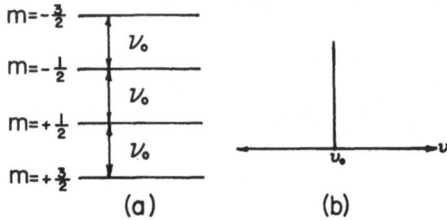

Figure 1.   (a) Energy levels and (b) NMR spectrum for a spin 3/2 nucleus in a magnetic field. (Only the Zeeman interaction is included. In a real system of spins the effect of the interaction between the spins would cause the spectrum to be broadened).

Structural Information

The energy levels (Fig. 1a) arising from the Zeeman interaction can be shifted by several other interactions of the nucleus with its environment. Through measurement of these interactions a variety of information about the local environment of the resonating nucleus is obtained. Three interactions relevant to this paper are the dipole-dipole, quadrupole, and chemical shift interactions. In the cases considered here, all of these are substantially smaller than the Zeeman interaction, so their effects can be calculated using perturbation theory.

Since a spin with a magnetic moment produces its own magnetic field, it will contribute to the total magnetic field experienced by a nearby magnetic moment. This is the origin of the dipole-dipole interaction. Its effect is to cause a distribution in resonance frequencies, since individual nuclei (of the species being studied by NMR) will experience slightly different total fields H_0 depending on the number, type, location, and orientation of neighboring magnetic moments. In glasses, this distribution generally produces a Gaussian NMR lineshape, but the line will narrow and become Lorentzian if the atoms begin to move. The strength of the dipolar interaction depends on the strength of the magnetic moments involved, is independent of H_0, has an angular dependence of $1 - 3\cos^2\theta$, and falls off as $1/r^3$;

2

here r is the distance from the resonating nucleus to the other magnetic moments, and θ is the angle between the field **H** and the position vector **r** going from the resonating nucleus to a neighboring magnetic moment.

Nuclei with spin 1 or greater also possess an electric quadrupole moment which will interact with an electric field gradient (EFG) present at the nuclear site. This will cause a shift in ν_0 which can be calculated and depends on H_0, θ, ϕ, Q_{cc}, and η. θ and ϕ are the polar angles of **H** with respect to the principal axes of the EFG tensor. Q_{cc}, the quadrupole coupling constant, is a measure of the strength of the interaction between the EFG and the nuclear quadrupole moment. η is the asymmetry parameter and is a measure of the departure of the EFG tensor from cylindrical symmetry. Figure 2 shows the spin-3/2 energy levels and the three transition frequencies for a single crystal when only first-order effects of the quadrupole interaction are dominant.

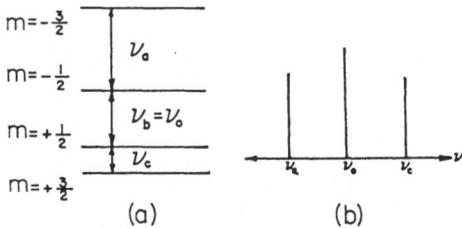

Figure 2.   Energy levels and NMR spectrum for a spin 3/2 nucleus in a magnetic field (Zeeman and quadrupolar interactions present).

In a powdered or glassy sample, all possible values of θ and ϕ will exist. When θ and ϕ are averaged over all possible orientations of **H**, the resulting spectrum is called a powder pattern. Figure 3 shows the powder pattern for a spin-3/2 system calculated to first order. Note the the central m = 1/2 ↔ -1/2 transition is unchanged from the case of no quadrupole interaction. This is because, to first order, the position of the central transition is unchanged and has no angular dependence. When the quadrupolar interaction is calculated to second-order, the central transition does have an angular dependence; and its powder pattern is shown in Fig. 4. Often Q_{cc} is large enough such that the powder pattern of Fig. 3 is spread over too wide a range of magnetic field to allow easy detection of the m = ±3/2 ↔ ±3/2 transitions. In these cases only the second-order powder pattern of the m = -1/2 ↔ 1/2 transition is observed.

3

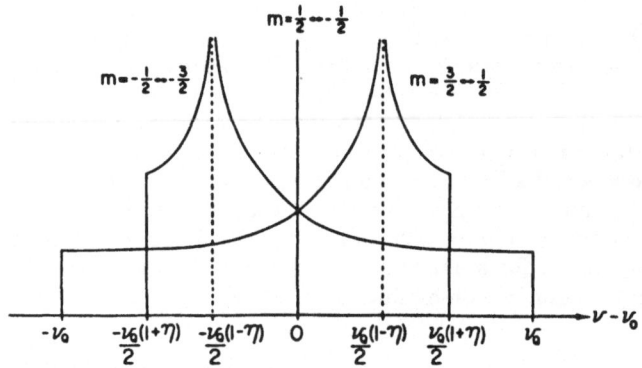

Figure 3.   First-order powder pattern for a random ensemble of spin 3/2 nuclei. Here $\nu_Q = 3Q_{cc}/2I(2I - 1)$.

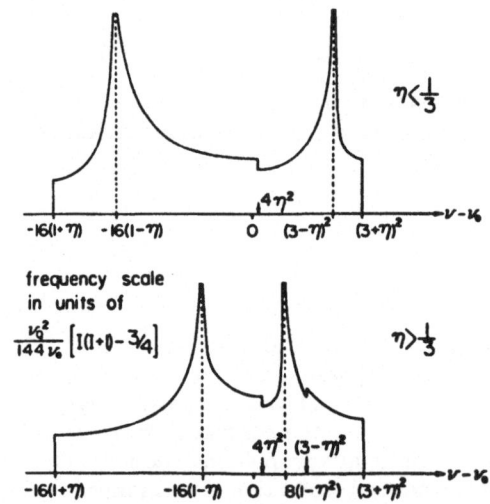

Figure 4.   Second-order powder pattern for the central transition of Fig. 3.

An experimental spectrum of ^{11}B (spin-3/2) in glassy and crystalline B_2O_3 is shown in Fig. 5a and Fig 5b respectively. In a lock-in experiment the field is swept

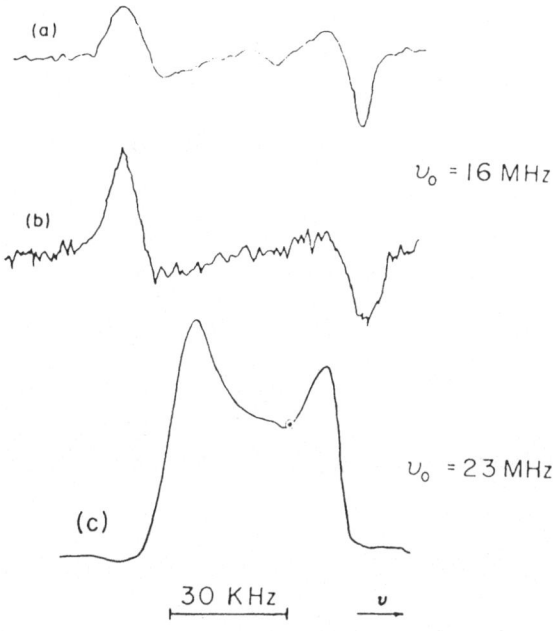

$\nu_0 = 16$ MHz

$\nu_0 = 23$ MHz

30 KHz ν

Figure 5.   The ^{11}B NMR spectrum in B_2O_3: (a) glassy B_2O_3 lock-in detection, (b) crystalline B_2O_3 lock-in detection and (c) glassy B_2O_3 fast passage detection.

while the derivative of the absorption spectrum is recorded. The absorption spectrum can be recorded directly by using the rapid fast passage technique. A rapid fast passage spectrum of B_2O_3 glass is shown in Fig. 5c. The spectrum in Fig 5c is narrower than the specra in Fig. 5a and Fig. 5b because the second-order quadrupole interaction is smaller in higher fields. The absorption spectrum is also obtained by integrating the derivative spectrum as shown in Fig. 6. The dipole-dipole interaction causes the sharp features of Fig. 4 to be rounded off in most actual experiments. The similarity of the two spectra in Fig. 5a shows that the short-range order about the boron atoms in B_2O_3 glass and crystal is identical. By simulating the experimental spectrum on a computer it is found that $Q_{cc} = 2.64 \pm 0.10$ MHz and $\eta = 0.12 \pm 0.04$. The small value of η indicates that the boron atom is trigonally bonded to three oxygen atoms with a slight distortion from perfect cylindrical symmetry[1].

5

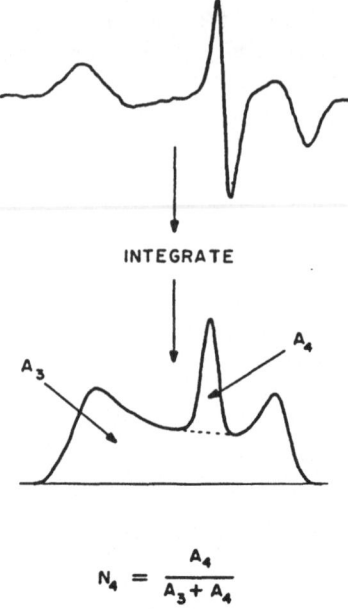

$$N_4 = \frac{A_4}{A_3 + A_4}$$

Figure 6.   Method for determining N_4 in a ^{11}B NMR spectrum.

The effect of **H** on the electrons near the nucleus may cause changes in the value of **H** experienced by the nucleus. This will cause the resonance frequency ν to be shifted from the value ν_0. This effect is sensitive to the chemical environment of the nucleus and is thus called the chemical shift interaction[2]. The strength of the chemical shift is proportional to H_0 and for a crystal depends on the orientation of **H** with respect to the crystal axes. For polycrystalline powders or glasses, this dependence can yield a powder pattern of some width and structure[3]. Figure 7 shows the powder pattern and derivative lineshape found in the case of an axially symmetric chemical shift tensor. Under conditions of rapid molecular motion or magic angle spinning, the angular dependence is averaged out and the chemical shift is reduced to its isotropic value which is proportional to H_0. The isotropic value of the chemical shift is different for different chemical environments; thus, by measuring the relative positions of resonance lines in an NMR spectrum, information about the different nuclear environments that exist in the sample can be obtained.

In NMR structural studies different atomic sites are distinquished by their different dipolar, quadrupolar and chemical shift parameters. But in vitreous materials the random nature of the atomic structure will often cause a site to have a distribution of some of its parameters. When a unique value of a parameter cannot be assigned to a spectral feature, then a distribution of that parameter is assumed in

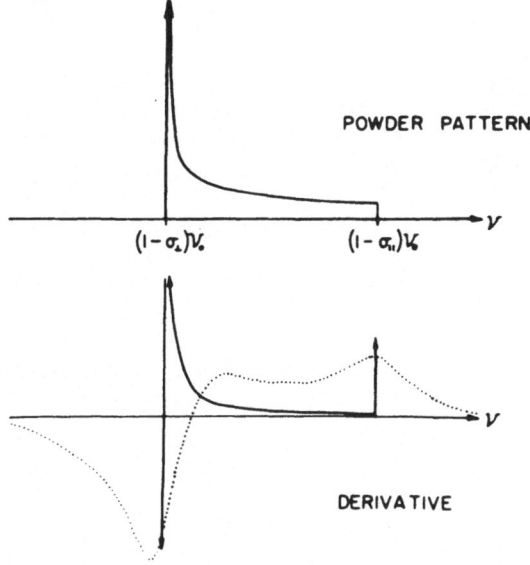

Figure 7.   NMR absorption powder pattern for a nuclear site with an anisotropic
axially symmetric chemical shift. Upper: absorption without dipolar
broadening. Lower: (solid line) first derivative; (dotted line) first
derivative including dipolar broadening.

the computer simualtion of that spectrum. Usually the distribution is assumed to be
Gaussian. In this case a site will be characterized by a mean value of a parameter
(i.e. Q_{cc}^0, η^0) and a standard deviation about that mean value (i. e. σ_{Qcc}, σ_η).

Dynamical Information

The reason that energy can be absorbed by the spin system is that the
population of spins in the lower energy states is greater than the populations of the
higher states. If the energy levels were to become equally populated then energy
would no longer be absorbed by the spin system. This is called saturation. The
process of spins returning to their equilibrium states is called relaxation. In a
magnetic field $\mathbf{H} = H_0\,\hat{z}$, equilibrium will yield a net bulk magnetization $\mathbf{M} = \sum \mu$
$= M_0\,\hat{z}$. If a large pulse of RF energy is applied such that the system is saturated,
then the magnetization will be zero. This is because all possible orientations of the
spins will occur with equal probability. The return of the magnetization to its
equilibrium value can be described as

$$\mathbf{M}(t) = M_0[1 - e^{-t/T}1]\,\hat{z} \tag{2}$$

where T_1 is called the spin-lattice relaxation time.

7

One of the main causes of spin-lattice relaxation is lattice motions which give rise to time-varying electromagnetic fields that can interact with nuclear magnetic dipole or electric quadrupole moments. The frequency dependence of the time-varying electromagnetic fields is called the spectral density $J(\omega)$, and the spin-lattice relaxation rate $1/T_1$ is directly related to the value of this function at the resonant frequency. $J(\omega)$ is the Fourier transform of the correlation function $G(t)$ which contains all the complicated dynamical information of the lattice.

A simple model of $G(t)$ is due to Bloembergen, Purcell and Pound[4] and is called the BPP model. The BPP model assumes

$$G(t) \propto e^{-t/\tau} \tag{3}$$

where τ is a typical time scale of the motion causing the relaxation (e. g. the time between jumps of an ion). The Fourier transform of equation 3 yields

$$1/T_1 \propto \tau/(1 + (\omega\tau)^2) \tag{4}$$

which implies that

$$1/T_1 \propto \tau \qquad \text{for } \omega\tau << 1 \tag{5}$$

$$1/T_1 \propto 1/\omega^2\tau \qquad \text{for } \omega\tau >> 1. \tag{6}$$

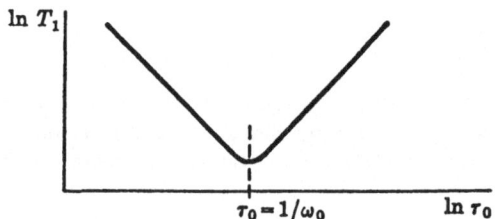

Figure 8.   Prediction of the BPP theory for $\ln T_1$ verses T.

A plot of $\ln(T_1)$ vs. $1/T$ for a system described by the BPP theory is shown in Fig. 8. The value of τ can be related to the diffusion constant D found in the Nernst-Einstein equation for DC conductivity: $\sigma(0)$

$$\sigma(0) = c(Ze)^2 D/kT \tag{7}$$

where kT is the Boltzman constant times the temperature, c is the carrier concentration, and Ze is the carrier's charge[5]. By assuming that the carrier undergoes uncorrelated jumps of length l at a rate ν, the diffusion constant can be expressed as $D = \nu l^2/2d$ where d is the dimensionality of the system. For ion motion, the hopping frequency $\nu = 1/\tau$ can often be described by an Arrhenius relation

$$1/\tau = \nu_0 e^{-E/kT} \tag{8}$$

where E is the activation energy of the hopping process and ν_0 is the attempt frequency. The activation energy is equal to the slope of the $\ln(T_1)$ vs. $1/T$ plot on either side of the minimum (see Fig. 8). Thus by measuring T_1 as a function of temperature, information about short-range motion of ions in glasses can be obtained. The theoretical treatment of NMR T_1 data is difficult and not yet complete. But for many systems the BPP theory outlined above is a very good starting point[6].

Information about the motion of ions can also be obtained from the behavior of the line shape as a function of temperature[7]. In a rigid solid the lineshape will be determined by the interactions discused above. But when the frequency of motion of the ion exceeds the line width of the rigid lattice resonance then the dipolar field experienced at the nuclear site will be averaged out leading to a narrowing of the lineshape. The extreme case of this occurs in liquids which, because of the very fast molecular motions found, can have extremely narrow linewidths in which only the isotropic chemical shift interaction is left (the EFG will also be averaged out).

STRUCTURAL STUDIES

NMR is useful in the study of glass structure because the interactions that perturb the Zeeman interaction depend mainly on the immediate nearest-neighbor environment of the nucleus under study. This is very valuable in glasses since they lack the long-range order which is necessary for detailed structural studies using x-ray diffraction.

Alkali Borate Glasses

In alkali borate glasses, three types of ^{11}B NMR line shapes caused by distinct types of boron-oxygen coordination exist (see Fig. 9). As is shown in Fig. 5, boron in pure B_2O_3 consists of boron trigonally coordinated to three oxygens (B_3), each of which is bonded to an additional boron atom (bridging oxygen)[8]. In this arrangement, an EFG would exist at the nucleus because of the charge asymmetry in the z-direction (perpendicular to the plane of the molecule), which would lead to a nonzero value of Q_{cc}. But η, which depends on the charge asymmetry in the xy-plane, would vanish. Thus the lineshape is as shown in Fig. 4 and Fig. 9b. As alkali oxide is added to the glass, the alkali ions enter the network interstitially, and the extra oxygen joins a B_3 unit to become part of a boron tetrahedrally coordinated to four bridging oxygens (B_4). Because of the tetrahedral charge symmetry, the EFG is near zero at the nuclear site. Thus, Q_{cc} and η would be near zero. In this case, the dipolar interaction would dominate and the spectrum would be a Gaussian lineshape (see Fig. 9a). The composition of alkali borate glasses is characterized by the value of R where the composition is $RX_2O \cdot B_2O_3$ (X is an alkali). When the amount of alkali exceeds $R = 0.5$, then B_4 units start to be replaced by boron atoms bonded to three oxygens with one of the oxygens being a nonbridging oxygen (B_3^-).

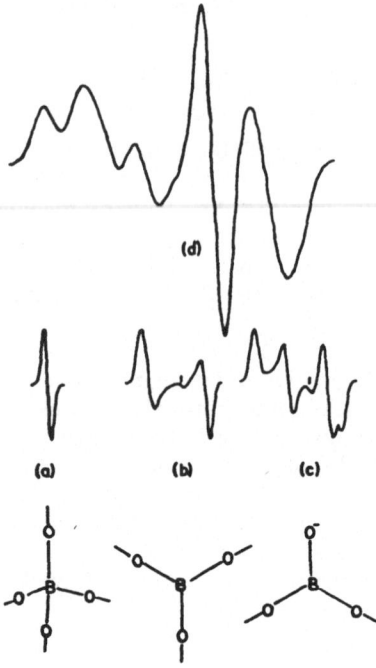

Figure 9.   The ^{11}B NMR spectrum of (a) a B_4 unit, (b) a B_3 unit, (c) a B_3^- unit and (d) a sample containing all three units.

This charge configuration would have a component of the EFG along the z-axis of the trigonal axis that would be similar to that of the B_3 unit, but it would have an asymmetric charge distribution in the xy-plane. This would lead to a similar value of Q_{cc} but a nonzero value of η giving rise to the spectrum shown in Fig. 9c. The fraction N_4 of borons that are in B_4 units can be easily obtained from the NMR spectrum by taking the ratio of the area under the narrow response to the area under the entire response[9] as shown in Fig. 6.

Figure 10 shows the value of N_4 as a function of R for several alkali borate glasses[10]. Many properties[11] (i. e. viscosity, thermal expansion) of alkali borate glasses show an extremum around the composition R = 0.5, where N_4 is maximum. In some cases, these changes in properties may be related to changes in N_4, but the so-called "borate anomaly" in alkali borate glasses (occuring around R = 0.2) is not associated with extrema or anomalies in N_4.

Additional details of the distributions and values of the quadrupole parameters in alkali borate glasses have been determined in the sodium and lithium borate system using ^{10}B NMR[12]. Figure 11 shows the ^{10}B NMR spectra for several

10

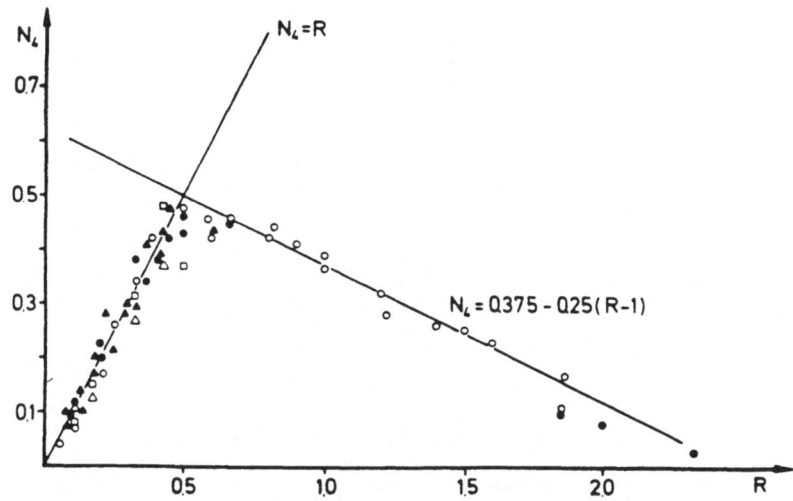

Figure 10. N_4 verse R in several alkali borate glasses: O lithium borate glasses, ● sodium borate glasses, ▲ potassium borate glasses, △ rubidium borate glasses, □ cesium borate glasses.

sodium borate glasses. A structural model based on the ideas of Krogh-Moe[13], was used to analyze this data. This model assumes that the glass is composed of a random arrangement of the structural groups found in the crystalline compounds of the glass-forming system. Figure 12 shows several of the structural groups used in the model. After all the various constrants (i. e. charge conservation, total boron conservation) of the model have been accounted for, only one parameter is left to fit the model to the NMR data. Figure 13 shows how closely the computer simulation fits the experimental spectrum for the main spectral feature. The concentrations of the different structural groups is thus obtained through this fitting procedure.

Sodium Borosilicate Glasses

The composition of sodium borosilicate glasses is characterized by the value of R and K where the composition is $RNa_2O \cdot B_2O_3 \cdot KSi_2O$. Through computer simulation of the ^{11}B NMR lineshapes the quantities N_4, N_{3S}, and N_{3A} (N_{3S} and N_{3A} are the fractions of borons that are in B_3 and B_3^- units respectively, see Fig. 7) have been measured through-out the glass forming range of this system[14]. Figures 14 and 15 show the N_4 and N_{3A} data for various K families as a function of R. The dashed lines in the figures represent the predictions of a model based on the the idea that the glasses are composed of various structural groupings found in the coresponding crystalline compounds.

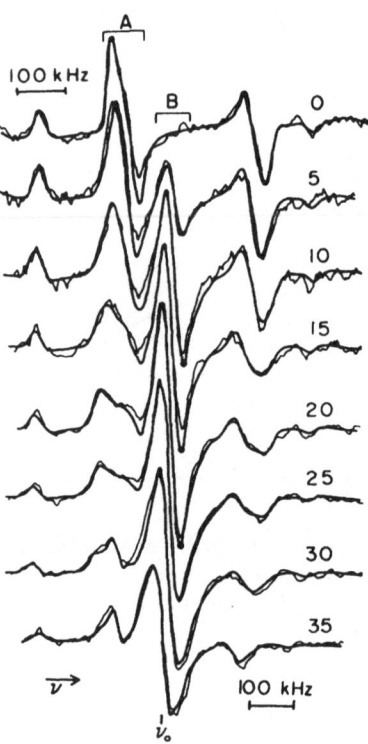

Figure 11. ^{10}B NMR spectra for eight $Na_2O \cdot B_2O_3$ glasses. The numbers on the right represent the molar concentrations of Na_2O. The calibration given in the lower right-hand corner is for the 35 mol% glass only. The calibration in the upper left-hand corner is for all the other glasses. The smooth lines are the computer-simulated spectra.

Oxygen in Silica Glass

^{17}O (I = 5/2) is the only stable isotope of oxygen with a magnetic moment. Since ^{17}O is only .037% abundant in nature, samples must be isotopically enriched. Figure 16 shows the ^{17}O spectra of SiO_2 glass[15]. Only one site was needed to simulate this spectrum. The quadrupolar parameters for this site are also given in Fig. 16. It is found that changing the value of η^0, $\sigma_{Q_{cc}}$ or σ_η by as much as 30% will still give an acceptable fit, but Q_{cc}^0 cannot be varied more than 1%. This can be related to a distribution in the value of the Si-O-Si bond angle a. It is found that $130° \lesssim a \le 180°$ which is consistent with an earlier x-ray investigation of SiO_2 glass[16].

Oxygen in Borate Glass

In vitreous SiO_2, only one site was needed to simulate the observed line shape. Figure 17 shows the NMR lineshape and computer simulation of ^{17}O in a sample of

12

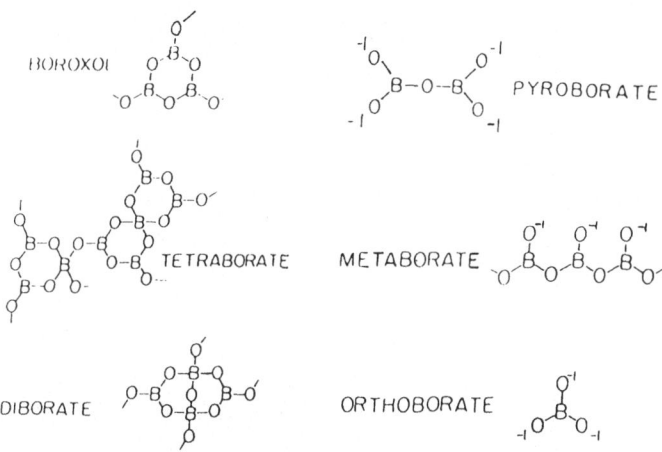

Figure 12. Structural groups found in alkali borate compounds and glasses.

B_2O_3 glass[17]. Two sites are needed to simulate this spectrum, indicating two types of oxygen environment. One has Q_{cc}^0 = 4.69 MHz and η^0 = 0.58 with a distribution σ_{Qcc} = 0.10 MHz: the other has Q_{cc}^0 = 5.75 MHz and η^0 = 0.40 with a distribution in η of σ_η = 0.20. The ratio of the two ^{17}O sites is 1.2. Previous x-ray work[18], and infra-red and Raman studies[19], indicate that vitreous B_2O_3 consists of a random three-dimensional network of B_3 units with a comparatively high fraction of six-membered boroxol rings. The NMR data support this model (see Fig. 18). Site one, with a very small distribution in the quadrupole parameters, corresponds to oxygen atoms in the relatively rigid boroxol rings (O(R)); site two is characterized by a larger coupling constant and a large distribution in η and corresponds to the oxygens O(C) which connect the boroxol rings and can have large variations in the B-O-B bond angle.

Silicon in Glass

^{29}Si is the only naturally occuring isotope of silicon with a magnetic moment. Until recently very few ^{29}Si NMR studies have been done because ^{29}Si has a natural abundance of only 4.7% and isotopically enriched ^{29}Si is prohibitively expensive. Also ^{29}Si with spin 1/2 is affected only by chemical shift and dipolar interactions. Without the high fields available with superconducting magnets, the dipolar interaction is larger than the chemical shift interaction prohibiting the detection of any chemical shift information.

By looking at ^{29}Si NMR in potassium silicate crystals and glasses, Harris and Bray were able to show that the chemical shift anisotropy is sensitive to the number

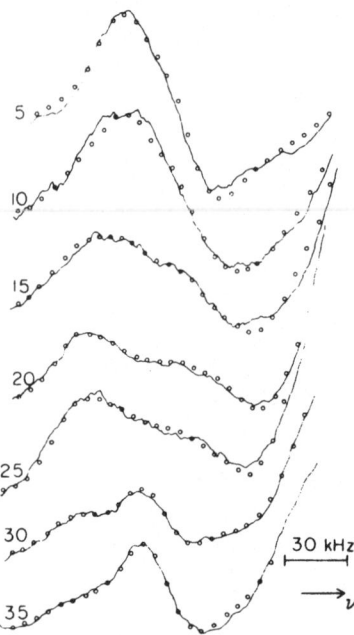

Figure 13.   [10]B NMR spectra of the main feature (labeled A in Fig. 11) in seven Na$_2$O·B$_2$O$_3$ glass (solid line) and the coresponding simulated spectra (circles).

of nonbridging oxygens surrounding the SiO$_4$ tetrahedra[20]. Because of potassium's low magnetic moment and the low abundance of [17]O and [29]Si, the dipolar broadening was sufficiently small to permit measurements of the [29]Si chemical shift anisotropy. By looking at crystalline potassium silicate they were able to distinguish between three different types of local symmetry due to the nonbridging oxygens surrounding the SiO$_4$ tetrahedra. Tetrahedra with zero or four nonbridging oxygens posses a spherical symmetry leading to a small anisotropy and narrow line shape. This is the line shape found in quartz and pure SiO$_2$ glass. Tetrahedra with one or three nonbridging oxygens show line shape structure due to a stronger anisotropy. Tetrahedra with two nonbridging oxygens may be distinguished by their non-zero asymmetry parameter, but this has not yet been observed. The spectra of glasses at compositions corresponding to crystalline compounds were found to resemble those of the crystals, indicating similar local structure. Figure 19 shows the comparison of crystalline and vitreous K$_2$O·2SiO$_2$.

Magic Angle Sample Spinning ^{29}Si NMR

The NMR signal strength is proportional to the strength of the magnetic field. Thus, by going to a higher field, one can overcome the problem of silicon's low

14

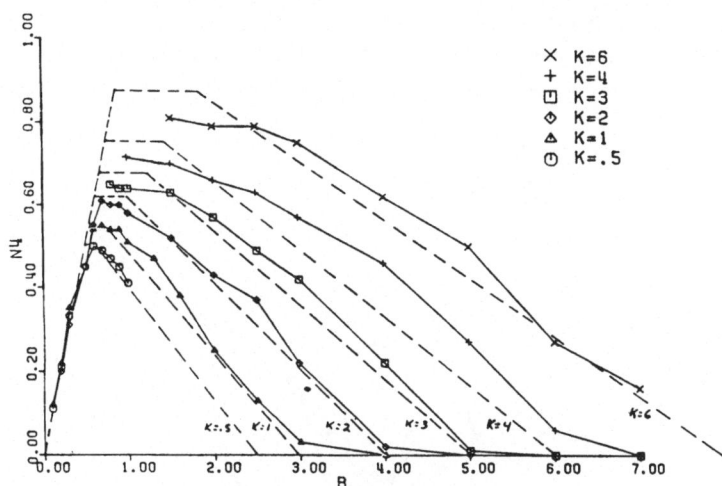

Figure 14. N_4 verses R for the different K families studied in $RNa_2O \cdot KSiO_2 \cdot B_2O_3$ glass.

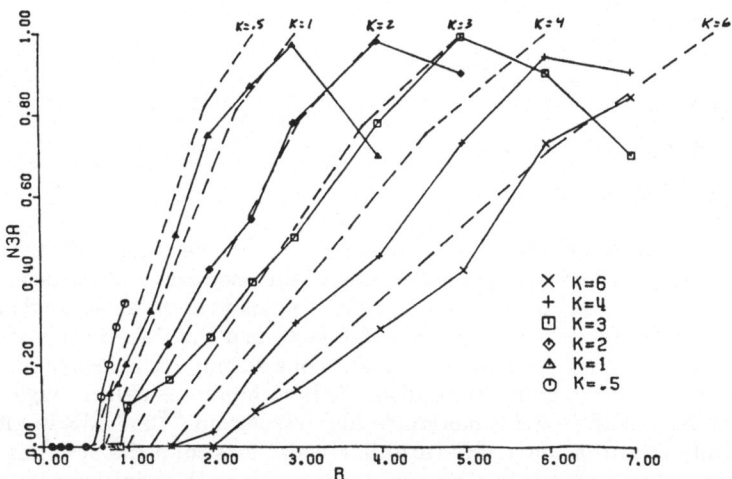

Figure 15. N_{3A} verses R for the different K families studied in $RNa_2O \cdot KSiO_2 \cdot B_2O_3$ glass.

natural abundance. The chemical shift is also proportional to the field strength so greater resolution is also gained by use of high magnetic fields. Even in high fields, however, the anisotropic chemical shift and dipolar broadening can obscure the isotropic chemical shift information. The dipolar broadening and chemical shift anisotropy can be made very small by using the fact that they both have an angular dependence of $1 - 3\cos^2\theta$. For the dipolar interaction, θ is the angle between the

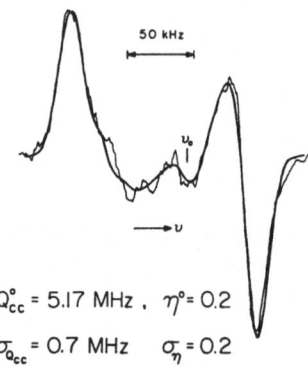

$$Q_{cc}^{o} = 5.17 \text{ MHz}, \quad \eta^{o} = 0.2$$

$$\sigma_{Q_{cc}} = 0.7 \text{ MHz} \quad \sigma_{\eta} = 0.2$$

Figure 16.   The ^{17}O NMR spectrum and the quadrupole parameters of the oxygen site in amorphous SiO_2. The smooth line is the computer simulated line-shape.

magnetic field $\mathbf{H}$ and the internuclear vector $\mathbf{r}$; for the chemical shift interaction, θ is the angle between $\mathbf{H}$ and the symmetry axis of the chemical shift tensor. When $\theta = 54.7°$ this term will vanish. When a sample of spins is spun at an angle of $54.7°$ the average angle between the magnetic moments and the field during one revolution is $54.7°$. Thus when a time scale longer than one revolution is looked at the dipolar and anisotropic chemical shift will vanish. This technique is called magic angle sample spinning (MASS) and it has made high resolution ^{29}Si NMR in glasses possible. Only lines that have a natural line-width less than the spinning frequency can be narrowed to a unique non-broadened line. Lines that are broader will have spinning sidebands on either side of the line at a distance of the spinning frequency. These sidebands will complicate the spectrum greatly if more than one chemical shift is present. The fastest speeds that can be consistently used are about four to five KHz. Higher speeds cause the spinner to burst due to the strong centripetal forces required.

16

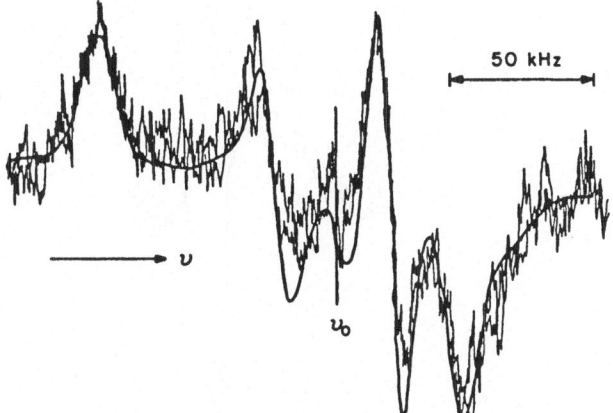

50 kHz

Figure 17. Two superimposed ^{17}O spectra from B_2O_3 glass and a computer simulation using the quadrupole parameters given in the text.

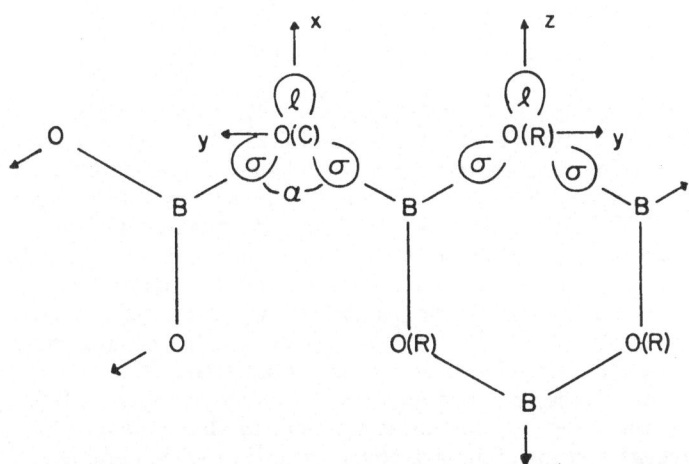

Figure 18. Boroxol model for B_2O_3 glass. The label σ refers to the oxygen σ bonds and the label l refers to the non-bridging oxygen lone pair orbitals.

17

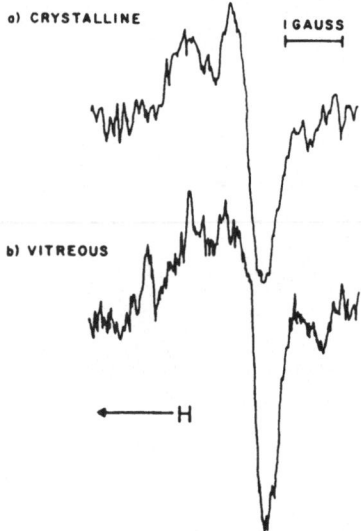

Figure 19. ^{29}Si NMR spectra for vitreous and crystalline $K_2O \cdot 2SiO_2$.

Through the use of MASS ^{29}Si, NMR it has been shown that silicon tetrahedra with different environments will give different chemical shifts for the ^{29}Si NMR frequency[21].

Silicon in Lead Silicate Glass During Heat Treatment

An interesting application of MASS ^{29}Si NMR in lead silicate glass shows several structural features of the evolution of lead silicate glass during heat treatment[22]. In Fig. 20, the spectrum of lead silicate glass after heat treatment at 550° C for 2, 4, 6, and 8 hours is shown. The spectrum with t = 24h was heat treated at 710° C for 24 hours and Fig. 20a is the spectrum of an untreated glass. The increased order present in the PbO-SiO$_2$ glass is evident in the decreased linewidth of the peak at -85 ppm (ν_0 = 49.7 Mhz) and the emergence of a more complicated spectrum. After 24 hours at 710° C, three well-separated lines emerge corresponding to those found in the ^{29}Si spectrum of the natural mineral alamosite. Since alamosite contains only three distinct silicon sites, there are probably three very similar sites in the annealed lead glass with the lead ions affecting the silicon chemical shift the same way the aluminum ions in alamosite do. The two broad lines that emerge and then fade are characteristic of silicon and lead clustering. The two lines at -77.7 ppm and -78.7 ppm are close to those found in the lead-rich 2PbO-SiO$_2$, and the broad line at about -97.0 ppm definitely corresponds to a silicon tetrahedron bonded to three other tetrahedra and one lead ion. Thus, during annealing PbO-SiO$_2$ goes through an interesting stage of lead and silicon clustering and finally ends up as a structure with silicon environments nearly indistinguishable from the natural mineral alamosite.

Aluminum in Glass

High field MASS ^{27}Al NMR has been used to distinguish four-and six-coordinated aluminum in $CaO \cdot Al_2O_3 \cdot P_2O_5$ glasses[23]. Figure 21 shows the MASS ^{27}Al NMR spectrum from crystalline $Al_4(P_4O_{12})_3$, $Al_2O_3 \cdot P_2O_5$ glass, and

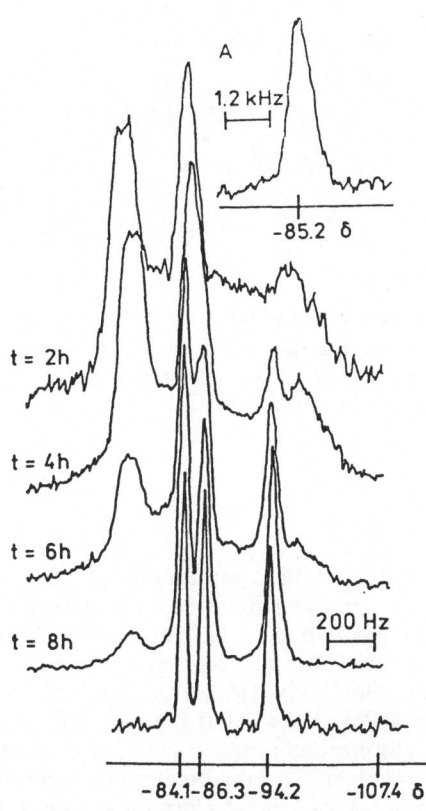

Figure 20.   The evolution of lead silicate glass during heat treatment. The bottom spectrum was treated at 710° C, the middle four were treated at 510° C and the top spectrum is from an untreated glass. (From reference 22)

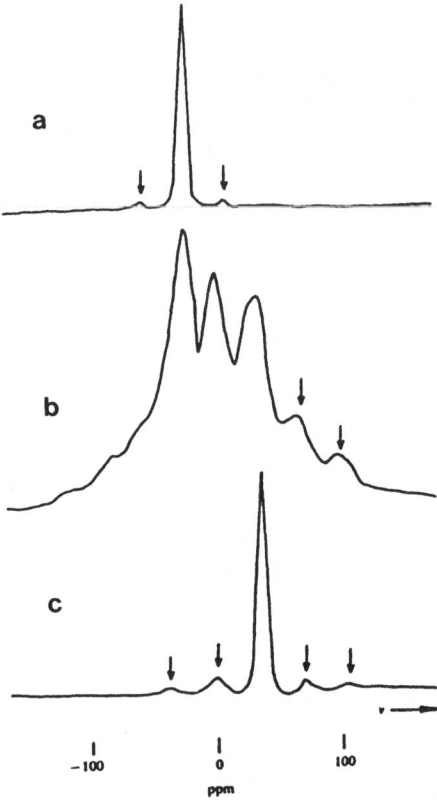

Figure 21.   Comparison of high field MAS ^{27}Al NMR spectra. (The marked peaks
are spinning sidebands). (a) Crystalline aluminium tetraphosphate,
$Al_4(P_4O_{12})_3$, (b) $Al_2O_3 \cdot P_2O_5$ glass (c) Crystalline α-tridymite $AlPO_4$
modification. (From reference 23).

crystalline α-tridymite $AlPO_4$. The crystalline $Al_4(P_4O_{12})_3$ consists of aluminum
octahedrally coordinated to six oxygen atoms with phosphorus atoms in the second
coordination sphere. The position of its NMR resonance identifies the resonance on
the left of the glass spectrum (Fig. 21b) as due to aluminums in a similar six
coordinated environment. Similarly, the resonance due to four-coordinated
aluminum in crystalline $AlPO_4$ shows that the resonance on the right of Fig. 21b is
due to four-coordinated aluminums in the glass. The identification of the central
peak in Fig. 21b is not yet clear. It may be due to six-coordinated aluminums which
are connected with other six-coordinated aluminums through an Al-O-Al bond which
would cause the second coordination sphere to differ from that of the aluminums that
cause the peak on the left. The broad background response is believed to be due to a
distribution of differently distorted sites causing a variation in the EFG for several of

the aluminum sites (The quadrupole interaction is not fully suppressed by MASS because of its angular dependence on both θ and ϕ).

Beryllium in Beryllium Fluoride Glasses

Figure 22 shows the NMR spectra[24] of ^{9}Be (spin 3/2) in a sample of BeF_2

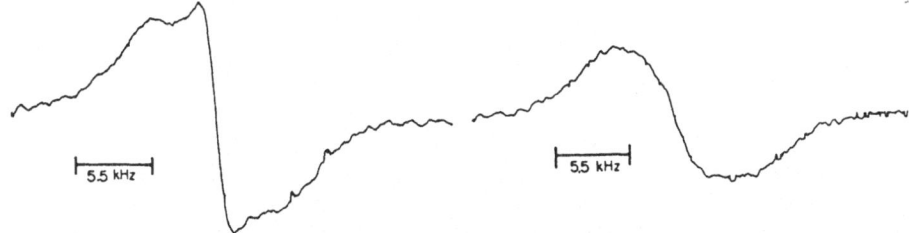

Figure 22. ^{9}Be NMR in pure beryllium fluoride glass and a sodium
fluoride--beryllium fluoride glass.

glass and a glass sample of molar composition $35NaF \cdot 65BeF_2$. Both glasses contain an 8.5 KHz broad line, but the pure BeF_2 glass also shows an anomalous narrow line with a width of 2.3 KHz superimposed on the wide feature. The broad line can be shown to arise from the dipolar broadening of a beryllium atom bonded to four fluorines. This linewidth agrees well with what is known from available x-ray data. The narrow line could be due to beryllium bonded to four oxygens; however, a chemical analysis of the sample did not reveal oxygen in sufficient quantities to account for the observed intensity of the narrow line which represented 4-8% of the total line intensity. A beryllium metal phase within the sample could also account for the narrow line; however, no electron-spin resonance signal with a g value of 2.0032, characteristic of beryllium metal[25], was found in a similarly prepared sample. It is believed that this narrow line is due to very small clusters of beryllium metal formed due to a fluorine deficiency that must be a result of the glass preparation process. (These glasses where prepared by melting the batch materials in a nitrogen atmosphere.) If the beryllium clusters are small enough, then the ESR spin-lattice relaxation time will become so small as to broaden the resonance so that it is undetectable. This fast relaxation is due to electron collisions with the cluster boundaries. The beryllium does not cluster in the binary glass because some of the fluorines available from the NaF are used to repair the BeF_2 network while the extra sodium ions exist interstitially in the network.

A wider field sweep performed on the a-BeF_2 sample revealed a spectrum typical of first order quadrupole broadening. The average coupling constant in this sample was found to be 100 KHz with a Gaussian distribution of 40 KHz about this value due to a distribution of bond angles in the BeF_4 tetrahedra.

The wide line in the binary sample was recorded at the low frequency of 1.17 MHz and was found to become asymmetricly broadened which is characteristic of the second-order interaction when a large distribution of EFG strength exists. The separation between the low-frequency divergence and the operating frequency allowed a determination of the coupling constant which was found to be 410 KHz with a distribution of 100 KHz about this value. The proposed existence of five-coordinated beryllium in alkali-modified vitreous BeF_2 is consistent with this larger value of Q_{cc}, since, as in ^{11}B NMR, changes of coordination number lead to large differences in Q_{cc}. Further evidence of a loss in tetrahedral symmetry in the binary sample is provided by the disappearance of the satellite transitions when recorded over a wide sweep range.

DYNAMICAL STUDIES

NMR is useful for determining the activation energies of ionic motion processes in glasses. The results from NMR are interesting in that, unlike other techniques which measure bulk properties, NMR measures motions which happen on the atomic scale. Thus the activation energies measured by NMR often are due to local motion of ions as opposed to long-range motions.

Fluorine Motion in Zirconium Fluoride Glasses

Fluorine motion in several heavy metal fluoride glasses based on ZrF_4 and HfF_4 have been studied using ^{19}F NMR[26]. In addition to having relatively high ionic conductivities[27], these glasses are of interest because of their potential application as ultra low loss infrared fibres.

Figure 23 shows a plot of $\ln T_1$ vs. 1000/T for a glass of molar composition $62ZrF_4 \cdot 33BaF_2 \cdot 5LaF_3$. Because these glasses crystallize at a relatively low temperature, T_1 could only be measured on the low-temperature side of the minimum in the $\ln T_1$ vs. 1/T plot. The frequency dependence of T_1 was found to obey the BPP relation in the temperature range studied. Through use of the BPP theory the activation energy E of the fluorine motion is found by taking the slope of the $\ln T_1$ vs. 1/T data. The activation energies were measured for several glasses and the results are summarized in Table 1. The binary system $57.7ZrF_4 \cdot 42.3BaF_2$ was found to have the lowest activation energy of 0.25 eV. The addition of glass stabilizing agents such as LaF_3 and AlF_3 to the binary glass raised the activation energies, indicating a suppression of fluorine motion in the more stable glasses. Substitution of hafnium for zirconium produced no effect on the activation energies measured in this study. Finally, the addition of 20 Mol% LiF was found to raise the activation energy to 0.7 eV while the addition of 20 Mol% NaF had little effect on the fluorine activation energy of the ternary $ZrF_4 \cdot BaF_2 \cdot LaF_3$ glass. This is consistent with earlier studies which indicated that lithium and sodium play different roles[28] within the vitreous network.

Lithium Motion in Fast Ionic Conduction Glasses--Motional Narrowing Study

The area of fast ionic conducting glasses is very important from a technological view. These materials have a good potential for use as solid electrolytes with a much higher energy density than conventional electrolyte batteries.

22

Ionic motion properties of the glass system $Li_2O \cdot Li_2Cl_2 \cdot B_2O_3$ where studied using ^7Li NMR[29]. The composition of these glasses is defined by A/B where

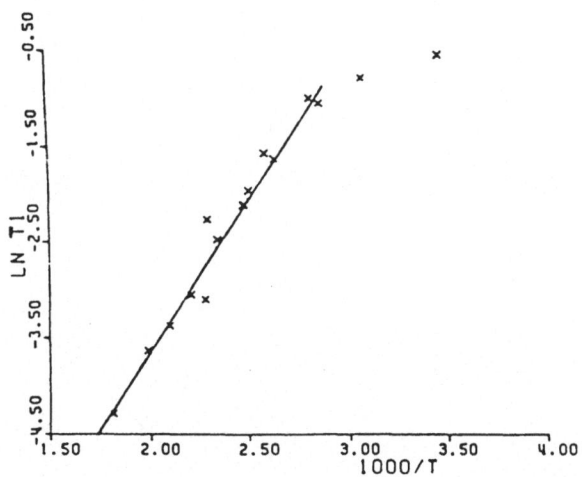

Figure 23. LnT_1 in seconds vs. 1000/T in K^{-1} for glass of molar composition $62ZrF_4 \cdot 33BaF_2 \cdot 5LaF_3$.

Table 1

Glass Composition Molar %	E_A(eV)
$57.7ZrF_4 \cdot 42.3BaF_2$	0.25
$63.5ZrF_4 \cdot 29.5BaF_4 \cdot 7LaF_3$	0.30
$62ZrF_4 \cdot 33BaF_2 \cdot 5LaF_3$	0.29
$57ZrF_4 \cdot 36BaF_2 \cdot 3LaF_3 \cdot 4AlF_3$	0.42
$57HfF_4 \cdot 36BaF_2 \cdot 3LaF_3 \cdot 4AlF_3$	0.44
$51ZrF_4 \cdot 20LiF \cdot 16BaF_2 \cdot 5LaF_3 \cdot 3AlF_3 \cdot 5PbF_2$	0.71
$51ZrF_4 \cdot 20NaF \cdot 16BaF_2 \cdot 5LaF_3 \cdot 3AlF_3 \cdot 5PbF_2$	0.31

23

$$A = mol\%Li_2O + mol\%Li_2Cl_2$$

$$B = mol\%Li_2O/(mol\%Li_2O + mol\%li_2Cl_2) \times 100$$

Four glasses of composition 50/10, 50/20, 50/30, and 50/45 where studied. Note that the total lithium content of the glasses is held constant, while the fraction of lithium added in the form LiCl is varied.

Figure 24 shows three of the spectra obtained for the 50/30 glass at various

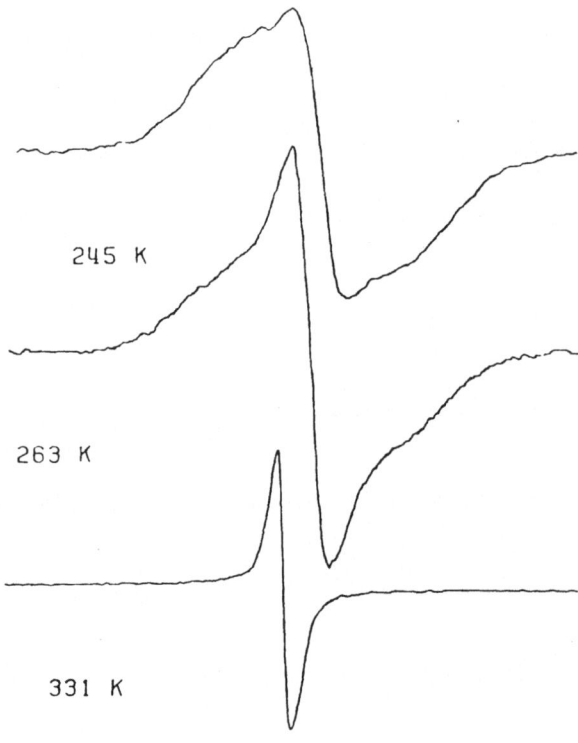

245 K

263 K

331 K

Figure 24. ^{7}Li NMR spectra of the 50/30 sample at various temperatures.

temperatures. As the temperature is increased, thermal motion causes the dipolar interaction to be averaged out. By computer simulation of the lineshape, the ratio of nuclei causing the narrow line shape to the total number of nuclei can be found. This is the fraction of lithium ions which are mobile (see Fig. 25).

This data cannot be explained by a model which assumes a thermally activated process with a unique activation energy (see Eq. 8). Because of the amorphous nature of glasses it is believed that there exists a range of activation energies. Several different distributions of activation energies can be used to fit the data in

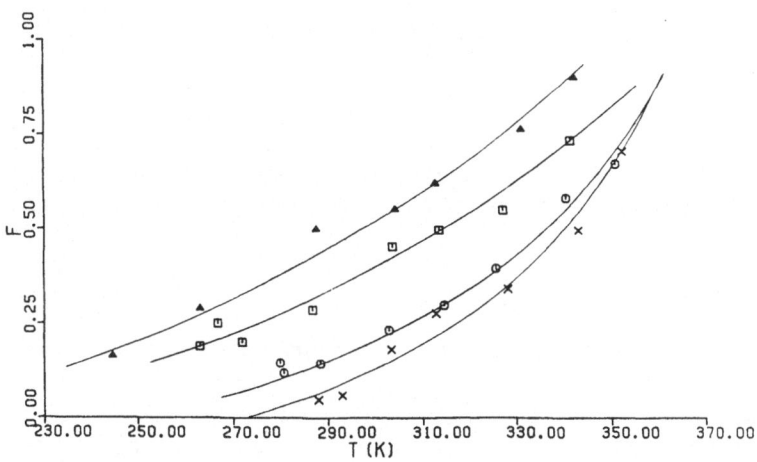

Figure 25. Fraction of mobile ions as a function of temperature. x -50/10,
O-50/20, ▢-50/30, △-50/45.

Fig. 25. One method assumes a unique activation energy for transition to the mobile
state (E_m) and a second unique activation energy (E_n) for the ion when it is in the
mobile state[30]. The sum of these two activation energies when added together
should give the total activation energy (E_A). These results are summarized in Table

Table 2

Sample	E_m(eV)	E_n(eV)	E_A(eV)	$<E>$(eV)	$\Delta E_{\frac{1}{2}}$(eV)	$E_A DC$(eV)
50/10	0.35	0.08	0.43	0.79	0.22	0.54
50/20	0.21	0.07	0.28	0.73	0.22	0.53
50/30	0.13	0.09	0.22	0.32	0.12	0.50
50/45	0.12	0.06	0.18	0.30	0.11	0.46

2. Another method assumes a Gaussian distribution of activation energies. The
mean activation energy ($<E>$) and half width ($\Delta E_{\frac{1}{2}}$) of the distribution which best
fit the data of Fig. 25 are also shown in Table 2. The lines drawn in Fig. 25
represent the prediction of the Gaussian fit model. Changing these parameters by
more than five percent will cause a noticable disagreement with the data. Since both
of these distributions will fit the data equally well, it cannot be said which if either is
an accurate model. The success of being able to fit a distribution of activation
energies to the data does indicate that the ions exist in some kind of distribution of

potential wells which the ions must overcome before they are mobile.

Also included in Table 2 are the activation energies measured through DC conductivity measurements ($E_A DC$). As can be seen, the activation energies, E_A, measured by NMR are consistenty lower than the activation energies measured by electrical conduction. This is the usual result for most materials. This is probably because NMR measures short-range motional processes while conduction is a bulk property.

Lithium Motion in Glass--T_1 Study

Shown in Fig. 26 is a plot of the ^{7}Li spin lattice relaxation rate ($1/T_1$) versus

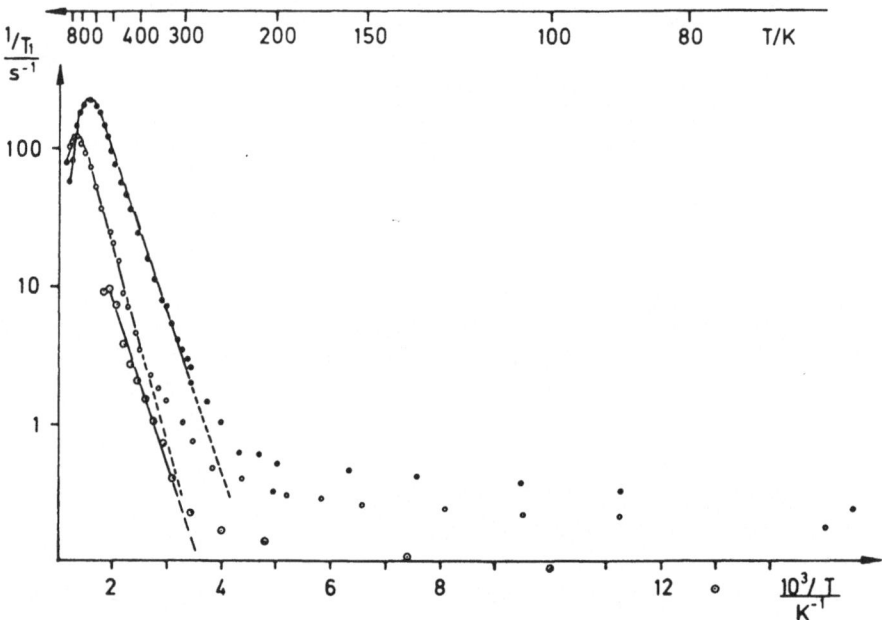

Figure 26. ^{7}Li spin-lattice relaxation rate versus reciprocal temperature for glasses containing 33 mol% Li_2O and 67% SiO_2 (O), B_2O_3 (●) and P_2O_5 (◉), respectively. (From reference 31).

1000/T for $33Li_2O \cdot 67SiO_2$, $33Li_2O \cdot 67B_2O_3$, and $33Li_2O \cdot P_2O_5$ glasses[31]. Above about 250 K the relaxation is dominated by coupling of the lattice to the quadrupole moment; the interaction is time-dependent because of motion of the lithium ion. Only a small part of the high temperature side of the curve was measurable because of the glass transformation temperature. The high asymmetry of these curves shows that the assumption of a unique τ in Eq. 3 is not valid. The data can be explained by assuming a distribution in τ. This is equivalent to a distribution in activation energies, which agrees with the result of the previous section.

The authors found that the activation energies for the silicate and borate glasses were similar. In the lithium phosphate glasses, there was a dependence of the activation energies on alkali concentration. When the lithium was partially substituted by another alkali, the activation energies were slightly enhanced consistent with the mixed alkali-effect[31].

Below 250 K, relaxation due to ionic diffusion becomes very small and another relaxation process becomes dominant. This is why the curve flattens out at low temperatures. Although there is still some controversy, and although complete understanding has not yet been obtained, it is believed that this relaxation is due to the existence of a two level potential energy well which is believed to be an intrinsic property of vitreous materials[32]

CONCLUSION

Clearly NMR has proved to be an invaluable tool in the study of glasses. Much information about both the structural and dynamical properties of glasses on the atomic scale is revealed through NMR. New and rapidly expanding NMR technology (i. e. superconducting magnets, MASS, computerized data handling) is greatly increasing the range of possible uses of NMR in the study of glasses.

ACKNOWLEDGMENTS

The research reported in this chapter was supported in part by the National Science Foundation through Grant No. DMR-8004488-02 and by the Materials Research Laboratory at Brown University.

REFERENCES

1. G. E. Jellison Jr., L. W. Panek, P. J. Bray and G. B, Rouse Jr., J. Chem. Phys. 66, 802 (1977).

2. C. P. Slichter, Principles of Magnetic Resonance, Springer Series in Solid State Sciences 1 (Springer, Berlin, 1978).

3. N. Bloembergen, and T. J. Rowland, Acta metall. 1, 739 (1953).

4. N. Bloenbergen, E. M. Purcell, P. V. Pound, Phys. Rev. 73, 679 (1948).

5. J. L. Bjorkstam and M Villa, Magnetic resonance review 6, 1 (1980).

6. W. Müller-Warmuth and H. Eckert, Physics Reports 88, 140 (1982).

7. J. R. Hendrickson and P. J. Bray, J. Mag. Res. 9, 341 (1973).

8. R. L. Mozzi and B. F. Warren, J. Appl. Crystallogr. 3, 251 (1970).

    J. Krogh-Moe, Phys. Chem. Glass 6, 46 (1975).

    G. E. Jellison Jr., L. W. Panek, P. J. Bray and G. B, Rouse Jr., J. Chem. Phys. 66, 802 (1977).

9. Y. H. Yun and P. J. Bray, J. Non-Cryst. Solids 27, 363 (1978).

10. Y. H. Yun and P. J. Bray, J. Noncryst. Solids 44, 227 (1981).

11. T. Abe, J. Am. Cer. Soc. 35, 284 (1952).

12. G. E. Jellison, Jr and P. J. Bray, J. Non-Crys. Sol. 29, 187 (1978).

    S. A. Feller, W. J. Dell and P. J. Bray, J. Non-Crys. Sol. 51, 21 (1982).

13. J.Krogh-Moe, Phys. Chem. Glasses 6, 46 (1965).

14. W. J. Dell and P. J. Bray, J. Non-Cryst. Solids 58, 1 (1983).

15. A. E. Geissberger and P. J. Bray, J. Non-Crys. Sol. 54, 121 (1983).

16. R. L. Mozzi and B. E. Warren, J. Appl. Cryst. 3, 251 (1970).

17. G. E. Jellison, Jr., L. W. Panek, P. J. Bray, and G. B. Rouse, Jr., J. Chem. Phys. 66, 802 (1977).

18. R. L. Mozzi and B. F. Warren, J. Appl. Crystallogr. 3, 251 (1970).

19. J. Krogh-Moe, Phys. Chem. Glass 6, 46 (1965).

20. I. A. Harris Jr. and P. J. Bray, Phys. Chem. Glasses 21, 156 (1980).

21. E. Lippmaa, M. Mägi, A. Samoson, G. Engelhardt, and A. R. Grimmer, J. Am. Chem. Soc. 102, 4889 (1980).

22. E. Lippmaa, and A. Samoson, Bruker Rep. 1, 6 (1982).

23. D. Müller, G. Berger, I. Grunze, G. Ladwig, E. Hallas and U. Haubenreisser, Physics and Chemistry of Glasses 24, 37 (1983).

24. P. J. Bray and W. J. Dell, Journal de Physique C9, No. 12, 131 (1982).

25. T. A. Kennedy and G. Seidel, Physics Letters 46A, 307 (1974).

26. P. J. Bray, D. E. Hintenlang, R. V. Mulkern,S. G. Greenbaum, D. C. Tran, and M. Drexhage, J. Non-Cryst. Sol. 56, 27 (1983).

27. D. Leroy, J. Lucas, M. Poulain, D. Ravaine Mat. Res. Bull. 13, 1125 (1978).

28.   A. Lecoq, M. Poulain, J. Non-Cryst. Sol, 34, 101 (1979).

29.   P. J. Bray, D. E. Hintenlang, R. V. Mulkern, S. G. Greenbaum, D. C. Tran, and M. Drexhage, J. Non-Crys. Sol. 56, 27 (1983).

30.   P. J. Bray, D. E. Hintenlang, M. L. Lui and R. V. Mulkern, Glastechnische Berichte, 56K, 892 (1983).

31.   E. Göbel, W. Müller-Warmuth, H. Olyschläger, and H. Dutz, J. Magn. Reson. 36, 371 (1979).

32.   W. Müller-Warmuth and H. Eckert, Physics Reports 88, 140 (1982) and the references sited therein.

APPLICATION OF INDUCTIVELY COUPLED PLASMA SPECTROMETRY FOR THE ANALYSIS OF CERAMICS AND GLASSES

Ramon M. Barnes

Department of Chemistry, GRC Towers
University of Massachusetts
Amherst, MA 01003-0035

INTRODUCTION

The analysis of ceramics, glasses, and other refractory materials by atomic spectroscopic techniques is a well - established analytical chemistry approach. During the past decade development of new spectrochemical plasma sources, in particular the inductively coupled plasma (ICP), direct current plasma jet (DCP), and microwave induced plasma (MIP), has revolutionized spectrochemical measurements of major, minor, and trace elements in diverse industrial materials. Owing to the very high temperature, stability, and low chemical interferences of the ICP discharge, ICP spectrometry is especially suited for the analysis of ceramics, glasses, and refractories. The objective of this article is to review the features and capabilities of ICP spectrometry for the analysis of ceramics, glasses, and refractory materials. The prospects for development are substantial, and maturity in the application of ICP spectrometry to these materials is expected to occur by the end of the decade.

Atomic spectroscopic methods for the analysis of ceramics, glasses, and refractories are documented in various sources. More than 20 methods for cement, glass, ceramic, and refractory materials are detailed in the American Society for Testing and Materials (ASTM) "Methods for Emission Spectrochemical Analysis" compilation[1]. These techniques primarily apply D-C arc spectrography, flame photometry, and atomic absorption spectrometry. X-ray fluorescence methods are also included. An excellent review by Wise et al.[2] on the "Analysis of Glasses and Ceramics by Atomic Spectroscopy" summarizes the procedures for (a) semiquantitative arc spectrography, (b) acid dissolution and fusion sample decomposition methods, (c) standards preparation techniques, and (d) element determinations. Although D-C

31

arc spectrography is excellent for semiquantitative estimation of as many as 65 elements in ceramics, glasses, and refractories, especially since solid and powder samples need not be dissolved and the measurement sensitivity is quite good, the technique lacks precision, is lengthy, and requires multiple data acquisition steps and special operator training and skill. Alternative methods based upon x-ray fluorescence, flame absorption and emission, and furnace absorption spectrometry have gained popularity during the 1970´s[1, 2]. The x-ray fluorescence approach generally requires sample fusion prior to simultaneous or sequential determination of major and minor elements, and the flame absorption and emission measurements are preferred for the alkali and alkaline earth elements after the sample has been dissolved. The flame technique generally requires hot flames (i.e. nitrous oxide - acetylene) to minimize chemical interferences of the refractory elements, but the analysis may not be sufficiently sensitive when refractory trace elements are sought. These limitations are generally overcome when plasma sources, like the ICP or DCP, are substituted for the flame. Since the preparation of the review by Wise et al.[2], plasma souces have become widespread in industrial analytical chemistry laboratories. Commercial ICP spectroscopy systems were introduced in 1973-4, but relatively few applications of the ICP for ceramics and glass analyses had been published before the Wise et al. review. However, ICP spectrometric analysis of refractories has grown steadily since then and today is probably the preferred analytical technique. The number of applications of ICP spectrometry for analysis of ceramics, glasses, and referactories in industrial laboratories is not represented proportionately by the volume of papers published, however. Tabulations of specific applications are compiled annually in the "Annual Reports on Analytical Atomic Spectroscopy"[3,4] in the section reviewing refractories, metal oxides, ceramics, slag, and cement. A biennial review on emission spectrometry also traces the growing role of plasma sources and their applications[5]. Since atomic spectroscopic analysis of rocks and minerals in many ways parallels the analysis of comminuted ceramic and glass samples, the recent "Handbook of Inductively Coupled Plasma Spectrometry" by Thompson and Walsh[6] provides an timely resource for ICP methodology.

ICP spectrometry is developing rapidly, and recent reviews follow this progress[7-12]. A unique publication devoted especially to plasma spectrochemical analysis, the ICP Information Newsletter, now in its tenth year provides another excellent resource[13]. A summary of ICP spectrometry and its applications are included in the following sections.

ICP SPECTROMETRY

The excellence of spectrochemical analysis in ICP spectrometry results from the high stability of the discharge, its unique annular shape of high temperature and electron density, its low noise and background, and its freedom from interferences when operated under appropriate experimental conditions. The discharge is generated by inducing a magnetic field in a flowing, electrically conducting gas, usually argon and sometimes argon and molecular gases, by means of an external, water-cooled coil that surrounds the quartz or boron nitride confinement tubes through which the gas flows. The plasma reaches temperatures on the order of 9000 K in an inert argon atmosphere without contact or contamination from electrodes. Introduction of the sample aerosol through the central channel formed in the discharge is a unique feature of the ICP and accounts for many of its favorable properties. Sample aerosol in the form of liquid droplets, solid powder, metal vapor, gaseous effluent, or fine powder slurries can be transported by the sample carrier gas into the central channel, where the liquids are evaporated, the solids vaporized, and the matrix and analytes atomized for suitable spectrochemical detection. Since the sample experiences temperatures greatly in excess of those available in combustion flames, the chemical interferences found in flame and furnace spectroscopy are absent in the ICP source. Similarly the effects of matrix elements, especially those with low ionization potential, in the ICP are significantly smaller than observed in the D-C arc, flames, and the DCP. Based upon these considerations, the ICP is the spectrochemical source preferred for refractory element determinations. Sample preparation and introduction techniques represent the major limitation to the universal application of the ICP source[7,14,15]. As the number of applications grow, ICP spectrometry tends to replace classical chemical analyses involving gravimetry, titrimetry, electrochemistry, and spectrophotometry. According to Thompson and Walsh[6] "ICP spectrometry is probably the first practical method for the simultaneous determination of large number of trace elements with the accuracy required for much petrological and geochemical research." This conclusion appears equally valid for the analysis of ceramics, glasses, and refractories.

Detection of the analyte once atomized and/or ionized in the ICP source occurs by atomic absorption, emission, or fluorescence spectrometry or by mass spectrometry. Although atomic absorption in the ICP is not common owing to the limitation of the single-element determination, commercial ICP instruments for atomic emission and fluorescence, and for mass spectrometry are available. ICP atomic emission spectrometry (ICP-AES) is certainly the most popular approach with more than 20 manufacturers producing simultaneous or sequential multielement analyzers. In the emission mode, all elements (except argon) can be measured theoretically, although in practice the limit is about 65-70. An ICP atomic fluorescence spectrometer (ICP-AFS), commercially available since 1981, is equipped with pulsed hollow cathode lamps for excitation of fluorescence radiation from atoms in the ICP discharge. Although significant progress has been made during the past year in HCL-ICP-AFS[16], the powers of detection for refractory elements nearly match those for conventional ICP-AES when a high-efficiency ultrasonic nebulizer is applied. However, the sensitivity of ICP-AFS can be improved with higher intensity excitation sources such as lasers[17],[18].

ICP source mass spectrometry (ICP-MS), first described in 1980 and introduced commercially in 1983, is a power extension of ICP-AES and one likely to become the major ICP measurement technique by the end of the decade[19-23]. Mass spectrometry is inherently more sensitive and the spectra are less complex than emission spectrometry. ICP-MS also permits isotopic ratio and isotope dilution analyses, and, with some development, may provide the direct determination of halogens especially F and Cl. Since sample introduction and ICP generation instrumentation are essentially identical for ICP-AES and ICP-MS, analysis methods for ceramics, glasses, and refractories developed for ICP-AES will be adaptable with minor changes to ICP-MS. As an example, the limits of detection for selected elements found in ceramics, glass, and refractories for ICP-AES, ICP-AFS, and ICP-MS techniques are as follows: Al AES 10 ng/mL, AFS (ultrasonic) 5 ng/mL, MS 0.2 ng/mL; B AES 3 ng/mL, AFS (ultrasonic) 60 ng/mL, MS 0.4 ng/mL; Co AES 3 ng/mL, AFS (ultrasonic) 1 ng/mL, MS 0.01 ng/mL; W AES 14 ng/mL, AFS (ultrasonic) 200 ng/mL, MS 0.05 ng/mL.

ICP spectrometry provides multielement analysis generally with minimal sample preparation and interferences. In their chapter on silicate rock analysis, Thompson and Walsh[6] explain ICP instrument calibration, wavelength selection, and practical realistic appraisal of the ICP-AES determination of major and trace elements. Among the limitations identified for ICP-AES which apply for the analysis of ceramics, glasses, and refractories include spectral line interferences and possible background effects arising from major concentrations of Al, Ca, and Mg[24]. Through matrix matched standards and appropriate background corrections, these limitations are minimized. These special interferences are absent with AFS and MS detection systems.

ANALYSIS OF CERAMICS, GLASS, AND REFRACTORIES

Applications of ICP spectrometry culled from the literature[3],[4] generally illustrate solution sample introduction of conventional ceramics, glasses, and refractories. Burdo and Su[25] recently reviewed fusion and acid decomposition methods for ceramics and glasses. Generally the large majority of ceramic, glass, and refractory samples encountered can be dissolved either by acid (HF) dissolution or suitable fusion procedures. Ceramics and glass industry laboratories equipped with ICP systems are able in practice to determine 15-20 major, minor, and trace elements in 10 - 100 mg of sample. High-technology ceramics[26], including composites, cermets, ceramic particles, fibers, and whiskers, have not been specifically described in the ICP spectrometry literature, although their analyses follow the analysis of bulk silicates, boron nitride, silicon carbide, aluminum oxide, or other refractory borides, carbides, nitrides, oxides, and sulfides. Furthermore, the direct analysis of ceramic, glass, or refractory powders, slurries, or solids by ICP spectrometry is less well developed than solution sample techniques although a number of new sampling techniques for these phases are now being explored[7]. Presently, solutions resulting from acid dissolution or fusion are preferred for ceramics and glasses, because calibration standards are readily prepared and the original sample is distributed homogeously.

Ceramics

The ICP analysis of ceramics has been reported by a limited number of authors although the extent of ICP anlayses in the ceramics industry in wider than the number of papers indicate. Debras-Guedon determined B in ceramic raw materials and products after samples were fused with sodium carbonate and ZnO[27]. Uchikawa et al. described the determinations of phosphorus[28] and cadmium[29] in ceramic materials and products. Limits of detection were 0.47 µg P/mL (P I 213.618 nm) and 0.6 ng Cd/mL (Cd I 228.802 nm) with analysis precision of 0.6 - 3.3% RSD for P and 1.9-26% RSD for Cd obtained for standard reference materials. Schroth determined Nd, Ti, Zr, and Ba in neodymium – titanium ceramic and Ti, Zr, Nb, Mg, Mn, and Ni in Pb-Zr-Ti ceramic after sample fusion[30]. A recent survey of instrumental and chemical methods for analysis of ferrites did not mention plasma spectrometry[31], but the application can be expected shortly.

Glass

The multielement analysis of small glass fragments by ICP-AES by Catterick and Hickman[32] has lead to elemental analysis and discrimination of glass samples for forensic characterization[32-35]. The ICP-AES results for Mn, Fe, Mg, Al, Sr, and Ba were especially useful for the discrimination of sheet glass samples with similar refractive indices and belonging to the same class. The ICP-AES technique has replaced both D-C arc emission spectrography and flame atomic absorption spectrometry. Small samples (0.2 - 0.5 mg) were dissolved in a mixture of HF and HCl, and limits of detection in a 0.5-mg sample were 10 µg Mn/g, 50 µg Fe/g, 500 µg Mg/g, 500 µg Al/g, 2000 µg Cr/g, and 5 µg Ba/g. Expansion of the ICP analysis to include a larger number of elements (e.g. 22) would enhance the discrimination among sheet glass samples. Hart and Adams[36] evaluated the ICP-AES results for 23 elements in ancient pottery samples as a method of assigning pottery to a kiln source. Romano-British pottery sherds were sampled with tungsten carbide tipped drill bits (300 mg), ground in an agate motar, and digested in hydrofluoric and perchloric acids. Differences were identified in pottery from two sites. These two examples illustrate the power of multielement ICP spectrometry combined with chemometric methods for anthropological and forensic needs. As chemometrics computer programs become more widely applied with ICP systems, ICP multielement analysis, especially for trace consistitutents, will probably lead to the replacement of atomic absorption by ICP spectrometry in archaeometry and forensics.

Weissman and Hallet[37] determined the phosphorus and silicon content of 600 - 1000 nm thick phosphosilicate glass films deposited on 5-cm diameter silicon wafers. After the thin film was removed from the wafer by etching with dilute HF, the glass film was found to contain 86 - 530 μg P and 920 - 1700 μg Si. Kojima et al.[38] measured Ca, Fe, Ge, Lu, Sm, and Y in Gd-Ga-bubble garnet films, and Grallath et al.[39] determined traces of boron in silicon and quartz after HF dissolution and boron distillation. Zil´bershtein et al[40]. analyzed monolithic quartz samples by continuously inserting 0.6 - 1 mm diameter rods directly into the ICP above the induction coil. The ICP spectrometric analysis of waveguide materials has not been described in the literature, although in industrial laboratories manufacturing waveguides, ICP-AES is applied in their analysis[3, 4].

Refractories

A variety of refractory materials have been successfully analyzed by ICP spectrometry. Rocks and minerals[6,19,41,42] and steel industry slags and other refractory oxides[43-46] are among the most popular applications. Hughes[46] survyed the analysis of oxide materials with ICP-AES and other techniques, and suggested the need for automated sample preparation methods. Analysis of slags and oxides in a steel mill by ICP-AES were described[43, 44].

Degré[47] described the ICP-AES analysis of cements and demonstrated 16-element determinations in clinkerized and non-clinkerized samples. Ward et al.[48] grouped rocks and cement together in their ICP-AES analysis of silicic materials for 12 major and trace elements. In·another example, Ru et al.[49] determined 14 rare earth elements in high-purity yttrium oxide by ICP spectrography. The determination of rare earth elements is simplified by the high temperature of the ICP source compared to flame spectrometry, although wavelength selection is more complicated owing to the line-rich spectra observed in the ICP. To extend the analysis of rare earth elements to lower concentration levels, ICP-MS is effective especially for geological materials[19, 23]. Barnes and Mahanti[50] compared acid digestion and NaOH fusion sample preparation methods for bauxite samples in their determination of 17 major and trace elements. Silicon in HF solution was determined by using an inert PTFE nebulizer and spray chamber along with an ICP torch with a graphite injector tube. Various HF-resistant arrangements now are available commercially[51]. Zamecheck et al.[52] analyzed pure aluminum and silicon and their oxides, and compared the ICP-AES results with D-C arc spectrography for sapphire. Saisho et al.[53] recently described the ICP-AES determination of 11 elements (Al, Ca, Cu, Fe, Hf, Mg, Na, Ni, Si, Ti, and Zr) in zirconia ceramics. Hulmston[54] determined 25 elements in boron by ICP emission spectrography after removal of the matrix with hydrofluoric acid. Typical of the photographic detection, the precision was 3-8% RSD, and the overall method precision was 10-15% RSD. The method can be extended to the analysis of boron carbide. Natansohn and Czuprya[55, 56] applied DCP spectrometry for the determination of Cr, Cu, Mn, Mo, Ni, B, Fe, and W at concentrations between 1-100 μg/g in silicon carbide and silicon nitride. Gremion and Duchoud[57] determined S, Ca, Mg, Sr, P, Na, and K in plaster and indicated that Si, Fe, Al, Ti, Ba, Mn, Cu, and Zn were also measured.

Mahanti and Barnes[58] determined Al, Cu, Fe, Si, Ti, and V in spectrographic grade graphite by ICP-AES. Samples were ashed with a purified magnesium nitrate ashing aid, and Cu, Fe, Ti, and V were chelated with a poly(dithiocarbamate) resin while Al was complexed with a poly(acrylamidoxime) resin. After liberation from the resins, the elements were determined in the absence of the magnesium which was not retained by the resins. The lowest quantitatively determinable concentrations ranged from 0.1 to 0.02 μg/g in the graphite. A similar procedure was applied in the analysis of industrial baked carbon and coke for Al, Ca, Fe, Ni, Ti, V, and Zn. One gram samples were ashed in a platinum crucible in the presence of pre-purified magnesium nitrate at 600°C and 750°C for 10 hours for the coke and graphite samples, respectively. The resulting ash was dissolved with 2 mL of aqua regia and 1 mL of HF in a PTFE bomb at 110°C for 3 hours. Boric acid was added to the cooled solution to complex the excess HF. The final volume was 25 mL, except for Al for which a 1:5 dilution was made. The lowest quantitatively determinable concentrations calculated for five times the limit of detection for 1 g sample in 25 mL of sample solution ranged from 3 μg Al/g, 2 μg Ni/g, and 1 μg Fe/g to 0.3 μg Ca/g, 0.4 μg Ti/g, 0.5 μg Zn/g, and 0.9 μg V/g. Precision of replication sample analyses in coke for major elements (Al, Fe, Ca, Ti) ranged from 1 to 4%, while for trace elements (Ni, V, Zn) was 2-12% RSD. Silicon was also determined in industrial carbon samples. For analysis, 0.2 - 0.4 g of sample was weighed into a platinum crucible, the sample ashed at 730°C, and the ash dissolved with 1 mL HCl and 0.5 mL HF. Boric acid was added, and the final solution was diluted to 25 mL. Recoveries for Si loss during ashing were 89-90%, and the detection limit was 4-5 μg/g.

Powder and Solids

The direct analysis of powder and solid samples by ICP spectrometry is not as well developed as solution analysis[6, 7, 15]. Difficulties in representative sampling, sample transport to the plasma, and calibration has limited progress although powder sampling was applied in 1967 for the quantitiative ICP analysis of lithium salts and alumina59. Powder introduction techniques include mechanically agitating59 or spark elutriating a powder bed, sparking briquetted powders or dispersed powders on tape, fluidizing a bed powder, nebulizing slurries, and laser ablating pressed pellets. In other approaches the powder or solid sample is inserted directly into the ICP40. For example, the powder sample is loaded into an arc-like carbon or metal electrode which is positioned in or near the bottom of the ICP discharge, where the electrode heats and the elements are evapored into the plasma. To date none of these systems has be made commercially effective as an ancillary sampling device for ceramics, glasses, or refractories, although a combined spark-sampling ICP arrangement is available60 but most often employed with metal solids. The direct ICP spectrometric analysis of non-conducting, refractory powders remains a major research challenge and one that merits considerably more attention than it has received.

PROSPECTS FOR FUTURE DEVELOPMENTS

ICP spectrometry is widely applied in the analysis of rocks and minerals, metals, agricultural materials and food products, and waters and wastes. Ceramics, glasses, and refractories can be analyzed after either acid digestion or fusion, and all metals, metalloides, and some non-metals, especially refractories, can be determined reliably by ICP spectrometry. The widespread adoption of ICP spectrometry in the ceramics and glass industry is expected to occur in the next few years as analysts become aware of its capabilities. ICP-MS offers a particularly valuable extension of ICP spectrometry owing to its almost uniform and high sensitivity for metals and metalloids. This capability should permit convenient analysis of extremely small ceramic and glass samples from the semiconductor manufacturing processes.

A number of significant needs exist in the ICP analysis of ceramics, glasses, and refractories. For example, a manual of standardized methods, including tested sample preparation, ICP operating parameters, and appropriate measurement conditions would significantly aid the industry. In plant utilization, a ruggedized, low-cost ICP instrument would be valuable. Finally, ICP systems analyze samples more rapidly than they can be prepared, so that the availability of an automated, commercially manufactured instrument for sample grinding, fusion/digestion, and dilution could provide a major increase in efficiency in ceramic, glass, and refractory materials laboratories. Total analysis automation in the ceramics and glass laboratory based upon ICP spectrometry and an automated sample preparation instrument is feasible before 1990.

38

ACKNOWLEDGEMENT

Support for preparation of this report was provided under
Department of Energy contract DE-AC02-77EV04320 and the ICP
Information Newsletter.

REFERENCES

1. "Methods for Emission Spectrochemical Analysis," American Society
   for Testing and Materials, 7th Edition, Philadelphia (1982).
2. W.M. Wise, R.A. Burdo, and J.S. Sterlace, Analysis of Glasses and
   Ceramics by Atomic Spectroscopy, Prog. Anal. Atom. Spectrosc.
   1:201 (1978).
3. M.S. Cresser and L. Ebdon, eds., "Annual Reports on Analytical
   Atomic Spectroscopy," Vol. 12, Royal Society of Chemistry,
   London (1983).
4. M.S. Cresser and B.L. Sharp, eds., "Annual Reports on Analytical
   Atomic Spectroscopy," Vol. 11, Royal Society of Chemistry,
   London (1982).
5. P.N. Keliher, W.J. Boyko, J.M. Patterson III, and J.W. Hershey,
   Emission Spectrometry, Anal. Chem. 56:133R (1984).
6. M. Thompson and J.N. Walsh, "A Handbook of Inductively Coupled
   Plasma Spectrometry," Blackie, Glasgow and London (1983).
7. R.M. Barnes, Progress in Inductively Coupled Plasma Analytical
   Spectroscopy, J. Testing Eval. 12:194 (1984).
8. R.M. Barnes, Frontiers in Inductively Coupled Plasma
   Spectroscopy, Chem. Anal. (Warsaw) 28:179 (1983).
9. R.M. Barnes, Recent Advances in Analytical Atomic Radiofrequency
   Emission Spectroscopy, Phil. Trans. R. Soc. London A 305:499
   (1982).
10. R.M. Barnes, Inductively Coupled Plasma Atomic Emission
    Spectroscopy: A Review, Trends Anal. Chem. 1:51 (1981).
11. R.M. Barnes, Recent Developments in Emission Spectroscopy with
    Inductively Coupled Plasma Discharge, Karl-Marx-Univ.
    (Leipzig) Wissensch. Z. 28:383 (1979).
12. R.M. Barnes, Recent Advances in Emission Spectroscopy:
    Inductively Coupled Plasma Discharges for Spectrochemical
    Analysis, Crit. Rev. Anal. Chem. 7:203 (1978).
13. R.M. Barnes, ed., ICP Information Newsletter, University of
    Massachusetts, Amherst, MA (1975-1984).
14. R.F. Browner and A.W. Boorn, Sample Introduction: The Achilles'
    Heel of Atomic Spectroscopy? Anal. Chem. 56:787A (1984).
15. R.F. Browner and A.W. Boorn, Sample Introduction Techniques for
    Atomic Spectroscopy, Anal. Chem. 56:875A (1984).
16. D.R. Demers, Hollow Cathode Lamp-Excited ICP Atomic Fluorescence
    Spectrometry -- An Update, Spectrochim. Acta 40B(1/2) (1985),
    in press.
17. B.D. Pollard, M.B. Blackburn, S. Nikdel, A. Massoumi, and J.D.
    Winefordner, Atomic Fluorescence Spectrometry in the

Inductively Coupled Plasma with a Continuous Wave Dye Laser, Appl. Spectrosc. 33:5 (1979).

18. J.D. Winefordner, Atomic Fluorescence Spectrometry, Past, Present, and Future, in: "Recent Advances in Analytical Spectrocopy," K. Fuwa, ed., Pergamon Press, Oxford (1982).

19. A.R. Date and A.L. Gray, Determination of Trace Elements in Geological Samples by Inductively Coupled Plasma Source Mass Spectrometry, Spectrochim. Acta, 40B(1/2) (1985), in press.

20. R.S. Houk, ICP Mass Spectrometry from the Eye of a Beholder, ICP Inf. Newsl. 10:194 (1984).

21. D.F. Douglas, ICP-MS at SCIEX, ICP Inf. Newsl. 10:196 (1984).

22. A.L. Gray, Continuing Development of ICP Source Mass Spectrometry at the University of Surrey, ICP Inf. Newsl. 10:200 (1984).

23. A.R. Date, ICP-MS Applications Development at the British Geological Survey, ICP Inf. Newsl. 10:202 (1984).

24. C. Trassy and J.M. Mermet, Interférences spectrales, in: "Les Applications Analytiques des Plasmas HF," Lavoisier, Paris (1984).

25. R.A. Burdo and Y.-S. Su, Fusion and Acid Decomposition of Silicate and Refractory Materials, Eastern Analytical Symposium, November (1983), ICP Inf. Newsl. 9:588 (1984).

26. H.J. Sanders, High-tech Ceramics, Chem. Engr. News 62(28):26 (1984).

27. J. Debras-Guedon, Le dosage du bore dans les matières premières et produits de l'industrie céramique par spectrométrie d'émission avec excitation par plasma induit par haute fréquence, Bull. Soc. Fr. Céram. 123:29 (1979).

28. H. Uchikawa, R. Furuta, and Y. Mihara, Determination of Phosphorus in Ceramic Materials and Ceramic Products by Inductively Coupled Plasma - Atomic Emission Spectrometry, Bunseki Kagaku 32:291 (1983).

29. H. Uchikawa, R. Furuta, and Y. Mihara, Determination of Cadmium in Ceramic Materials and Ceraminc Products by Inductively Coupled Plasma - Atomic Emission Spectroscopy, Bunseki Kagaku 32:673 (1983).

30. H. Schroth, Quantitative Emissionspektranalyse von der Keraminkmassen mit induktiv gekoppelter Plasmaanregung, Z. Anal. Chem. 269:286 (1979).

31. C. McCrory-Joy and D.C. Joy, Chemical and Instrumental Analysis of Ferrites, Talanta 30:299 (1983).

32. T. Catterick and D.A. Hickman, Sequential Multi-element Analysis of Small Fragments of Glass by Atomic-emission Spectrometry Using an Inductively Coupled Radiofrequency Argon Plasma Source, Analyst 104:516 (1979).

33. D.A. Hickman, G. Harbottle, and E.V. Sayre, The Selection of the Best Elemental Variable for the Classification of Glass Samples, Forensic Sci. Int. 23:189 (1983).

34. D.A. Hickman, Elemental Analysis and the Discrimination of Sheet Glass Samples, Forensic Sci. Int. 23:213 (1983).

35. D.A. Hickman, Linking Criminals to the Scene of the Crime with

Glass Analysis, <u>Anal</u>. <u>Chem</u>. 56:844A (1984).

36. F.A. Hart and S.J. Adams, The Chemical Analysis of Romano-British Pottery from the Alice Holt Forest, Hampshire, by Means of Inductively - Coupled Plasma Emission Spectrometry, <u>Archaeom</u>. 25:179 (1983).

37. S.H. Weissman and S.G. Hallet, Quantitative Analysis of Phosphosilicate Glass Films on Silicon Wafers, <u>Sandia</u> <u>Report</u> <u>SAND82-0039</u> (1982).

38. H. Kojima, E. Kitazume, F. Nagata, and M. Ezawa, Analysis of Elements (Ca, Fe, Ge, Lu, Sm, Y) in Bubble Garnet Films by Inductively Coupled Plasma Atomic Emission Spectrometry, <u>Bunseki</u> <u>Kagaki</u> 30:667 (1981).

39. E. Grallath, P. Tshöpel, G. Kölblin, U. Stix, and G. Tölg, Zur Spektralphotometrie und Emissionssekmetrie mit CMP, ICP von Bor-Spuren in Metallen, Silicium und Quartz nach HF-Ausschluss und Abtrennung durch BF_3-Destillation bzs ausschuettein von BF_4-Ionen Assoziaten, <u>Fresenius</u> <u>Z</u>. <u>Anal</u>. <u>Chem</u>. 302:40 (1980).

40. Kh. I. Zil´bershtein, Recent Soviet ICP Studies, <u>ICP</u> <u>Inf</u>. <u>Newsl</u>. 8:445 (1983).

41. S.E. Church, Multielement Analysis of Fifty-four Geochemical Reference Samples Using Inductively Coupled Plasma - Atomic Emission Spectrometry, <u>Geostandars</u> <u>Newsl</u>. 5:133 (1981).

42. S.E. Church, Trace Element Determination in Geological Reference Materials - An Evaluation of the ICP-AES Method for Geochemistry Applications, <u>in</u>: "Developments in Atomic Plasma Spectrochemical Analysis," R.M. Barnes, ed., Wiley, London (1981).

43. A. Wittman, J. Hancroft, W. Hughes, and K. Ohls, Application of ICP-OES in Steelworks Laboratories, <u>in</u>: "Developments in Atomic Plasma Spectrochemical Analysis," R.M. Barnes, ed., Wiley, London (1981).

44. J.O. Burman, ICP-OES Applications in Steel Industry. Steel and Slag Analysis, <u>in</u>: "Developments in Atomic Plasma Spectrochemical Analysis," R.M. Barnes, ed., Wiley, London (1981).

45. G.M. Russel and A.E. Watson, The Spectrometric Analysis of Chromium Bearing Materials with Particular Reference to Ferrochromium Slags and Chromite Ores," <u>NIM</u> <u>Report</u> <u>1907</u> National Institute of Metallurgy, Randburg, South Africa (1977).

46. H. Hughes, Analysis Survey 1: Oxide Materials, <u>Iron</u> <u>Steel</u> <u>Int</u>. 53:13 (1980).

47. J.P. Degre, Plasma Emission Spectrometry for Analysis of Cements, <u>ICP</u> <u>Inf</u>. <u>Newsl</u>., 7:384 (1982).

48. A.F. Ward, V.J. Luciano, and L.F. Marciello, Development of an Analytical Procedure for Elemental Determinations in Silicic Materials Using the ICAP, <u>in</u>: "Applications of Plasma Emission Spectrochemistry," R.M. Barnes, ed., Heyden, Philadelphia (1979).

49. N.-C. Ru, W.-M. Chang, Z.-C. Jiang, and Y.-E. Zeug, The

Determination of Trace Rare Earth Elements in High-Purity
Yttrium Oxide by Inductively Coupled Plasma - Atomic Emission
Spectroscopy, Spectrochim. Acta 38B:175 (1983).

50. R.M. Barnes and H.S. Mahanti, Analysis of Bauxite by Inductively
Coupled Plasma - Atomic Emission Spectroscopy, Spectrochim.
Acta 38B:193 (1983).

51. G.F. Wallace, V.V. Pirc, and R.D. Ediger, A Hydrofluoric Acid
Resistant Sample Introduction System for ICP Atomic Emission
Spectroscopy, Can. J. Spectrosc. 27:46 (1982).

52. W. Zamechek, R.J. Ledwandowski, R.G. Parkhurst, and A.J. Ellgren,
Trace Metal Analysis in Silicon and Aluminum Metals by
Inductively Coupled Plasma, in: "Applications of Inductively
Coupled Plasmas to Emission Spectroscopy," R.M. Barnes, ed.,
Franklin Institute Press, Philadelphia (1978).

53. H. Saisho, M. Tanaka, K. Sushida, and K. Nakayama, Accurate
Analysis of Zirconia Ceramics by Inductively Coupled Plasma
Emission Spectrometry, in: "The Pittsburgh Conference and
Exposition on Analytical Chemistry and Applied Spectroscopy.
1984 Abstracts," Atlantic City (1984).

54. P. Hulmston, The Application of Inductively-Coupled Plasma
Emission Spectrometry to the Determination of Impurities in
Boron and Boron Compounds, Anal. Chim. Acta 155:247 (1983).

55. G. Czupryna and S. Natansohn, Analysis of Silicon Nitride, in:
"Advances in Materials Characterization," D.R. Rossington,
R.A. Condrate, and R.L. Snyder, eds., Plenum, New York (1983).

56. S. Natansohn and G. Czupryna, Determination of Impurities in
Industrial Products by D.-C. Plasma Emission Spectrometry,
Spectrochim. Acta, 38B:317 (1983).

57. A. Gremion and J.M. Duchoud, A New Conception of Laboratory
Analysis in the Plaster Industry, Ciments, Bétons, Platres,
Chaux No. 742 -3 (1983).

58. H.S. Mahanti and R.M. Barnes, Analysis of High-Purity Graphite
for Trace Elements by Inductively Coupled Plasma Atomic
Emission Spectrometry after Chelating Resin Preconcentration,
Anal. Chem. 55:403 (1983).

59. H.C. Hoare and R.A. Mostyn, Emission Spectrometry of Solutions
and Powders in a High-Frequency Plasma Source, Anal. Chem.
39:1153 (1967).

60. J.Y. Marks, D.E. Fornwalt, and R.E. Yungk, Application of a Solid
Sampling Device to the Analysis of High Temperature Alloys by
ICP-AES, Spectrochim. Acta 38B:107 (1982).

42

RAMAN SPECTRA AND STRUCTURE OF FLUORINE- AND

WATER-BEARING SILICATE GLASSES AND MELTS

Björn O. Mysen and David Virgo

Geophysical Laboratory
Carnegie Institution of Washington
Washington, D.C., 20008

INTRODUCTION

Transport and thermodynamic properties of water- and
fluorine-bearing silicate melts differ significantly from those of
their F- and H_2O-free equivalents (e.g., Hirayama and Camp, 1969;
Kogarko and Kriegman, 1973; Burnham, 1979). For example, their
viscosities and densities decrease dramatically as only a few percent
of these components are added. Furthermore, the magnitudes of such
changes caused by either water or fluorine resemble each other.
Liquidus phase equilibria of fluorine- or water-bearing silicate
systems also differ dramatically from those of F- and H_2O-free
systems. From these observations it can be inferred that F and H_2O
profoundly affect the structure of silicate melts. In order to
understand the solution mechanisms it is necessary to determine the
individual complexing mechanisms of F and H_2O with the silicate
materials.

EXPERIMENTAL METHODS

Compositions studied were in the systems SiO_2-H_2O, SiO_2-NaF and
SiO_2-AlF_3. The water-bearing samples were equilibrated at 15 kbar
and 1550°C for 20 hr or more in solid-media, high-pressure apparatus
(Boyd and England, 1960). Under these conditions, the water
solubility in SiO_2 melt is greater than the maximum water content
used (15 wt%) in the present experiments. The quenched glasses with
15 wt% H_2O did, however, appear milky white as a result of evenly
distributed micrometer and submicrometer bubbles in the glass. It
is, therefore, unlikely that the 15 wt% H_2O could be quenched in the
sample. The fluorine-bearing samples were formed by melting at

43

1550°C and 1 atm in sealed 3 mm I.D. and 7 mm O.D. Pt containers. Quenched melts (glasses) were formed by isobaric temperature-quenching at 250-500°C/sec.

Raman spectra were obtained with an automated Raman system whose main feature is an LSI 11 minicomputer interfaced with a photon counter and the slit- and wavelength drives of the Raman spectrometer. The digitized output is stored on floppy disks. These data were subsequently transferred to a VAX 11/780 for numerical manipulations. In the present study, 0.5-1.0 W of the 514-nm line of an Ar^+ ion laser and the 532-nm line of a Nd:YAG laser were used for sample excitation (see Mysen et al., 1982a; Seifert et al., 1982, for additional details of the system).

Statistical Analysis of Spectra

The spectrum of silica glass (Fig. 1) will be used to illustrate the statistical deconvolution procedures used in this study. Before curves are fitted to the spectra, the raw data are corrected for temperature and frequency dependence of scattering intensity (Long, 1977). An expression of the form:

$$I_{corr} = I\{\nu_0{}^3[-exp(-hc/kT) + 1]\nu/(\nu_0 + \nu)^4\}, \tag{1}$$

where ν_0 and ν are the frequencies of the exciting line and Raman shift, respectively, have been used for this correction. The I and I_{corr} are the raw and corrected Raman intensities, respectively. The Raman intensities in portions or all of the resulting spectrum are normalized to the highest intensity in the spectrum and are then deconvoluted into individual lines with a least-squares minimization routine based on the techniques described by Powell and Fletcher (1963) and Powell (1964a,b) (e.g., Fig. 1B). The line shape, number of lines, and line parameters (half-width, position, and intensity) are independent variables. The principal statistical measures of the quality of the fits are χ^2 and the distribution of the residuals.

The frequency range between 700 and 1300 cm^{-1} for SiO_2 glass has been fitted with mixtures of Gaussian and Lorentzian lines and with a different number of bands (Figs. 1 and 2). Notably, the χ^2 for a fixed number of lines (6) increases by about 200% as the line shape is changed from Gaussian to Lorentzian (Fig. 2). The decreased statistical quality with increasing Lorentzian component is also seen as a pronounced systematic periodicity of the residual distribution (Fig. 1). The χ^2 also increases rapidly as the number of lines in the broad band centered near 1200 cm^{-1} is decreased from two to one and as the band centered near 800 cm^{-1} is deconvoluted into one rather than two bands. The relative intensities of individual Raman bands both in the 800 cm^{-1} and in the 1000-1300 cm^{-1} range are sensitive to the choice of line shape (Fig. 2).

From a statistical point of view, the Raman spectrum of an amorphous material such as SiO_2 glass is best fitted with Gaussian lines, whose line widths typically vary between 25 and 60 cm^{-1}. The

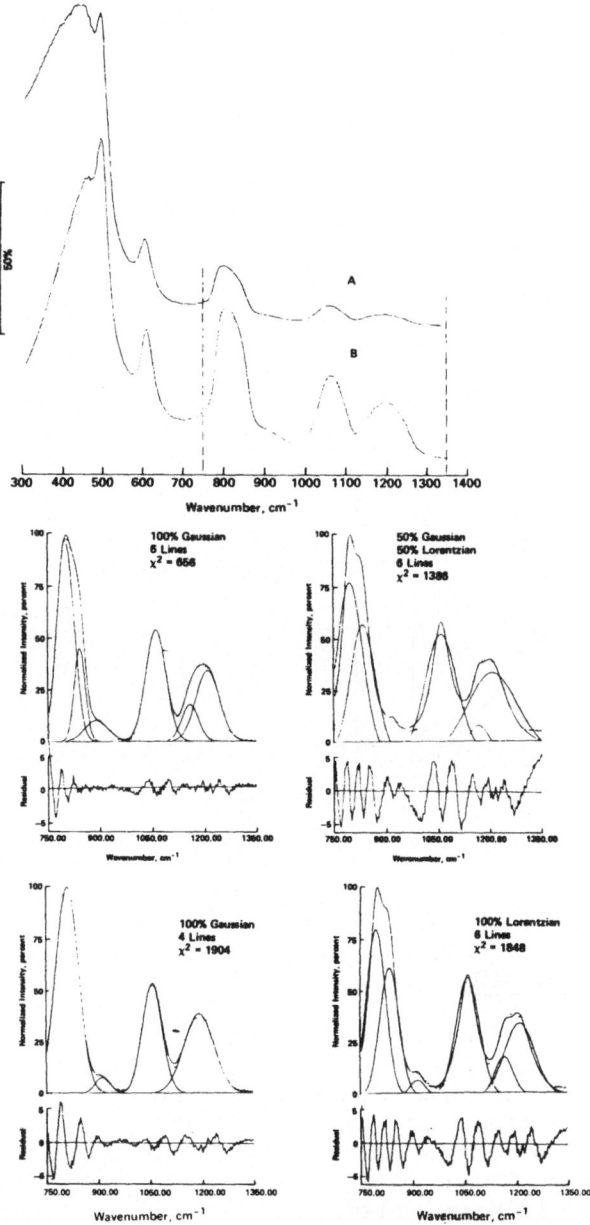

Fig. 1.   (A) Uncorrected and (B) temperature- and frequency-corrected
Raman spectrum of vitreous SiO_2. Dashed lines denote
portions deconvoluted by Gaussian and Lorentizian lines and
with different numbers of lines as shown in examples below.

spectrum of SiO_2 glass fitted in this fashion has been interpreted to
reflect the existence of two three-dimensionally interconnected
structures, differing mainly in that the average Si-O-Si angle
differs by 5 to 10° (Seifert et al., 1982).

45

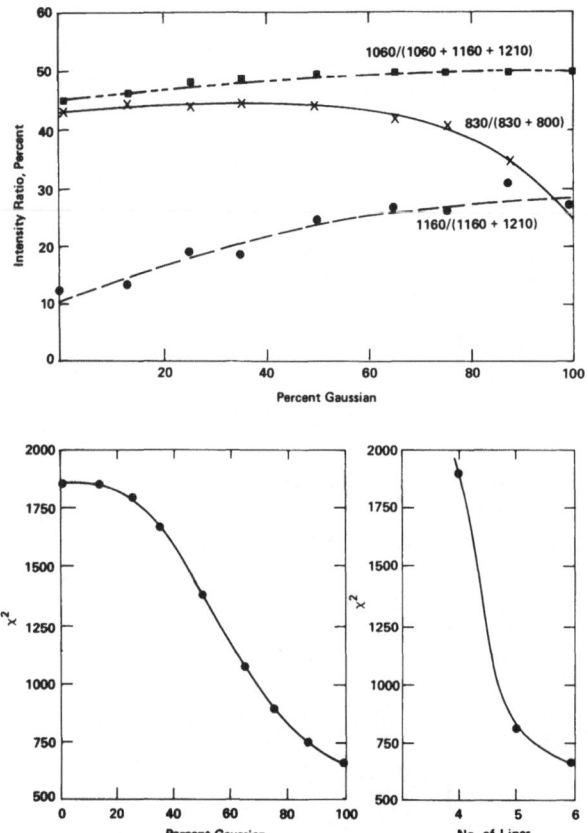

Fig. 2. X^2 of fits such as shown in Fig. 1 as a function of number of lines, and proportion of Gaussian and Lorentzian component with calculated intensity ratios obtained with similar mixtures of line shapes.

RESULTS

Fluorine-Bearing SiO_2 Melts

Temperature- and frequency-normalized Raman spectra of glasses in the systems SiO_2-NaF and SiO_2-AlF$_3$ as a function of $F/(F + Si)$ in the range 0.14-0.50 show two major differences from the spectrum of fluorine-free vitreous silica (Figs. 1 and 3). There is a polarized band near 935 cm^{-1}. The Raman intensity in the high-frequency envelope between 900 and 1300 cm^{-1} shows a sharp intensity increase near 1100 cm^{-1} in both systems.

The spectra in Figure 3 have been deconvoluted into individual bands of Gaussian line shape (Fig. 4). The high-frequency regions characteristically contain the 935-cm^{-1} band from Si-F stretching most probably from an SiO_3F^{3-} entity in the quenched melts (e.g., Takusagawa, 1980; Dumas et al., 1982). The simplest stoichiometry of such a structural unit is $Si_4O_7F_2$. There is also a sharp (30-cm^{-1} half-width at half-height) polarized band near 1100 cm^{-1}. Its relative intensity increases with increasing fluorine content, as does that of the 935-cm^{-1} band in the SiO_2-AlF$_3$ quenched melts

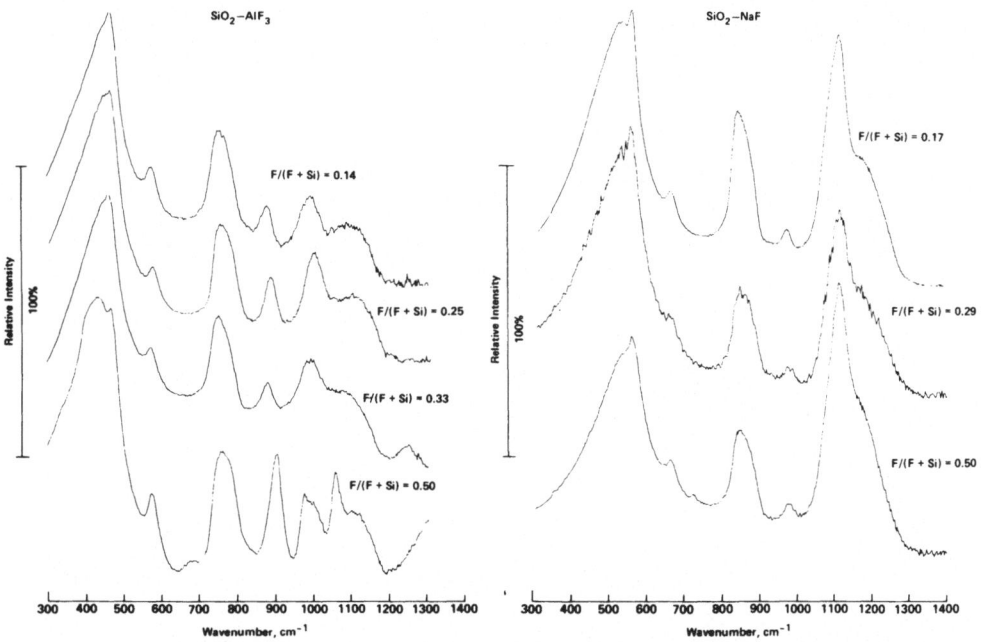

Fig. 3. Temperature- and frequency-corrected Raman spectra of quenched melts on the joins SiO_2-AlF_3 and SiO_2-NaF.

(Figs. 4 and 5). In the SiO_2-NaF quenched melts, the intensity of the 935-cm^{-1} band is insensitive to fluorine content and is significantly less than the intensity of the analogous band in the SiO_2-AlF_3 melts (about 30% of the intensity). The 1100-cm^{-1} intensity increases, and at the highest AlF_3 content, an additional polarized band near 960 cm^{-1} is evident. The latter two bands stem from antisymmetric Si-O$^-$ stretch vibration (where O$^-$ represents a nonbridging oxygen) in $Si_2O_5^{2-}$ and SiO_3^{2-} units in the fluorine-rich silicate melt (Furukawa et al., 1981; Mysen et al., 1982b). The addition of fluorine results, therefore, in significant depolymerization of SiO_2 melt.

The relative intensities derived from the deconvoluted spectra are a measure of relative abundance of fluorine- and oxygen-bearing structural units in the melt. The only signature from the fluorine-bearing samples is the Si-F stretch vibration appearing near 935 cm^{-1}. With up to at least F/(F + Si) = 0.33, the intensity of the 935-cm^{-1} band in the system SiO_2-Al_3 is a linear function of the total fluorine content of the sample (Fig. 5). This line can be extrapolated through the origin. Fluorine in SiO_2-AlF_3 glasses appears, therefore, dissolved in only SiO_3F^{3-} tetrahedral units (neutral formulation: $Si_4O_7F_2$). With F/(F + Si) > 0.33, the Raman spectra show weak shoulders near 360 and 400 cm^{-1}, shoulders that may

47

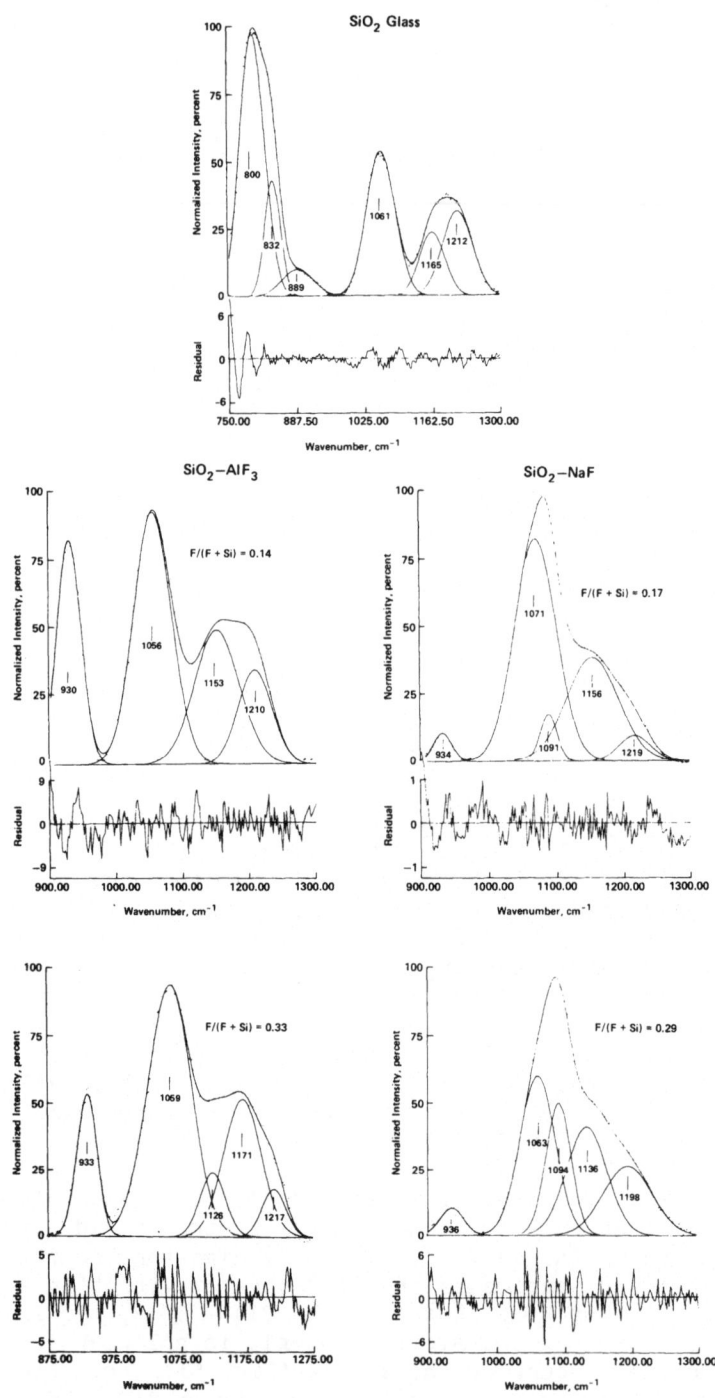

Fig. 4. Examples of deconvoluted spectra of fluorine-bearing SiO_2 in the frequency range 900-1300 cm^{-1}.

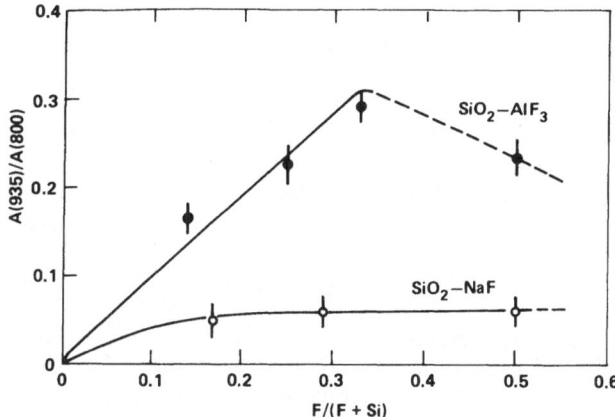

Fig. 5. Intensities of Si-F stretch band at 935 cm^{-1} relative to Si-O° symmetric stretch bands near 800 cm-1.

indicate the existence of Al-F bonds in the melts (Gilbert et al., 1975). Within the experimental sensitivity of the method there is no spectroscopic evidence of AlF_4^- in any of the samples.

The intensity correlation in Figure 5 can be used as a calibration curve for Raman intensity vs. $Si_4O_7F_2$ content of the glasses (Fig. 5). It is seen (Fig. 5) that the proportion of $Si_4O_7F_2$ in the SiO_2-NaF glasses is significantly less than the total amount of fluorine present. The remaining fluorine most likely is present as NaF complexes. One may then write equations that express the solubility mechanisms of NaF and AlF_3, respectively, in SiO_2 melt:

$$6SiO_2 + 2NaF = Si_4O_7F_2 + Na_2Si_2O_5, \tag{2}$$

and

$$18SiO_2 + 2AlF_3 = 3Si_4O_7F_2 + Al_2(Si_2O_5)_3, \tag{3}$$

with the equilibrium constants

$$K_1 = [Si_4O_7F_2][Na_2Si_2O_5]/[SiO_2]^6[NaF] \tag{4}$$

and

$$K_2 = [Si_4O_7F_2]^3[Al_2(Si_2O_5)_3]/[SiO_2]^{18}[AlF_3]^2. \tag{5}$$

With the possible exception of the SiO_2-AlF_3 sample with F/(F + Si) = 0.5, only $Si_4O_7F_2$ units occur in the SiO_2-AlF_3 quenched melts. Equilibrium (3) is, therefore, dramatically shifted to the right, and the value of K_2 cannot be obtained with any accuracy from the Raman data. The value of K_1 can be evaluated under the assumption that the mole fractions of the structural units in the melts can be used to approximate the activities. The NaF concentration is obtained by difference between the amount of fluorine in $Si_4O_7F_2$ and the total amount of fluorine added. The abundance of the $Na_2Si_2O_5$ complex can be obtained by using the Raman cross sections for antisymmetric Si-O$^-$ stretching in the $Si_2O_5^{2-}$ unit (Mysen et al., 1982b). The

49

proportions of structural units in the SiO_2-AlF_3 melts were obtained in a similar manner. These proportions are shown in Figure 6.

The equilibrium constant for the SiO_2-NaF system [eq. (4)] is 67 ± 8, which yields a standard-state free energy for the reaction of −15 kcal/mole. For the SiO_2-AlF_3 system the reaction has proceeded almost completely to the right, reflecting, therefore, the much greater stability of NaF compared with AlF_3 complexes in the melts.

Water-Bearing SiO_2 Melts

Temperature- and frequency-corrected Raman spectra of H_2O- and D_2O-bearing quenched SiO_2 melt (quenched from 1550°C at 15 kbar) with up to 15 wt% H_2O or D_2O (Figs. 7 and 8) show several features resulting from dissolved water. In the frequency range up to about 1300 cm^{-1}, there are two bands between 930 and 970 cm^{-1} that derive from Si-OH stretching. In all H_2O- and D_2O-bearing samples, only one band near 970 cm^{-1}, rather than two near 930 and 970 cm^{-1}, respectively, results in χ^2 values that are 600 to 1000% higher. The distribution of the residuals also becomes distinctly periodic. The two bands near 930 and 970 cm^{-1} are, therefore, necessary on statistical grounds. A simple analysis of the spectra of H_2O- and

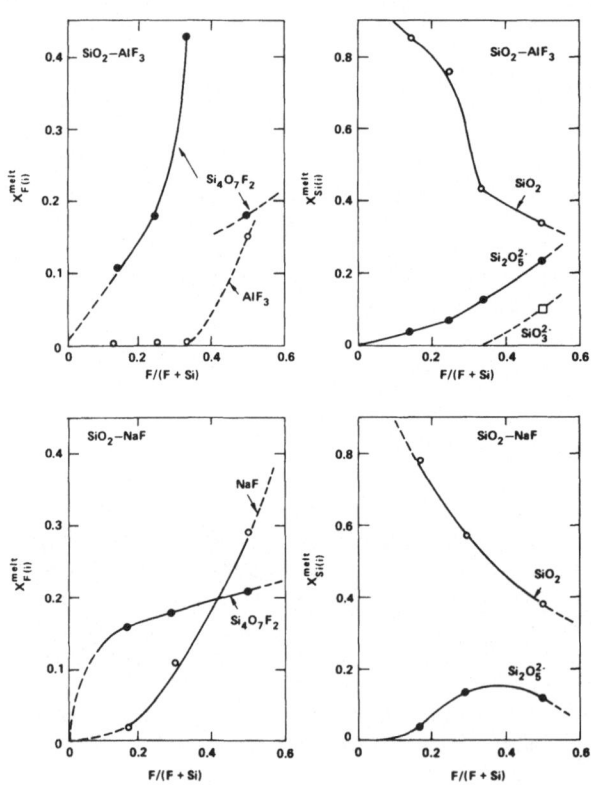

Fig. 6. Relative proportions (mole fractions) of structural units in quenched melts in the systems SiO_2-NaF as a function of F/(F + Si) of the systems.

50

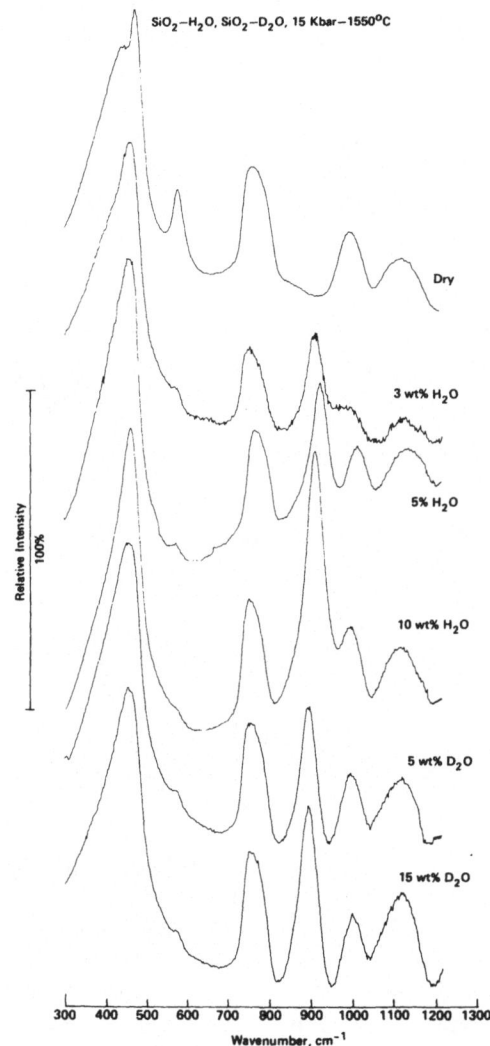

SiO$_2$-H$_2$O, SiO$_2$-D$_2$O, 15 Kbar-1550°C

Dry

3 wt% H$_2$O

5% H$_2$O

10 wt% H$_2$O

5 wt% D$_2$O

15 wt% D$_2$O

Relative Intensity

100%

300 400 500 600 700 800 900 1000 1100 1200 1300

Wavenumber, cm^{-1}

Fig. 7. Temp-
erature - and
frequency-
corrected spectra
of quenched melts
on the joins SiO$_2$-
H$_2$O and SiO$_2$-
D$_2$O as a function
of wt% H$_2$O (and
D$_2$O) added.

D$_2$O-bearing quenched SiO$_2$ melts shows that both bands reflect the
existence of two structurally distinct Si-OH bonds. These vibrations
are principally those of Si vibrating against OH, and substitution of
OH with an OD group results in a reduction of frequency (Freund,
1982):

$$\nu(SiOH)/\nu(SiOD) = \sqrt{\mu(SiOH)/\mu(SiOD)} = 1.02, \qquad (6)$$

where μ is the reduced mass of the oscillator. Both the 970- and the
930-cm^{-1} band in the spectra from the sample with 10 wt% H$_2$O have
shifted to 955 and 916 cm^{-1} by substitution of D for H (Fig. 8). The
frequencies predicted from the simple expression in eq. (6) are 951
and 912 cm^{-1}, respectively. These frequencies are identical within
the positional uncertainties of the fitted bands (±5 cm^{-1}). The
proportion of the Si-OH bonds resulting in the lower-frequency band

51

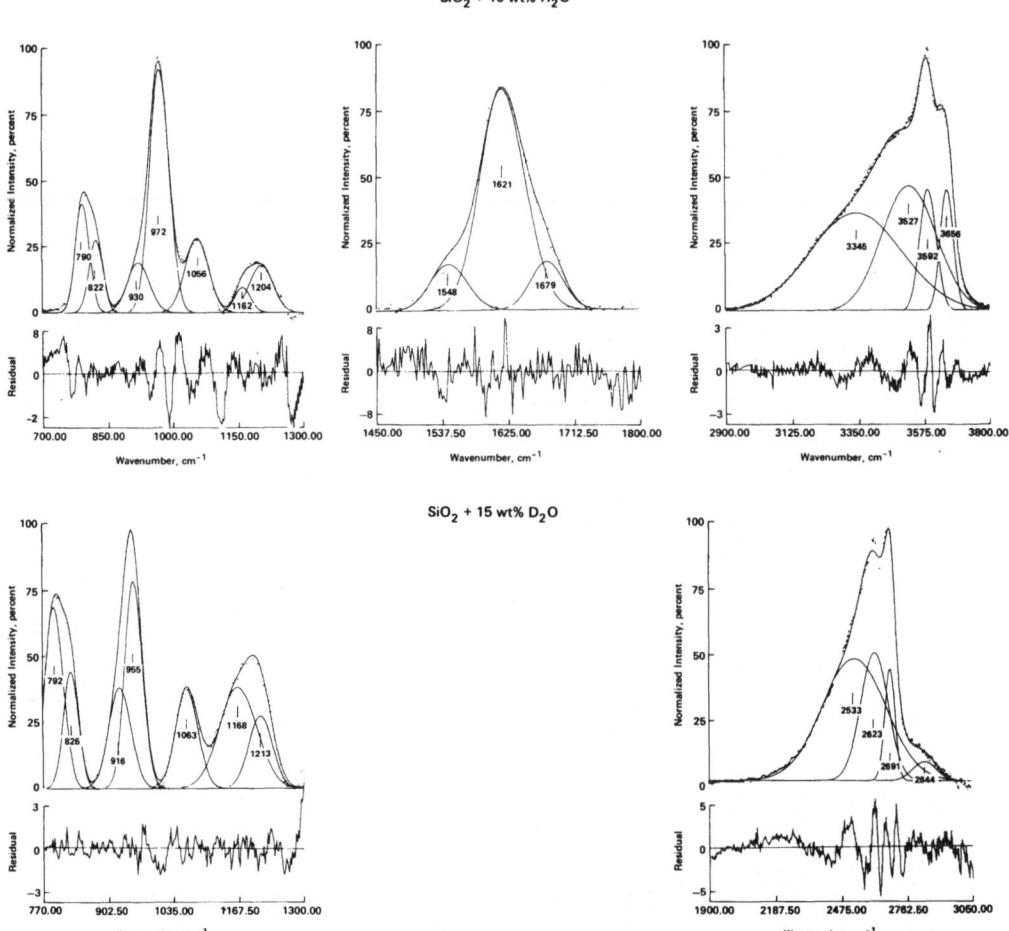

SiO₂ + 10 wt% H₂O

SiO₂ + 15 wt% D₂O

Fig. 8. Examples of deconvoluted spectra in the 800-1300, 1400-1800, and 3000-4000 cm⁻¹ regions of SiO₂-H₂O quenched melts and in the corresponding region between 2000 and 3000 cm⁻¹ for deuterated silica. Note the absence of DOD bending bands in SiO₂-D₂O.

increases with increasing total water content (Fig. 9). The total amount of hydroxylated silicon in water-bearing SiO_2 also increases systematically with increasing water content (Fig. 9).

The band near 1600 cm⁻¹ (HOH bending) (Fig. 8) shows that molecular water is present in all samples studied. Its intensity relative to the 800-cm⁻¹ bands does increase slightly with increasing total water content (Fig. 9). One may, therefore, write a simple expression that describes the solution of H_2O in SiO_2 melt:

$$2Si_4O_8 + 2H_2O = Si_4O_7(OH)_2[1] + Si_4O_7(OH)_2[2], \qquad (7)$$

where $Si_4O_7(OH)_2[1]$ and $Si_4O_7(OH)_2[2]$ represent the two hydroxylated silicate units in the melt. This expression differs from the common expression of this equilibrium only in that the two OH-bearing units are described as separate entities. The lower frequency of the band from the Si-OH[1] stretch vibration may be the result of a longer Si-OH bond length in this unit compared with that of the Si-OH[2] bond. The equilibrium constant

$$K_3 = [Si_4O_7(OH)_2[1]][Si_4O_7(OH)_2[2]]/[Si_4O_8]^2[H_2O]^2, \qquad (8)$$

may be evaluated from the Raman spectra. The proportion of molecular H_2O relative to water dissolved as OH can be estimated by comparing the relative intensities of the $920+970$-cm^{-1} and 1600-cm^{-1} bands as a function of total water content with the method suggested by Seifert et al. (1981) to obtain relevant Raman cross sections. It can be seen from the results (Table 1), that the proportion $OH/(OH + H_2O)$ ranges between 0.25 and 0.20, where, however, the absolute amount of H_2O is significantly less sensitive to total water content in the sample than that of OH (calculated as H_2O). By using the mole fractions obtained (Table 1), the value of K_3 (and ΔG) appears to increase rapidly with increasing water content. This effect either results from real changes in the equilibria or stems from the fact that the activity coefficient ratio, $\gamma Si_4O_7(OH)_2[1]\gamma Si_4O_7(OH)_2[2]/(\gamma Si_4O_8)^2(\gamma H_2O)^2$, is a function of the water content of the sample.

Table 1. Structural Information on Quenched SiO_2-H_2O Melts

Mole % H_2O[a]	OH[1][b]	OH[2][b]	H_2O[b]	SiO_2[b]	K_3[c]	ΔG[d]
29.2	0.022	0.192	0.078	0.708	1.4	-1.2
41.2	0.052	0.256	0.104	0.588	3.6	-3.6
59.7	0.093	0.387	0.117	0.403	16.2	-10.1

[a] Amount of water added.
[b] Mole fraction of the two hydroxylated silicate units, molecular water, and SiO_2.
[c] Equilibrium constant [eq. (8)].
[d] Standard-state free energy of reaction (7) (kcal/mole).

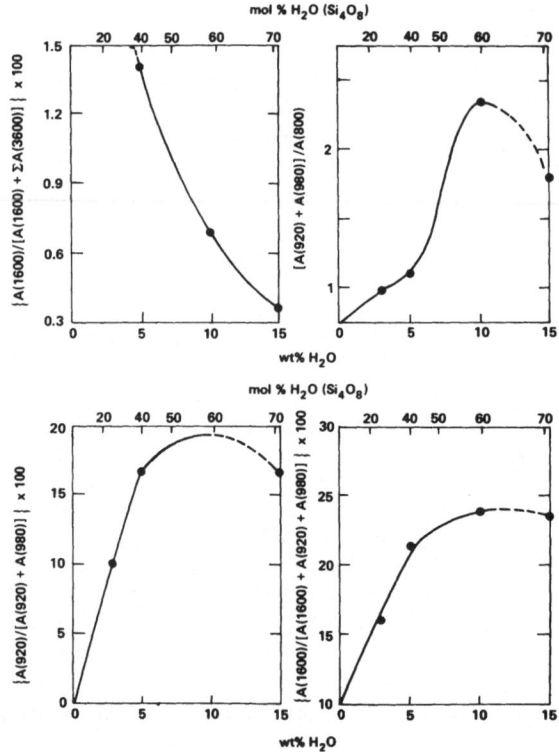

Fig. 9. Relevant intensity ratios of Raman bands from Si-O, Si-OH, HOH, and OH vibrations in the system SiO_2-H_2O as a function of water content.

SUMMARY

Fluorine and water, dissolved in SiO_2 under equilibrium conditions, attack the $SiO_4{}^{4-}$ tetrahedra by replacing bridging oxygen bonds with either F^- or OH. Fluorine replaces one bridging oxygen to produce an Si-F and a nonbridging oxygen bond, with additional fluorine dissolved as sodium and aluminum fluoride complexes. The sodium fluoride complex is significantly more stable than the aluminum fluoride complex. Water forms two OH bonds per broken oxygen bridge. Additional water dissolves in a molecular form.

REFERENCES

Boyd, F. R., and England, J. L., 1960, Apparatus for phase equilibrium measurements at pressures to 50 kilobars and temperatures up to 1750°C, J. Geophys. Res., 65:741.

Burnham, C. W., 1979, The importance of volatile constituents, in: "The Evolution of the Igneous Rocks--Fiftieth Anniversary Perspectives," H. S. Yoder, Jr. ed., Princeton Univ. Press, Princeton.

Dumas, P., Corset, J., Cavalho, W., Levy, Y., and Neuman, Y., 1982, Fluorine-doped vitreous silica analysis of fiber optics performed by vibrational spectroscopy, J. Non-Cryst. Solids, 47:239.

Freund, F., 1982, Solubility mechanisms of H_2O in silicate melts at high pressures and temperatures: a Raman spectroscopic study: discussion, Am. Mineral., 67:153.

Furukawa, F., Fox, K. E., and White, W. B., 1981, Raman spectroscopic investigation of the structure of silicate glasses. III. Raman intensities and structural units in sodium silicate glasses, J. Chem. Phys., 75:3226.

Gilbert, B. G., Mamantov, G., and Begun, A., 1975, Raman spectra of aluminum fluoride containing melts and the ionic equilibrium in molten cryolite type mixtures, J. Chem. Phys., 62:950.

Hirayama, C., and Camp, F. E., The effect of fluorine and chlorine substitution and fining of soda-lime and potassium-barium silicate glass, Glass Technol., 10:123.

Kogarko, L. N., and Kriegman, L. D., 1973, Structural position of fluorine in silicate melts (according to melting curves), Geochem. Int., 9:34.

Long, D. A., 1977, "Raman Spectroscopy", McGraw-Hill, New York.

Mysen, B. O., Finger, L. W., Seifert, F. A., and Virgo, D., 1982a, Curve-fitting of Raman spectra of amorphous materials, Am. Mineral., 67:686.

Mysen, B. O., Virgo, D., and Seifert, F. A., 1982b, The structure of silicate melts: implications for chemical and physical properties of natural magma, Rev. Geophys., 20:353.

Powell, M. J. D., 1964a, An efficient method for finding a minimum as a function of several variables without calculating derivatives, Computer J., 7:155.

Powell, M. J. D., 1964b, A method for minimizing a sum of non-linear functions without calculating derivatives, Computer J., 7:303.

Powell, M. J. D., and Fletcher, R., 1963, A rapidly converging descent method for minimization, Computer J., 6:163.

Seifert, F. A., Mysen, B. O., and Virgo, D., 1981, Quantitative determination of proportions of anionic units in silicate melts, Carnegie Inst. Washington Year Book, 80:301.

Seifert, F. A., Mysen, B. O., and Virgo, D., 1982, Three-dimensional network structure in the systems $SiO_2-NaAlO_2$, $SiO_2-CaAl_2O_4$ and $SiO_2-MgAl_2O_4$, Am. Mineral., 67:696.

Takusagawa, N., 1980, Infrared absorption spectra and structure of fluorine-containing alkali silicate glasses, J. Non-Cryst. Solids, 42:35.

SIMULTANEOUS CRYSTALLITE SIZE, STRAIN AND STRUCTURE

ANALYSIS FROM X-RAY POWDER DIFFRACTION PATTERNS

Scott A. Howard[*] and Robert L. Snyder

New York State College of Ceramics

Alfred University Alfred, NY 14802

INTRODUCTION

Methods for determining crystallite size and strain and crystal structure parameters, from X-ray powder diffraction patterns have occupied the attention of crystallographers for many years. The inherent asymmetry of X-ray powder diffraction profiles has been one of the principal difficulties limiting this work. The resurgence of developments in this area has primarily resulted from the computer automation of diffractometers. The ability to collect digitized representations of the line profiles and apply numerical methods to their analysis has led to a number of new and exciting developments.

The most exciting of these has been the development of a whole pattern refinement method to obtain crystal structure information from neutron diffraction data by Rietveld (1969). The mostly symmetric Gaussian nature of neutron diffraction lines has allowed this method to develop to its current rather mature state.

The application of the neutron Rietveld method to X-ray patterns has however, never been entirely satisfactory because of the asymmetric nature of X-ray diffraction profiles, which result from the convolution of eight separate functions. In many refinements, the least squares residual is determined by the profile mismatch between the model and observed data rather than being determined by the fit of the structural parameters. The various successes reported to date have often been the result of defect disorder which tends to obliterate the inherent peak asymmetry and produce a more Gaussian shape.

_____

[*]Present Address: The Department of Ceramics University of Missouri-Rolla, Rolla, Mo 65401

57

These have not allowed for the Profile Shape Function (PSF) contributions from either the instrument or the sample.

The observed diffraction profile, as stated by Jones (1938) and applied to powder diffraction systems by Taupin (1973) and Parrish (1976), is a result of the convolution of a specimen profile (S) and a function modeling the aberrations introduced by the diffractometer. Taupin and Parrish have identified the instrumental function itself to be a convolution of a function representing the diffractometer's optics (G) and a function representing the wavelength distribution of the X-rays (W). The line profile (F) can be expressed as:

$$F = (W*G)*S + background \qquad (1)$$

where the (*) represents the convolution operation. Since both W and G are fixed for a particular instrument/target system, (W*G) may be regarded as a single entity.

The specimen function S for a sample with no defect broadening is a delta function (i.e., a line with height but no width). Using a delta function for S in (1) yields:

$$F = (W*G) + background \qquad (2)$$

Hence, for a pattern of an ideal sample, with background appropriately accounted for, the profiles of F are identical to the profiles of (W*G).

The purposes of the current study are four fold. 1. To establish a method based on our previous profile fitting experience, for establishing (W*G). 2. To develop deconvolution procedures for isolating the S function. 3. To analyze the S function for the size and strain origin of its broadening and 4. To incorporate these exact profile modeling techniques into the Rietveld crystal structure analysis method.

Description of Background

The description of the background in a powder diffraction pattern is critical to profile refinement because any background function must correlate strongly with the PSF. Two methods commonly used to describe background involve selecting points between peaks and interpolating between them (Rietveld, 1969), or refining the coefficients of a polynomial along with the the profiles' parameters (Wiles and Young, 1981). None of the methods proposed to date (Snyder, 1983) will routinely produce an accurate description of X-ray background due to the fact that the true background, in low symmetry materials where the powder pattern presents a continuum of peaks at high diffraction angle, is never sampled.

In this study we have incorporated an empirical approach to background determination analogous to the empirical (W*G) approach to profile fitting. Background was carefully measured from a quartz crystal cut along an extinct reflection. Thus neither amorphous nor crystalline scattering contributions are present. This background curve is a measure of the (W*G) background, in that there are no sample contributions present. Although there is no theoretical reason for sample contributions to background to cause a uniform vertical displacement of this function, we find that this assumption adequately describes the background in the diffraction patterns used in this study.

Visual examination of the quartz pattern at lowest values of 2θ suggested a functional form of the type:

$$I(i) = A \quad 2\theta(i) + B \quad (2\theta(i) - C)^D + O \qquad (3)$$

Where A, B, C, and D are constants obtained from a least squares fit to the quartz pattern. The offset O is the only background variable used during profile refinement. This approach greatly minimizes the correlations between profile and background parameters during profile refinement.

The Modeling of Diffraction Profiles

Our previous work (Howard and Snyder, 1983, 1985b,c) evaluated seven profile models and three regression techniques and concluded that the split-Pearson VII function (Brown and Edmonds, 1980) combined with a Gauss-Newton or a Marquardt least-squares optimization algorithm gave excellent fits to diffraction lines obtained under a wide variety of conditions. The Pearson VII function has the form:

$$I = I_0 / (1 + kx^2)^m \qquad (4)$$

where $k = (2^{1/m}-1) / (FWHM/2)^2$ and m is the shape factor whose value determines the rate at which the tails fall. FWHM (or H_k) is the full width at half maximum. The split-Pearson VII, as illustrated in Figure 1, was shown to consistently refine against observed lines with the lowest residual error.

The spectrum from a Cu X-ray tube target has the $K\alpha_1, \alpha_2$ and α_3 wavelengths as the dominant components. We may take advantage of knowing their respective wavelengths and relative intensities by calculating the positions and intensities for the diffracted lines from the α_2 and α_3 based on the parameters of the α_1. If the PSF has these three components "built-in", then its shape will allow it to fit only those areas of the pattern that exhibit the intrinsic α_1, α_2, α_3 triplet. In addition, constraining the α_2 and α_3 components greatly reduces the number of variables in the refinement thereby lowering the number of false minima likely to slow down or entrap the refinement algorithm.

The fixing of the α_2 and α_3 positions and intensities in the split-Pearson VII function reduces the number of parameters in the model from 18 to 14 as illustrated in Figure 2. However, if the shape factors, m, and the H_k's of the α_2 and α_3 are also fixed to be the same as the α_1, then the number of variable parameters can be reduced to six.

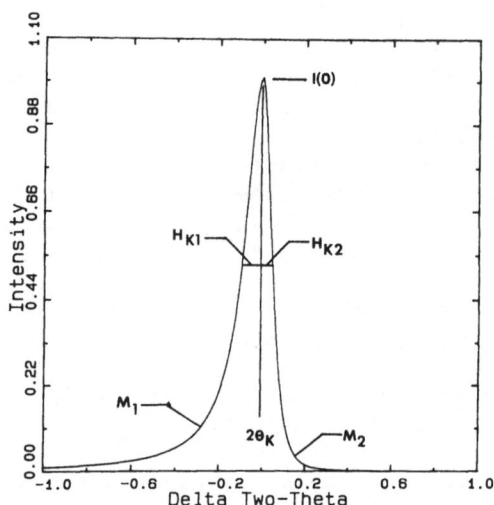

Figure 1. The Split-Pearson VII function uses two halves of the Pearson VII function with a common peak angle and intensity. The Full width at half maximum H_k and the exponents m for each side are varied independently.

Establishing the (W*G) Calibration Curves

In order to establish the W*G function for later evaluation, the profiles of a standard which shows no sample broadening must be fit with the split-Pearson VII function. Ideally one would like a uniform particle size of about 5 μm, for an unstrained, high-symmetry material to use as a standard. Although the U. S. National

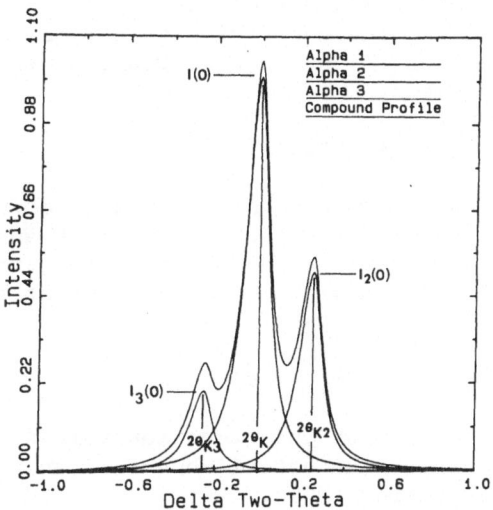

Figure 2. The split-Pearson VII profile shape function with the α_2 and α_3 components "built-in." For a particular set of $2\theta_k$ and I_0 values, $2\theta_{k2}$, $2\theta_{k3}$, I_{02} and I_{03} can be calculated. This leaves six values for m and six for H_k or 14 total to be determined.

Bureau of Standards SRM640A Si sample has a small amount of crystallite size broadening, it was chosen as the standard for want of a
better material.

The 11 lines of this reference material lie in the region of
27° to 140° 2θ for Cu radiation. This region was step scanned at
0.01° 2θ increments with a 10 second count time at each step. A
Siemens D500 Automated diffractometer operating at 40kV, 30 mA was
used. This unit was configured with 1° divergence and anti-scatter
slits and a diffracted beam graphite monochromator with a 0.05°
focusing slit.

The 11 profiles were refined using the constrained split-
Pearson function described above. Each profile was refined
separately in a region with enough points to allow an accurate
description of the profile. In each case the background was refined
along with the profile parameters. The program SHADOW (Howard and
Snyder 1985a) was developed in FORTRAN 77 to carry out these
refinements. SHADOW is a very general profile or pattern fitting
program which permits the use of a wide variety of PSFs and either a
Gauss-Newton or Marquardt optimization algorithm.

The FWHM values obtained from the refinement were used to
determine the value of the coefficients in a polynomial expression.
The expression derived for neutron diffraction (Caglioti, Paoletti,
Ricci, 1958) was chosen:

$$(H_k)_{2\theta} = a\ \tan^2\theta_k + b\ \tan\theta_k + c \tag{5}$$

Two equations of this type are established by doing a least-squares
regression for both the low angle and the high angle sides of the
split-Pearson VII functions used to fit to the standard's profiles.
A similar type of polynomial was used for the shape factors. The
quadratic expression used was of the form:

$$m = a'(2\theta_k)^2 + b'(2\theta_k) + c' \tag{6}$$

Least squares regression of this function versus the two sets of
shape factors completes the establishment of an analytical
expression for evaluating (W*G).

Modeling the Specimen Broadening Contribution of an X-ray Profile

It is generally accepted that (S) may be represented by a
Lorentz (or Cauchy) function when profile broadening is caused by
small crystallite size. When strain is responsible, (S) is
typically assumed to be represented by a Gaussian function. To test
these two convolutes, the program SHADOW was modified to generate
the required profiles. Diffraction patterns were obtained from
three materials. Two commercial aluminas were used: Linde A and C

62

with a nominal particle size of 0.3 μm and 1.0 μm respectively. A sample of tungsten with a nearly mono-particle size distribution of approximately 5 μm was also examined. The tungsten sample exhibited a high degree of line broadening. The analysis progressed as follows:

Generation of the (W*G) Function. Since the (W*G) model includes the α_2 and α_3 lines, only estimates for the α_1 line positions need be obtained from the pattern. These positions are used to generate profiles from the (W*G) function, which is expressed as four polynomials in 2θ. One for each of the split-Pearson variables. (W*G) for each line is generated by using one split-Pearson VII function for each wavelength component. To limit the range over which the profiles are generated, I_0 is set to 100.0 for all peaks and the tails were generated until they fell below 0.001 counts per second.

Generation of the Specimen Profile. Lorentzian and Gaussian profile functions were generated from their equations, normalized to obtain unit integrated area.

Numerical Convolution of (W*G) and (S). In the absence of an analytical convolution function for the profile models, numerical techniques were employed. The (W*G) profiles were generated at discreet values, i.e., at the same values of 2 that the observed pattern was measured. Since S is symmetric, j values are generated from -n to n, where n is the number of points in the tails. (S) is also generated at intervals corresponding to that of the pattern.

The convoluted profile is obtained by:

$$(W*G)*S(i) = \sum_{j=-n}^{+n} (W*G)(i-j)\ S(j) \tag{7}$$

The convolute gathers intensity contributions from all points in both (W*G) and S. The only approximation in the numerical convolution comes from the fact that the n limits are finite. The broader the FWHM of (S), the greater the smearing of the profiles and loss of apparent resolution.

Profile analysis. Patterns of the tungsten and alumina samples were collected on an instrument previously calibrated for (W*G). Each pattern was step-scanned from 20-140° 2θ with an angular increment of 0.0° and count time of 20 seconds per step.

The lines in each pattern were fit using the two models for S. $2\theta_k$ and I_0 were refined for the (W*G) component, and the FWHM for (S). The least-squares error criterion was:

$$Rwp = \frac{\Sigma \; w_i[I_{i(obs)} - I_{i(calc)}]^2}{\Sigma \; w_i[I_{i(obs)}]^2} \qquad (8)$$

Summation is over all points in the segment being refined, and the weights are the reciprocal variances of the observations.

Broadening as a Function of Angle: Particle Size and Strain Contributions. Figures 3 and 4 show the results of modeling S with a Cauchy and Gaussian profile. It is clear that for our samples the Cauchy correctly models the broadening, independent of its origin.

Since all non-specimen related broadening terms, like the tan spectral broadening, have been accounted for in (W*G), pure crystallite size and strain effects were simply modeled by the Scherrer equation:

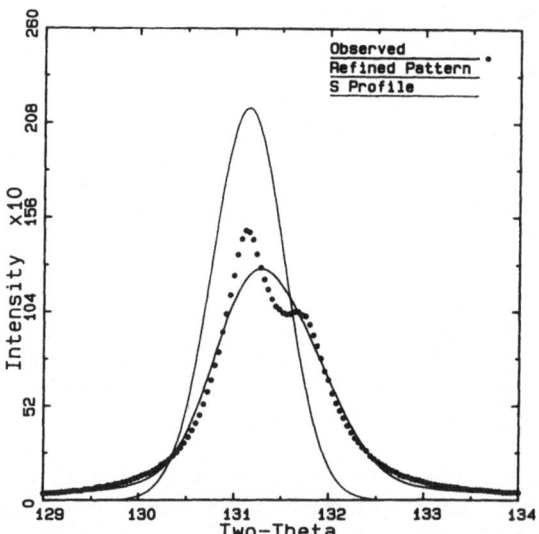

Figure 3.   Shown above is the fit of (W*G)*S to a line from the tungsten specimen. A Gaussian specimen profile was convolved with the instrument profile during refinement. The poor fit as a result of using a Gaussian profile for S is obvious.

64

$$B_{\tau} = \frac{\lambda}{\tau \cos \theta_k} \tag{9}$$

and the strain broadening was modeled by:

$$B_\varepsilon = 4 \varepsilon \tan \theta_k \tag{10}$$

Where: B_τ = the integral breadth of the (S) profile (radians),
τ = crystallite thickness,
B_ε = the integral breadth of the (S) profile (degrees), and
ε = the X-ray crystallite strain.

The slope of the curves B vs. $\lambda/\cos \theta_k$ and B vs. $\tan \theta_k$ give $1/\tau$ and ε respectively. A non-linear least-squares was used to determine values for these parameters and to generate the curves in figures 5 and 6. The results of the regression analysis indicate

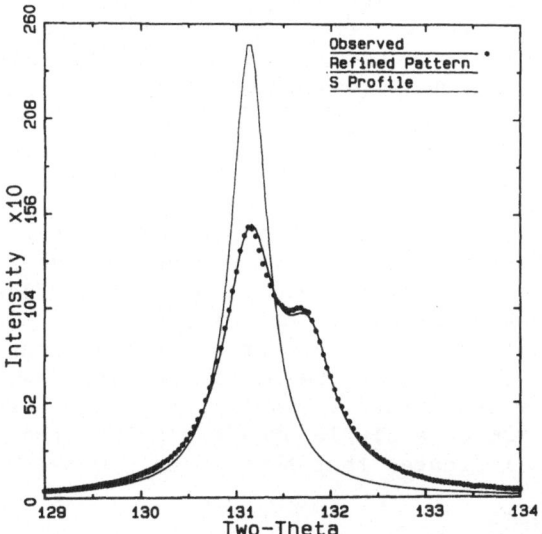

Figure 4. The quality of fit obtained using a Lorentzian function for S was far superior than using a Gaussian function for this tungsten specimen. The cause of this broadening was determined to be solely due to strain.

65

that both alumina samples exhibited pure particle size broadening while the tungsten exhibited pure strain broadening.

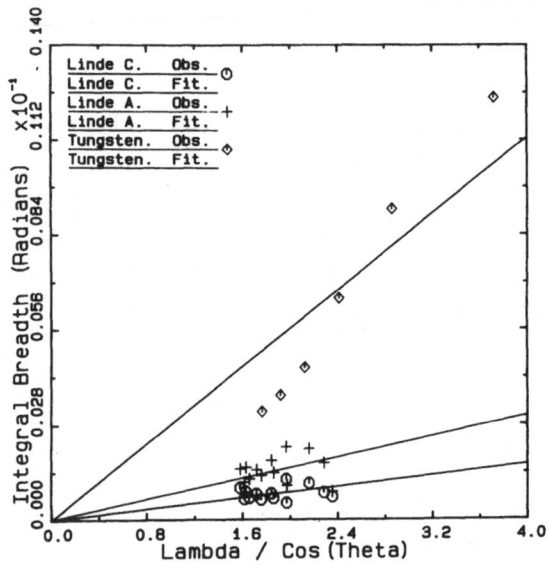

Figure 5. Characterization of the specimen broadening based on crystallite size. Linde aluminas A and C follow the trend expected from Equation 9 for particle size determined broadening.

Program SHADOW uses these simple models to allow for the simultaneous determination and refinement of τ and/or ε. The addition of these two parameters is more than offset by eliminating the FWHM parameter for each line being refined. SHADOW was then used to refine those segments over which the B data was taken. The refinement of τ and ε from the non-linear least squares and SHADOW refinements are summarized in Table 1. 'Individual' refinement refers to the use of a single broadening function, while 'simultaneous' indicates that both τ and ε were refined.

Rietveld Analysis

With the (W*G) and S functions properly modeled for X-ray diffraction profiles, this "Parrish Method" can now be applied to the Pattern-Fitting-Structure-Refinement procedure of Rietveld. The formalism developed above allows for the discarding of a number of

empirical parameters in the Rietveld method. These include the
asymmetry parameter (P) and the profile half-width parameters U,V,W.

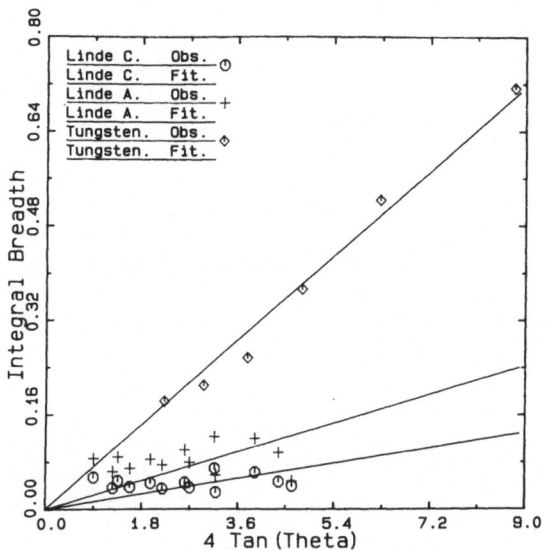

Figure 6.   Characterization of the specimen profile broadening based
on crystallite strain. Tungsten follows the strain relation of
Equation 10 while the alumina samples do not.

The replacement of the P,U,V and W parameters by τ and ε
removes all empirical parameters from the refinement, leaving in
their place parameters associated with known physical processes.
All empirical parameters associated with the peak asymmetry are
incorporated into the (W*G) function and are fixed during
refinement.

When both strain and crystallite size effects are present, the
FWHM of the (S) profile was assumed to be a linear addition of the
two components. That is, $B = B_\tau + B_\varepsilon$. The basis for this assumption is that the convolution of two Lorentzian functions yields
another Lorentzian function. Thus,

$$L(strain) * L(size) * (W*G) = L(combined) * (W*G) \qquad (11)$$

The value for the FWHM of the new convolute profile is simply a sum
of the component FWHMs. We also assume that the order of convolution does not matter.

TABLE 1. Results from Particle Size and Strain Evaluation

Parameters obtained from least-squares

Specimen	ε	R%	τ (nm)	R%	ε	τ (nm)	R%
Tungsten	0.0794	5.9	35.5	20.5	0.079	----	5.9
Linde A	0.0270	44.2	130.9	26.9	---	129.5	26.9
Linde C	0.0144	43.4	244.7	25.3	---	243.4	25.3
		individual				simultaneous	

Parameters obtained from program SHADOW

Specimen	ε	Rwp%	τ (nm)	Rwp%	ε	τ (nm)	Rwp%
Tungsten	0.0763	7.8	41.0	11.4	0.0763	---	7.8
Linde A	0.0353	17.7	116.5	17.7	---	117.3	18.8
Linde C	0.0169	17.6	248.9	17.6	---	219.9	17.8
		individual				simultaneous	

Modification to the program of Wiles and Young

The X-ray version of the Rietveld procedure as developed by
Young, Mackie and von Dreele (1977) and Wiles and Young (1981) was
modified to incorporate the (W*G) deconvolution and modeling of S
described above. The existing program architecture was preserved so
that the deconvolution procedures can be "switched" on and off. The
following four options are allowed when deconvolution is "on":

0) No broadening present, i.e., use the (W*G) profiles alone,
1) the broadening of (S) is considered being due to particle size,
2) (S) broadening is due to strain effects, and
3) allow for broadening from both crystallite size and strain.

Turning 'off' the (W*G) deconvolution and allows the program to
function as originally designed.

To generate the derivatives necessary for the optimization
algorithm, some numerical differentiation was required. The
convolution process added to this program was the same as previously
employed by these authors in program SHADOW [Howard and Snyder,
1985a]. Refinements were carried out on each of the specimens by
varying the same structure parameters. However, the first
refinement was performed in the original program context, i.e.,
using P,U,V, and W. The second refinement used the (W*G)
deconvolution process.

DISCUSSION OF RESULTS

Regarding (W*G) Deconvolution in Profile Fitting

For each of the samples, the broadening of the specimen profile B was determined by using both a Gaussian and Lorentzian specimen function convolved with the (W*G) function. To determine the quality of fit obtained from the refinement, the residual errors were compared, and the differences between refined and observed profiles were visually inspected. Figure 7 shows the fitting of selected lines for two of the samples.

In all cases, the use of the Lorentzian distribution function yielded a lower value for the residual errors. Visual comparison showed the Gaussian function to produce distorted representations of the observed lines. For these reasons, further use of the Gaussian function was not considered. Brown and Edmonds (1980) found that for Guinier data, without deconvoluting S, strain broadening was Gaussian and particle size broadening was Lorentzian. In our samples, both broadening terms are Lorentzian in nature.

Incorporation of (W*G) deconvolution into the Rietveld Method

Broadening in these samples could be modeled by the size and strain functions (9) and (10). The alumina samples exhibited almost pure size-determined broadening. Linde A exhibited a higher degree of broadening than that of C. Linde C showed a perceptible degree of broadening over the Si (SRM640a) profiles used to determine (W*G).

The broadening of the tungsten profiles was entirely a result of a strain in the sample. This is supported by SEM micrographs and the automatic elimination of crystallite size dependencies by the refinement algorithms. Most disconcerting was the quality of fit using a Cauchy specimen PSF. Strain broadening is routinely assumed to be Gaussian in nature.

Finally, the introduction of an analytical expression describing (S) broadening simplifies profile refinement. This constrained broadening reduces the number of parameters undergoing refinement while characterizing and quantifying the source of the broadening. However the cost is a considerable increase in the execution time of the program.

Table 2 lists the results from the six refinements performed. As verified by the data in the table, the inclusion of the (W*G) deconvolution has lowered the residual errors after refinement. Crystallite size for the Linde A and C specimens showed a larger degree of broadening in A than in C. A segment of the refined patterns are show in Figure 7. However, attempts to refine both size and strain failed. Since no constraints were available in this version of the Rietveld program, there was a tendency to make the crystallite strain parameters (ε) negative. The accompanied lowering of the residual error was done at the expense of introducing a value for strain that was physically unrealistic.

The fit of the tungsten pattern again indicated that (W*G) deconvolution lowered residual error compared to the other method of refinement. Again, trouble was encountered refining both crystallite size and strain parameters. In this case, unrealistically large values for the crystallite size hindered refinement and was therefore removed from the processing.

TABLE 2. Results from the Rietveld Structure Refinement

Specimen	with (W*G)			without (W*G)		
	ε	τ(nm)	R%	ε	τ(nm)	R%
Tungsten	0.0315	NA	9.2	NA	NA	7.7
Linde A	NA	134.9	15.4	NA	NA	18.7
Linde C	NA	257.8	15.4	NA	NA	19.6

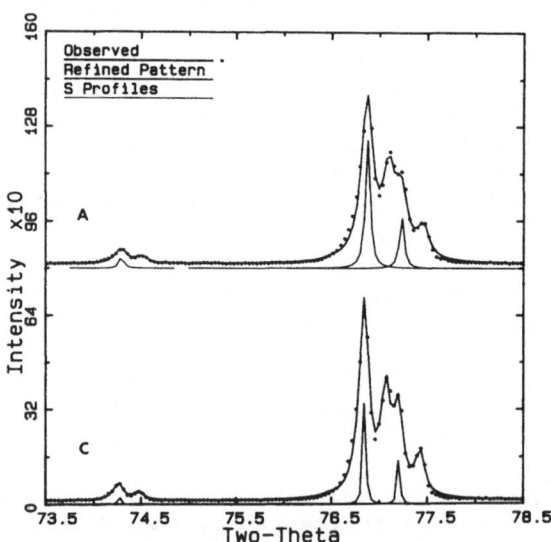

Figure 7. Both Linde a and C aluminas show significant line broadening. The inner curves are the deconvoluted S functions.

The FWHM of these S values show that the X-ray crystallite size of linde A is about half that of Linde C.

Application of (W*G) deconvolution in the Rietveld structure refinement aids in the description of the profile intensity distributions. Replacement of the numerical parameters P,U,V and W with analytical parameters characterizing the specimen crystallite size and strain was effective in describing the broadening of the specimen profiles as a function of angle.

Acknowledgement

The authors gratefully acknowledge the support of the The National Science Foundation (Grant No. DMR-8115242) and of the NYS College of Ceramics.

References

1. Brown, A. and Edmonds, J. W., (1980). Adv. X-ray Anal., 23 361.
2. Caglioti, G., Paoletti, A. and Ricci, F. P., (1958) Nuclear Instruments and Methods, 3, 223-226
3. Howard, S. A. and Snyder, R. L. (1983). Adv. in X-ray Anal., 26 73-81.
4. Howard, S. A. and Snyder, R. L. (1985a). NYS College of Ceramics Technical Publication.
5. Howard, S. A. and Snyder, R. L. (1985b) submitted to J. Appl. Cryst.
6. Howard, S. A. and Snyder, R. L. (1985c) submitted to J. Appl. Cryst.
7. Jones,F. W. (1938). Proc. R. Soc. London Ser. A, 166 16-43.
8. Parrish, William, Huang, T. C. and Ayers, G. L. (1976) Am. Cryst. Assoc. Monograph, 12 55-73.
9. Rietveld, H. M. (1969). J. Appl. Cryst., 2 65-71.
10. Snyder, R. L. (1983) Adv. X-ray Anal., 26 1-11.
11. Taupin, Daniel (1973). J. Appl. Cryst., 6 266-73.
12. Wiles, D. B. and Young, R. A. (1981). J. Appl. Cryst., 14 149-151.
13. Young, R. A. and Wiles, D. B. (1982). J. Appl. Cryst., 15 430-8.

THE INFRARED AND RAMAN SPECTRA OF PHOSPHOROSILICATE PHASES

WITH 1:1 SiO_2/P_2O_5 MOLAR RATIOS

I.N. Chakraborty and R.A. Condrate, Sr.

Institute of Glass Science and Engineering
NYS College of Ceramics, Alfred University
Alfred, New York 14802

INTRODUCTION

The chemical and physical properties of glass and crystalline phases intimately depend upon their network structures. Phosphorosilicate phases possessing 1:1 SiO_2/P_2O_5 molar ratios may contain six-coordinated silicon atoms in their solid state structures. For instance, X-ray diffraction analyses indicate that the crystal structures of the various SiP_2O_7 polymorphs do possess six-coordinated silicon atoms.[1,2] This paper will investigate the vibrational spectra of various phosphorosilicate phases with 1:1 $SiO_2P_2O_5$ molar ratios in order to obtain structural information. First, the infrared and Raman spectra will be investigated for various crystalline SiP_2O_7 polymorphs using normal coordinate analysis. This analysis will lead both to structural parameters such as force constants and to band assignments for phosphorosilicate phases with set crystalline structures possessing SiO_6-units in their network. Then, the vibrational spectra of SiP_2O_7 glass will be analyzed to determine whether such structural arrangements with six-coordinated silicon atoms exist in the glasses. The structural and spectral changes which occur upon addition of Na_2O to this glass will also be discussed. Finally, the effects of devitrification on vibrational spectra due to heat treatment of these glasses will be analyzed.

EXPERIMENTAL PROCEDURE

Four crystalline polymorphic forms (cubic (Pa3), monoclinic ($P2_1/C$), tetragonal and monoclinic ($P2_1/n$)) were prepared by the procedures described by Makart.[2] The investigated glasses were prepared from mixtures of ammonium dihydrogen phosphate, silica and sodium carbonate. The method used in this study has been described elsewhere[3] in detail. These glasses were melted in a covered silica crucible in a globar furnace at 1200-1500°c for 4-6 hours. an excess of P_2O_5 was added to each batch of glass to compensate for the P_2O_5 loss during melting. The glasses were poured onto a cold metal plate and converted later to suitable experimental specimens for spectral analysis by cutting and grinding. All glasses except for those containing 25% and 30% Na_2O concentrations were clear. The latter glasses were translucent due to liquid-liquid immiscibility. No crystalline phases were detected for these glasses by X-ray diffraction. Glasses were also heated in an electric furnace before spectral investigation in order to obtain desired thermal histories.

The infrared absorption spectra (4000-300 cm^{-1}) were measured on a double beam grating spectrophotometer. A purging device was attached to the instrument to maintain a moisture- and CO_2- free atmosphere in the instrument. The instrument was calibrated using a standard polystyrene sheet. The infrared spectra of glasses and crystals were measured using the KBr pellet technique. The comparison of the OH stretching bands suggests that the water contents in the glasses were less than 0.01%.

The Raman spectra were measured using a double grating spectrometer with an argon ion laser. Scattered radiation was collected at 90° to the incident beam. The instrument was calibrated using distilled Indene. The spectra of polycrystalline samples were measured with the samples in capillary tubes. The glass samples were polished so that they had two parallel flat sides and one perpendicular flat side. The direction of the incident laser beam was perpendicular to the two parallel sides. Raman spectra were measured perpendicularly to the third flat side. The glass samples were positioned such that the laser beam passed very near the perpendicular side to avoid internal depolarization effects.

RESULTS AND DISCUSSION

A) Analysis of the Vibrational Spectra of Several Crystalline SiP_2O_7 Polymorphs

Cubic SiP_2O_7 belongs to space group T_h^6 (Pa3) with the cell parameter a = 7.46 Å and four formula units per unit cell. The atomic arrangement for this material is illustrated in Fig. 1. Its Bravais unit cell consists of four SiO_6-octahedral and eight PO_4-tetrahedral units. All Si-O bond distances in the SiO_6- octahedral units are 1.835 Å. The PO_4-tetrahedral units contain two different types of phosphorus oxygen bonds. Each PO_4-tetrahedral unit possesses three short phosphorus-oxygen bonds ($P-O_I$) with a bond length of 1.38 Å and a longer bond ($P-O_{II}$) with a bond length of 1.41 Å. The six oxygen atoms of the octahedral units are connected to

74

different PO_4-units. Also, only three oxygen atoms (O_I-type) of the PO_4-units are shared by different SiO_6-units. The other oxygen atom (O_{II}-type) is shared by another PO_4- unit to form a P–O_{II}–P bridge. Factor group analysis predicts the following selection selection rules for this structure:

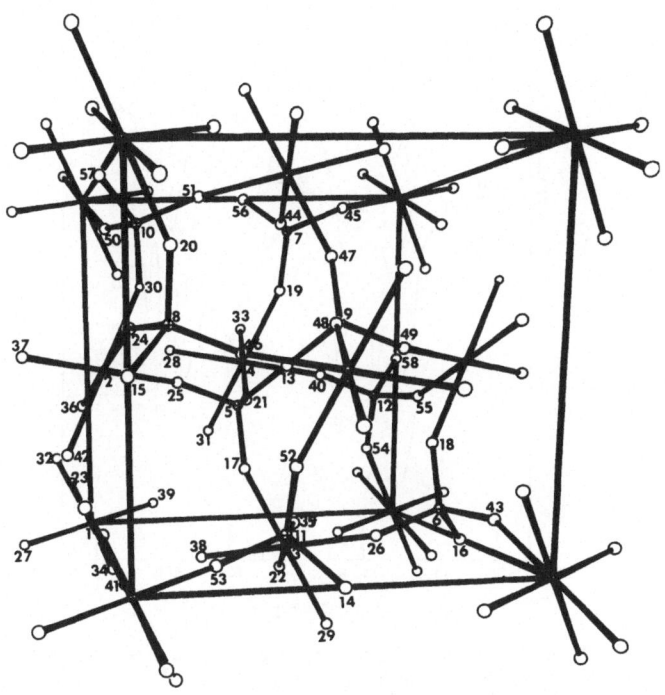

Fig. 1   Atomic Arrangement for Cubic SiP_2O_7

$$4A_g(R) + 6A_u(i.a) + 4E_g(R) + 6E_u(i.a.) + 12F_g(R) + 17F_u(IR).$$
Only the vibrations with F_u symmetry are infrared active, and the vibrations with A_g-, E_g- and F_g-symmetry are Raman active. The selection rules predict 17 infrared-active and 20 Raman-active bands. No coincidences are expected on the basis of the mutual exclusion principle.

The infrared and Raman spectra are illustrated for cubic SiP_2O_7 crystals in Fig. 2 & 3, respectively. The far infrared spectrum is not illustrated. Due to the overlapping of bands and weak intensities, a smaller number of bands are observed in the infrared and Raman spectra than predicted by factor group analysis.

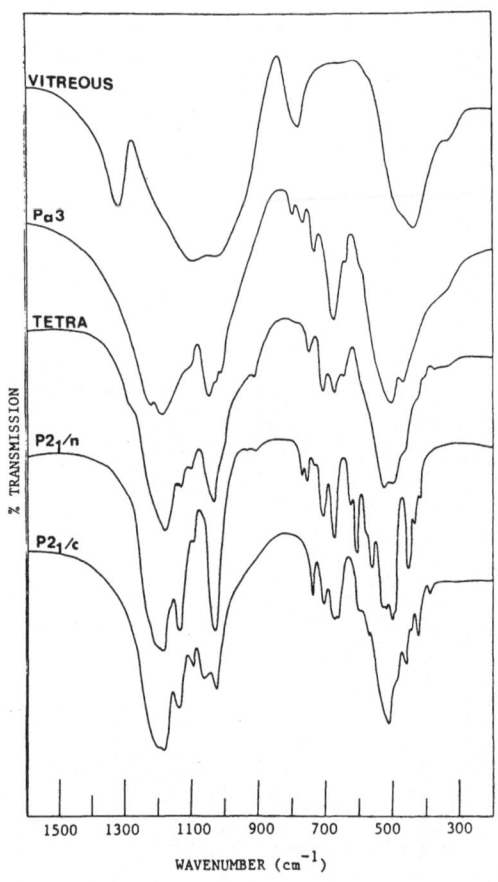

Fig. 2 The Infrared Spectra of SiP_2O_7 Glass and Related
Crystalline Polymorphs.

Vibrational spectra were analyzed by normal coordinate analysis
using programs modified by S. C. Cherukuri.[4] A Born von Korman correction was incorporated into the programs for the k=0 selection
rules for crystals. A modified valence force field has been used
for the calculation of the first order wavenumbers. Three stretching force constants, f_{ra}, f_{rb} and f_{rc}, are associated with Si-O,
P-O_I and P-O_{II} stretching coordinates, respectively. Six bending
force constants, f_{aa}, f_{ab}, f_{ac}, f_{ad}, f_{ae} and f_{af}, are associated
with O-Si-O (107.7°), O-Si-O (72.2°), O_I-P-O_I, O_I-P-O_{II}, Si-O_I-P and
P-O_{II}-P bending coordinates, respectively. In order to obtain a
better fit between the observed and calculated wavenumbers, five
stretch-stretch and stretching-bending interactions were also
incorporated. The force constants, f_{rara}, fr_{aaa}, f_{raab}, f_{rbrb} and
f_{rbac}, are associated with the (Si-O)(Si-O), (Si-O)(O-Si-O large

76

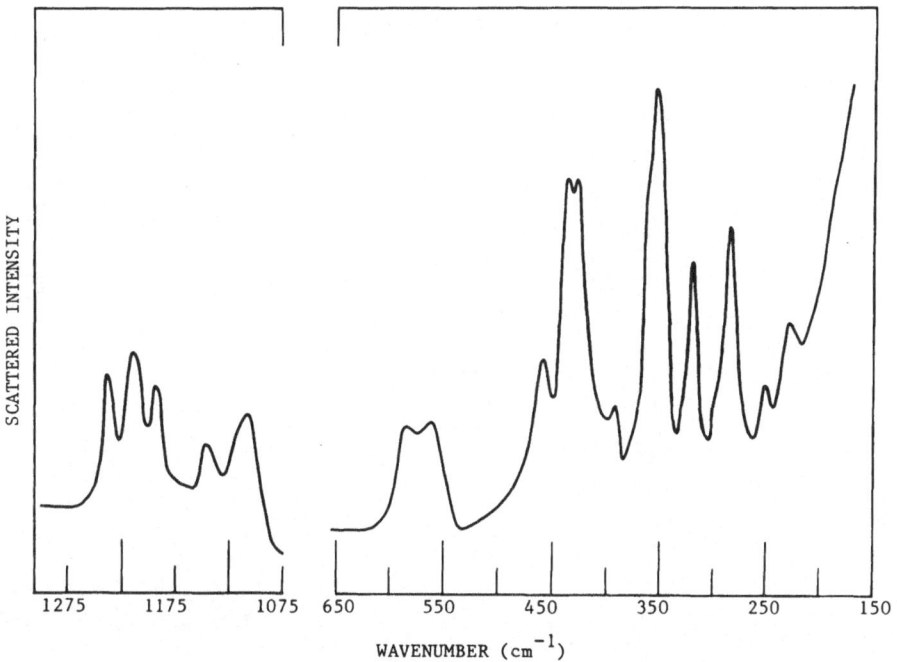

Fig. 3   The Raman Spectrum of Cubic SiP_2O_7.

angle), (Si–O) (O–Si–O small angle), (P–O) (P–O) and (P–O) (O–P–O)
interactions, respectively. The interactions between types of (P–O)
stretching coordinates were approximated to be same. A similar
approximation was made for all interactions between P–O stretching
and O–P–O bending coordinates.

Table 1 compares the calculated and the observed wavenumbers
Table 1. Observed and Calculated Wavenumbers (cm^{-1}) for Cubic
SiP_2O_7.

Symmetry species	Wavenumbers		Major
F_u	observed	calculated	contributions
(Infrared)	1230	1228	f_{rc}
	1195	1216	f_{rc}
	1100	1090	f_{rb}, f_{ra}
	1052	1052	f_{rb}, f_{ra}
	1015	1034	f_{rc}, f_{ra}
	798	804	f_{ra}, f_{aa}, f_{ab}
	765	764	f_{ra}, f_{rb}
	732	740	f_{rb}, f_{ac}
	675	679	f_{rb}, f_{rc}, f_{ac}
	639	623	f_{ra}, f_{rb}, f_{rc}, f_{ac}
	595	604	f_{ra}, f_{rb}, f_{ac}
	510	515	f_{ra}, f_{rb}, f_{ac}, f_{ad}

(continued)

77

Table 1. (Continued)

Symmetry species	Wavenumbers		Major contributions
	observed	calculated	
	473	488	f_{ra}, f_{ac}, f_{ad}, f_{af}
	212.7	221.6	f_{ra}, f_{aa}, f_{ab}, f_{ad}, f_{ae}
	183.9	184.0	f_{ra}, f_{aa}, f_{ab}, f_{ad}, f_{ae}
	176.6	174.4	f_{ra}, f_{aa}, f_{ab}, f_{ad}, f_{ae}
	150.0	147.7	f_{ra}, f_{aa}, f_{ab}, f_{ac}, f_{ae}, f_{af}
A_g (Raman)	1234	1253	f_{ra}, f_{rb}, f_{re}
	1107	1095	f_{ra}, f_{rb}
	462	479	f_{ra}, f_{ab}, f_{ac}
	353	366	f_{ra}, f_{rb}, f_{ab}, f_{ac}, f_{ad}
E_g (Raman)	1192	1210	f_{rb}, f_{rc}
	--	716	f_{rb}, f_{ac}
	589	597	f_{ra}, f_{rb}, f_{rc}, f_{ac}
	251.8	251.5	f_{rb}, f_{aa}, f_{ab}, f_{ac}, f_{ad}, f_{ae}, f_{af}
F_g (Raman)	1212	1217	f_{rc}
	1143	1125	f_{ra}, f_{rb}
	--	795	f_{ra}, f_{rb}
	564	577	f_{ra}, f_{vb}, f_{rc}, f_{ac}
	438	451	f_{ra}, f_{rb}, f_{ab}, f_{ac}, f_{ad}
	429	442	f_{ra}, f_{rb}, f_{ab}, f_{ac}, f_{ad}, f_{af}
	396	403	f_{ra}, f_{ab}, f_{ac}, f_{ad}, f_{ae}, f_{af}
	--	336	f_{rb}, f_{aa}, f_{ab}, f_{ac}, f_{ad}
	318	328	f_{rb}, f_{aa}, f_{ab}, f_{ac}, f_{ad}
	283.6	299.7	f_{rb}, f_{ac}, f_{ad}, f_{af}
	235.0	231.4	f_{ra}, f_{aa}, f_{ab}, f_{ad}, f_{ae}
	189.2	193.0	f_{ra}, f_{aa}, f_{ab}, f_{ad}

for cubic SiP_2O_7. This Table also indicates the symmetry species associated with the particular wavenumbers and the major contributors to the corresponding modes of vibration. Band assignments will not be discussed in detail because Table 1 is self explanatory. However, one may note that on the basis of the normal coordinate analysis, observation of infrared bands around 650 cm^{-1} for a phosphorosilicate solid with a 1:1 SiO_2/P_2O_5 molar ratio would suggest that it contains SiO_6- and PO_4- units linked similarly to cubic SiP_2O_7. The force constants for cubic SiP_2O_7 are listed in Table 2. Their structural interpretation will be discussed later.

Table 2. Refined Set of Force Constants for Cubic SiP_2O_7

Force constant	Coordinate or	Interaction Value
f_{ra}	Si–O	3.4586
f_{rb}	P–O (short bond)	5.5788
f_{rc}	P–O (long bond)	5.3560
f_{aa}	O–Si–O (small angle)	0.6451
f_{ab}	O–Si–O (large angle)	0.5137
f_{ac}	O–P–O (large angle)	0.8323
f_{ad}	O–P–O (small angle)	0.7239
f_{ae}	Si–O–P	0.1612
f_{af}	P–O–P	0.2290
f_{rara}	(Si–O) (Si–O	0.1677
f_{raaa}	(Si–O (O–Si–O large angle)	0.1555
f_{raab}	(Si–O) (O–Si–O small angle)	0.1801
f_{rbrb}	(P–O) (P–O) (average)	0.2926
f_{rbac}	(P–O) (O–P–O) (average)	0.1820

*Units: Stretch: mdyne/A°
       Bend: mdyne A°/rad^2
       Stretch/bend: mdyne/rad

The unit cell of monoclinic SiP_2O_7 with space group C_{2h}^5 consists of four formula units per unit cell. The Bravais unit cell of this phase consists of four distorted SiO_6–octahedral and eight distorted PO_4–tetrahedral units. The Si–O bond length for the octahedral unit ranges from 1.732 to 1.792 Å. The P–O bond length in the PO_4–tetrahedral unit ranges from 1.493 to 1.581 Å. The interconnection of PO_4 and SiO_6– units in the monoclinic polymorph is similar to that in the cubic form. The six equatorial oxygen atoms of each SiO_6–octahedral unit are connected to different PO_4–tetrahedral units. On the other hand, only three oxygen atoms

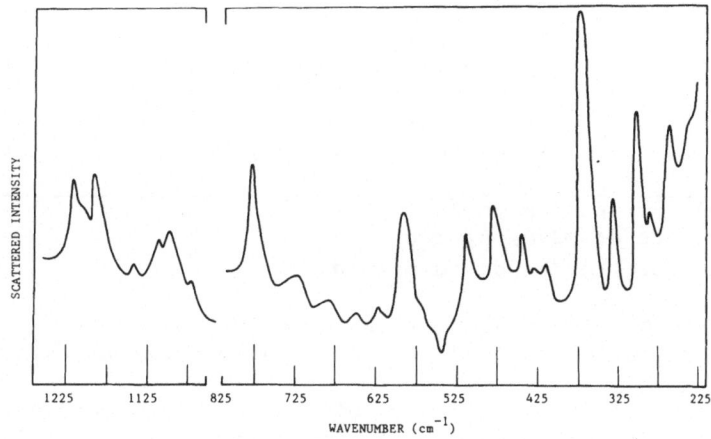

Fig. 4. The Raman Spectrum of Monoclinic SiP_2O_7.

of PO_4-tetrahedral units are connected to different SiO_6-octahedral units. Again, two PO_4-tetrahedral units are connected to each other through the longest P–O bond. Factor group analysis predicts the following selection rules for monoclinic SiP_2O_7:

$$30A_g (R) + 30B_g (R) + 29A_u (IR) + 28B_u (IR).$$

Due to the presence of a center of symmetry in the Bravais unit cell, no Raman band is active in the infrared spectrum and vice versa. According to selection rules, 60 bands are active in the Raman spectrum while 57 bands are active in the infrared spectrum.

The infrared spectrum of monoclinic SiP_2O_7 was illustrated in Fig. 2. The Raman spectrum is illustrated in Fig. 4. Again, normal coordinate analysis generates a close fit between observed and calculated wavenumbers for the first order modes of monoclinic SiP_2O_7. Bands are found in the same major regions of the spectra as for cubic SiP_2O_7. For instance, infrared bands are found in the 625–700 cm^{-1} region, indicating similarities in the local symmetry and structure involving SiO_6– and P_2O_7– units. Differences in band splittings are due to differences in the factor groups.

The force constants calculated by normal coordinate analysis for monoclinic SiP_2O_7 are listed in Table 3. The interpretation of

Table 3.   Refine Set Force of Constants for Monoclinic SiP_2O_7.*

Force constant	Coordinate or Interaction	Value
f_{ra}	Si–O (short bond)	3.5037
f_{rb}	Si–O (long bond)	3.1931
f_{rc}	P–O (short bond)	5.6579
f_{rd}	P–O (long bond)	5.6280
f_{aa}	O–Si–O (average)	0.6448
f_{ab}	O–P–O (large angle)	0.9016
f_{ac}	O–P–O (small angle)	0.8299
f_{ad}	Si–O–P	0.2388
f_{ae}	P–O–P	0.2595
f_{rara}	(Si–O) (Si–O) (average)	0.1317
f_{rcac}	(P–O) (O–P–O) (average)	0.1794
f_{raaa}	(Si–O) (O–Si–O) (average)	0.2508
f_{rcrc}	(P–O) (O–P–O) (average)	0.3723

*Units: Stretch: mdyne/A° $_2$
    Bend: mdyne A°/rad^2
    Stretch/bend: mdyne/rad

these force constants along with those for cubic SiP_2O_7 is reasonable on the basis of trends in the related chemical parameters.

The force constant for the Si–O stretching coordinates for SiO_6-octahedral units have been calculated for the SiP_2O_7 phases in

the range of 3.1-3.6 mdyne/A. These force constants are lower than
the force constants obtained in earlier studies for Si—O stretching
coordinates involving SiO_4-tetrahedral units, because the bonds
involving the same A and O atoms in an AO_6-unit should be weaker
than that those in AO_4-units. However, this force constant is lower
for stishovite (SiO_2) which also possesses SiO_6-units. One can
account for this difference by the fact that each oxygen atom in
stishovite is coordinated by three silicon atoms, while the oxygen
atoms in SiP_2O_7 polymorphs coordinate to only two atoms. Such
structural differences should have substantial effect on the
magnitude of the Si—O stretching force constants.

The P—O stretching force constants for P_2O_7-groups in SiP_2O_7
crystals were found in the range of 5.3-5.7 mdyne/A depending on the
length of the P—O bond. In each case, the long bond has a smaller
force constant than the short bond. P—O stretching force constants
have been reported in the literature for NCA treatments using a
model of isolated $P_2O_7^{4-}$-ions for $M^{IV}P_2O_7$ type of materials.[5,6] Both
the P—O stretching and bending force constants obtained in this
study agree well with the values reported by Hezel et al.[5] The
values reported by Mooney et al.[6] are higher than the calculated
values in the present work. However, no consideration was taken in
the earlier work concerning the effect of the crystalline matrix on
the vibrational modes of the $P_2O_7^{4-}$-ion. This consideration considerably alters the calculated force constants. The factor group effect
will also split the bands observed for isolated $P_2O_7^{4-}$-ions and thus
the full symmetry associated with different vibrations of isolated
$P_2O_7^{4-}$-ions can not be directly compared with those observed for
SiP_2O_7 polymorphs. However, as predicted earlier by both workers[49,50]
many of the P—O stretching vibrations occur in the range of
900-1200 cm^{-1}. Similar predictions have been made regarding the P—O
stretching vibrations of cubic and monoclinic SiP_2O_7--crystals. The
O—P—O bending vibrations for free $P_2O_7^{4-}$-ions appear in the 400-750
cm^{-1}. The results obtained for cubic and monoclinic SiP_2O_7 crystals
are consistent with these mentioned observations for other
materials. The bands noted in the above ranges are associated with
the relevant bending vibrations along with minor coupled
contributions from other stretching and bending vibrations.

B) Analysis of the Vibrational Spectra of Untreated Glasses

The infrared spectra of $50SiO_2 \cdot 50P_2O_5$ glass along with those of
the different crystalline polymorphs are illustrated in Fig. 2. A
major difference in the spectra of the crystals and glasses is the
presence of the band at 1330 cm^{-1} for the glass which is not
observed for the related crystals. Also, the band observed at
ca. 650 cm^{-1} for the crystals is absent for the related glass. The
latter band involved a coupled stretching vibration involving SiO_6
-units and PO_4-units with no P=O bonds. These two spectral observations seem to exclude the possibility of six fold coordination for

silicon atoms in the SiP_2O_7 glass network. Thus, the expected coincidence between the structure of SiP_2O_7 glass and its corresponding crystalline polymorphs does not appear to occur.

Structural interpretation of the vibrational spectra for SiP_2O_7 glass can be conducted best by comparing its spectra with those of related crystalline and glass phases. This investigation compares the spectra of pure P_2O_5 and SiO_2 glasses along with selected glass compositions in the $Na_2O-P_2O_5$, Na_2O-SiO_2, $P_2O_5-SiO_2$ and $Na_2O-P_2O_5-SiO_2$ systems to those of the above-mentioned glass. The vibrational spectra of $xNa_2O.(50-x/2)P_2O_5.(50-x/2)SiO_2$ glasses are illustrated in Fig. 5 & 6.

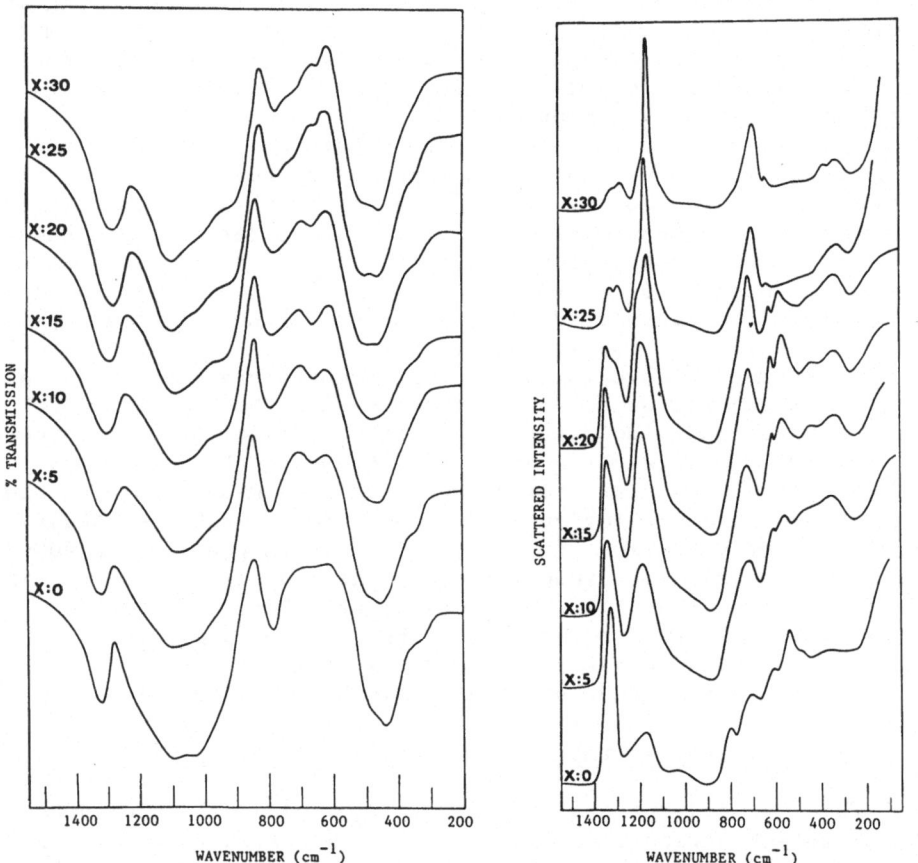

Fig. 5. Raman Spectra of xNa_2O. Fig. 6. Infrared Spectra of xNa $(50-x/2)SiO_2$. $(50-x/2)P_2O_5$ Glasses. $xNa_2O.(50-x/2)SiO_2.(50-x/2)P_2O_5$.

The network of vitreous P_2O_5 consists of PO_4-tetrahedral units in which one oxygen atom is doubly bonded to phosphorous and each of the other three oxygen atoms is bonded to another phosphorous atom. The addition of alkali oxide leads to depoly-merization of the three dimensional network of P_2O_5 with the formation of $(PO_2-O)_n^{n-}$ chains. At the metaphosphate composition the glass structure is usually composed of infinite chains.

The vibrational spectral bands for pure vitreous P_2O_5 in the 1350-1200 cm^{-1}, 1200-600 cm^{-1} regions and the region below 600 cm^{-1} are associated with P=O stretching, P-O stretching and O-P-O bending vibrations, respectively.[8,9] The Raman spectra of a $SiO_2-P_2O_5$ glass with 30 mole% P_2O_5 has been studied by Shibata et al.[10] Based on their assignments, the Raman band at 1330 cm^{-1} for $50SiO_2 \cdot 50P_2O_5$ glass can be assigned to a P=O stretching vibration. The band at 1180 and 800 cm^{-1} in these glasses are assigned to coupled vibra-tions of P-O and Si-O stretching coordinates of P-O-Si linkages, and their presence indicate that P-O-Si linkages are present in these glasses. The band at 708 cm^{-1} is assigned to an O=P-O bending vibra-tion. The band at 537 cm^{-1} is assigned to a O-P-O bending vibration. The shoulder at ca. 430 cm^{-1} can be assigned to Si-O-Si and P-O-Si bending vibrations. The weak bands at lower wavenumbers are associ-ated with internal vibrations of the phosphate units.

With the increase in alkali concentration in these glasses, the intensity of the P=O stretching band decreases while the band at ca. 1180 cm^{-1} increases. At 15 mole% Na_2O concentration, a shoulder appears at ca. 1300 cm^{-1}. With further increase in alkali concentration, this shoulder resolves into a band whose intensity decreases. A similar effect has been observed by Bobovich[11] for alkali phosphate glasses. The band at ca. 1300 cm^{-1} is attributed to symmetric P-O stretching vibrations of delocalized (PO_2) -units. This delocalization leads to a decrease in the bond order of the P=O bond shifting the related band to a lower wavenumber. The decrease in the intensity of this band with higher alkali concentration may be attributed to a change in dipole moment change during vibration associated with this band due to structural changes. The spectra of the glass with 30 mole% Na_2O resembles that of the sodium metaphos-phate glass[12]. The sharp band at 1168 cm^{-1} has been assigned to the PO_2 symmetric stretching vibration and the band at ca. 695 cm^{-1} is assigned to the P-O-P stretching vibration in metaphosphate chains. The low wavenumber bands at 375 and 325 cm^{-1} is assigned to the internal vibrations of metaphosphate chains. The weak band at 1330 cm^{-1} indicate the presence of residual P=O bonds associated with $(O=PO_{3/2})$ -units The absence of Si-O$^-$ stretching and bending bands at ca. 1050 cm^{-1} and 550 cm^{-1}, respectively, excludes the associa-tion of Na_2O with SiO_2.[13] Thus, all of the Na_2O appears to combine with P_2O_5 to form metaphosphate chains while residual P_2O_5 combines with SiO_2. The network in this system primarily consists of a metaphosphate-rich regions and silica-rich regions.

Transmission electron micrographs measured for glasses containing alkali oxide indicate liquid-liquid immiscibility.[7] The association of Na_2O with P_2O_5 is also indicated by the emergence of a band at ca. 420 cm^{-1} for the intermediate Na_2O concentrations. This band is associated with Si-O rocking vibrations involving SiO_4 tetrahedral units. The appearance of this band at higher alkali concentrations can be attributed to the lower scattering intensity of silicate units compared to the metaphosphate units. The lowering in the intensity of the band at 537 cm^{-1} (the O-P-O bending vibration of $(O=PO_{3/2})$-units) with an increase in alkali concentration also indicates the lowering in the concentration of the corresponding phosphate units. This observation indirectly supports the formation of metaphosphate units. Thus, at an intermediate Na_2O concentration, the band at ca. 1180 cm^{-1} can be assigned to a coupled combination of P-O and Si-O stretching vibrations for the P-O-Si linkages and a PO_2-symmetric stretching vibration of the metaphosphate groups. The decrease in intensity of the band at 800 cm^{-1} with an increase in alkali concentration indicates the lowering in concentration of P-O-Si linkages. Thus, with the increase in Na_2O concentration the band at 1180 cm^{-1} becomes primarily associated with the PO_2-symmetric stretching vibration of metaphosphate units. The sharpening of this band indicates that PO_2-stretching vibration becomes more localized with increasing Na_2O or metaphosphate concentration.

The glass with no Na_2O possesses strong infrared bands at 1331, 1100, 1040, 794, 487 and 453 cm^{-1} (See Fig 6). While the infrared spectra of pure vitreous P_2O_5 possesses bands at 1285, 1150, 950, 780, 650 and 475 cm^{-1}, vitreous SiO_2 possesses a strong band at 1060 cm^{-1} and a shoulder at 1180 cm^{-1}. These latter bands have been assigned to Si-O stretching vibrations. The bands at 810 and 440 cm^{-1} for SiO_2 have been assigned to O-Si-O and Si-O-Si bending motions. Based on these observations, band assignments may be made for SiP_2O_7 glass. The band at 1331 cm^{-1} in phosphosilicate glasses may be assigned to a P=O stretching vibration. Also, the band at 1100 cm^{-1} is assigned to a combination of P-O stretching vibration of the P-O-P and P-O-Si bridging units. The band at 1040 cm^{-1} is assigned to Si-O stretching vibrations of the Si-O-Si and Si-O-P bridged units. The band at 794 cm^{-1} can be assigned to the bending motions of P-O-P, Si-O-P and O-P-O units. The low wavenumber bands at 487 cm^{-1} is assigned to O-P-O bending vibrations. The band at 453 cm^{-1} can be assigned to a combination of Si-O-P and Si-O-Si bending vibrations, respectively. The weak shoulder at ca. 350 cm^{-1} can be assigned to the skeletal deformation of phosphate units.

With the addition of alkali to phosphosilicate glasses, the P=O stretching band shifts continuously to lower wavenumber. This shift can be attributed to the lowering in the bond order of P=O bonds due to delocalization of P=O and P-O bonds in the metaphosphate units. New bands at 655 and ca. 950 cm^{-1} become prominent in the spectra with the addition of Na_2O to phosphosilicate glass. The band at 665 cm^{-1} has been assigned to O=P-O bending vibrations, and the band at 930 cm^{-1} can be assigned to asymmetric stretching vibrations of the P-O-P units in the metaphosphate chains.

As noted earlier, the Raman spectra indicate that the addition of Na_2O to the phosphosilicate network leads to the formation of metaphosphate chains. Thus, the band at 1100 cm^{-1} for 30 mole% Na_2O glass can be assigned primarily to the symmetric stretching vibration of PO_2-units of the metaphosphate group. The shoulder at 1050 cm^{-1} is assigned to Si-O stretching vibrations of SiO_4-tetrahedral units. At 25 and 30 mole% Na_2O concentrations, the shoulder at ca. 760 cm^{-1} can be assigned to symmetric stretching of the metaphosphate chains. For intermediate Na_2O concentrations, the band at ca. 1100 cm^{-1} can be assigned to a combination of symmetric PO_2-stretching, P-O stretching of PO_4-tetrahedral units and Si-O stretching of SiO_4-tetrahedral units. Similarly, for the intermediate Na_2O concentration, the band at 780 cm^{-1} can be assigned to a combination of Si-O-Si stretching and symmetric stretching of POP units in metaphosphate chains. Due to the formation of metaphosphate chains, the intensity of infrared band associated with PO_2-units should increase while a reverse effect would be expected for terminal PO_3-units. This effect has been observed earlier in the infrared spectra of alkali phosphate glasses.[15] Thus, based on the infrared and Raman spectral data, we may conclude that sodium ions in $Na_2O-P_2O_5-SiO_2$ system associate primarily with metaphosphate chains, and the silicate network remains unaltered.

C)   Analysis of the Vibrational Spectra of Heat-Treated Glasses

The glasses studied in this system were heat-treated in two stages. The glasses were first treated at their transformation temperature(T_g) for 24 hours to induce nucleation and later treated at 20°C below the dilatometric softening point for different periods of time to develop crystal growth. Only the glasses with compositions containing between 0 and 5 mole% Na_2O could be devitrified homogeneously. Glasses with higher than 5 mole% Na_2O concentration showed only surface devitrification. The effect of heat treatment on vibrational spectra and structure were studied in detail only for glasses with 0 and 5 mole% Na_2O concentration. The 1:1 $SiO_2-P_2O_5$ glass was first treated at 760°C for 24 hours and then treated at 820°C for different periods of time. The glass containing 5% Na_2O concentration was first treated at 520°C for 24 hours and then treated at 560°C for different periods of time.

Analyses of the powder diffraction patterns of heat treated glasses containing 0% Na_2O indicate that the devitrified products for glasses contain only cubic SiP_2O_7 crystals. A similar analysis of the X-ray patterns for glasses containing 5% Na_2O indicates that their devitrified products also contain some $3SiO_2 \cdot 2P_2O_5$ and $Na_4P_2O_7$ crystals. Thus, the devitrified products for all of these glasses consist mainly of crystals with 6-fold coordination for the silicon atoms, while as noted earlier the glasses with related compositions possess silicon atoms with four-fold coordination.

The infrared spectra for the series of heat treated SiP_2O_7 glasses are illustrated in Fig. 7. A new band appears at ca. 650 cm^{-1} in the infrared spectra of these glasses along with those of heat treated glasses containing 5% Na_2O when they were heat treated at their corresponding transformation temperatures. A similar effect

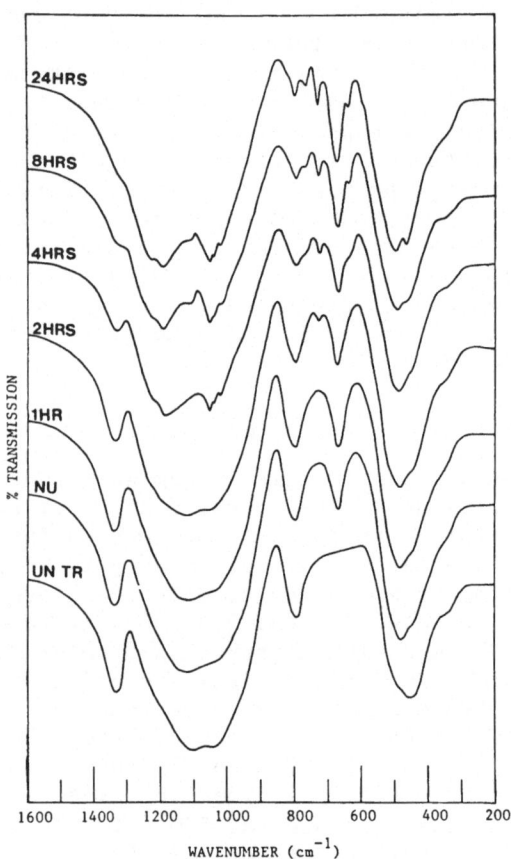

Fig. 7.   Infrared Spectra of SiP_2O_7 Glasses Heat Treated at 820°C for Different Periods of Time.

has also been observed when the glasses were treated 40°C below the T_g for five days. This band was not observed in the spectrum of untreated glasses. The band at ca. 650 cm^{-1} persists, becoming sharper with further treatment. As mentioned earlier, this band has been assigned to a fundamental mode of cubic SiP_2O_7 by normal coordinate analysis treatment. Another major change occurs in the infrared spectra of these glasses in the high wavenumber region. The band

86

at 1330 cm^{-1} gradually decreases with the increase in heat treatment time. The latter band had been assigned to a $P=O$ stretching vibration of the phosphate units in the glass.

During the devitrification of these glasses, the broad band centered at 1100 cm^{-1} splits into two broad groups. Bands centered at 1200 cm^{-1} are assigned to $P=O$ stretching vibrations while bands centered at 1050 cm^{-1} are primarily due to stretching vibrations of $P-O$ bonds in cubic SiP_2O_7 crystals. The bands in the 700-800 cm^{-1} region are primarily associated with $Si-O$ stretching vibrations involving SiO_6-octahedral units. A few extra bands appear in this region for devitrified samples containing 5% Na_2O. These bands are associated with crystalline $3SiO_2.2P_2O_5$ and $Na_4P_2O_7$ phases. The lower wavenumber bands in both the devitrified and untreated glasses can be assigned to different bending vibrations of $O-P-O$ and $O-Si-O$ units.

The Raman spectra of heat-treated 5% Na_2O glasses also reveal information similar to that observed from the infrared spectra. With an increase in heat treatment time, the intensity of the $P=O$ stretching mode decreases indicating a decrease in concentration of this type of bond in the material. After heating the glass for 24 hours at 560°C, there is only a trace of $P=O$ bonds in the network. This spectra mainly consists of characteristic bands for cubic SiP_2O_7 at 353, 429 and 438 cm^{-1} along with those for the other phases such as $3SiO_2.2P_2O_5$ and $Na_4P_2O_7$.

The major spectral changes during the heat treatment of the glasses were the disappearance of $P=O$ bands in the infrared spectra as well as the Raman spectra, and the emergence of a band at ca. 650 cm^{-1} in the infrared spectra associated with six fold coordinated silicon atoms. The development of six fold coordination for Si atoms and the loss of $P=O$ bonds during devitrification can possibly be explained by the following mechanism. The nonbridging oxygen atoms in the phosphate units involving $P=O$ bonds become bridging oxygen atoms with the formation of $P-O-Si$ linkages.

Comparing vibrational spectra indicates that the equimolar $SiO_2-P_2O_5$ glass composition does not appear to possess any structural similarities with that of related SiP_2O_7 crystalline polymorphs. Analysis of the spectra of the SiP_2O_7 glass suggests that its glass network possesses PO_4-units with $P=O$ bonds and SiO_4-units. The silicon coordination changes from 4 to 6 upon devitrification of the SiP_2O_7 glasses, generating cubic SiP_2O_7 crystals. The change in the coordination of the Si atoms occurs due to the formation of more $P-O-Si$ linkages at the expense of $P=O$ bonds. Na_2O associates in $Na_2O-P_2O_5-SiO_2$ glasses mainly with P_2O_5 to form a sodium metaphosphate network portion while the silicate network portion retains its bridged structure. The nonbridging oxygen atoms in these latter glasses are associated mainly with phosphate units.

REFERENCES

1. V. F. Liebau, G. Bissert, and N. Koppen, (1968). Z. Anorg. Allg. Chem., $\underline{359}$, 113.

2. V. H. Makart, (1967). Helv. Chim. Acta., $\underline{50}$ (47), 399.

3. E. M. Rabinovich and M. I. A. Kisilev (1980). J. Mater. Sci., $\underline{15}$, 2027.

4. S. C. Cherukuri (1983). PhD. Thesis, Alfred University.

5. A. Hezel and S. D. Ross (1967). Spectrochim Acta, Part A, $\underline{23}$, 1583.

6. R. W. Mooney and R. L. Goldsmith (1969). J. Inorg. Nucl. Chem., $\underline{31}$, 937.

7. I. N. Chakraborty, (1984). Ph.D.Thesis, Alfred University.

8. J. Wong, (1976).J. Non-Cryst. Solids, $\underline{20}$, 83.

9. F. L. Galeener and J. C. Mikkelsen (1979). Solid State Commun. ,$\underline{30}$, 505.

10. N. Shibata, M. Noridudhi and T. Edahiro (1981). J. Non-Cryst. Solids,$\underline{45}$, 115.

11. Ya. S. Bobovich (1962). Opt. Spektrosk. (Engl. Trans.), $\underline{13}$, 274.

12. V. Fawcett, D.A. Long and L.H. Taylor (1977). Proc.of the 5th Int.Conf.on Raman Spectrosc., Universitat Freiberg, Edited by E. D. Schmid, J. Brandmuller, W. Keifer, B. Schraderand H. W. Schrotter, Hans Ferdinand Schulz Verlag, Breisgau.

13. S. A. Brawer and W. B. White (1975). J. Chem. Phys., $\underline{63}$ (6), 2421.

14. R. J. Bell, N. F. Bird and P. Dean (1968). J.Phys.C., $\underline{1}$, 299.

15. G. Kh. Cherches, V. V. Pechkovskii,V.V., M. I. Kuz'menkove, and Barranikova,T.I.(1978). Sov. J.Glass Phys.Chem. (Engl. Transl.), $\underline{4}$ (2), 200.

MEASUREMENT OF NONSTOICHIOMETRY IN BaTiO$_3$

USING RAMAN SPECTROSCOPY

Marek W. Urban and Bahne C. Cornilsen

Department of Chemistry and Chemical Engineering
Michigan Technological University
Houghton, MI 49931

INTRODUCTION

The nonstoichiometry and point defect accommodation in low temperature (900°C) BaTiO$_3$ influences the Raman spectrum. Therefore it is possible to monitor this nonstoichiometry and to use this technique to study the point defect chemistry of BaTiO$_3$. Earlier research by Eror and Loehr indicated that the half-band width of a 525 cm^{-1} vibrational band broadened in Ba-excess or Ti-excess material.[1] This observation is confirmed. We also report variation of Raman band intensities with changes in oxygen nonstoichiometry as well as Ba/Ti ratio. These dependencies are expected to be indicative of the defect species which influence the band shapes. The origin of the spectral parameter changes is not completely understood, although it is believed to be related to disorder introduced by the point defects,[1,2] or changes in polarizability induced by electronic structure changes near point defects, vacancies or dopants. Vibrational spectra can be more sensitive than neutron scattering for the study of low level defect concentrations (< 1%).[2] For example, between 10 and 550 ppm of nickel vacancies have been detected in NiO using disorder activated Raman scattering.[3] The presence of doubly ionized nickel vacancies was indicated by the oxygen partial pressure dependence of the data.

EXPERIMENTAL

The BaTiO$_3$ powders (MTU sample set) were prepared using a modification of Pechini's liquid mix technique.[4] This technique allows precise control of cation composition (Ba/Ti ratio), and the cations are homogeneously mixed at the atomic level. BaCO$_3$* and titanium triethanolamine chelate in isopropyl alcohol solution# (Tyzor ET) were dissolved along with citric acid+ in distilled/deionized water. Use of this titanium source eliminates the need to use ethylene glycol, minimizing impurities and improving precision. A second set of samples (OGC sample set) were prepared from powders supplied by Professor N. G. Eror of the Oregon Graduate Center (O.G.C.).**

Pressed pellets of BaTiO$_3$ have been sintered in platinum for six hours at 900°C in a controlled oxygen partial pressure. CO/CO$_2$ or Ar/O$_2$ mixtures were used, and the P$_{O_2}$ was confirmed using an Y:ZrO$_2$ oxygen sensor. The use of sintered pellets reduces the influence of adsorbed carbonate on the spectra and eliminates any hexagonal phase material that can form at lower temperatures (700°C).[5] This made more precise measurement of the spectral parameters possible. The samples were quenched to room temperature from 900°C, and the spectra were recorded at room temperature.

The Raman spectra are multi-scan spectra recorded on a signal averager using a four-slit, double monochromator.[3] These provide improved signal-to-noise ratios allowing precise measurement of the spectral parameters. Four scans were summed for each sample at 20 cm^{-1}/min and 1 cm^{-1}/address, with 2 cm^{-1} slits. A 514.5 nm argon ion laser exciting line was used.

RESULTS AND DISCUSSION

The Raman spectrum of BaTiO$_3$ (Figure 1) is sensitive to both the cation ratio, Ba/Ti, and the oxygen nonstoichiometry. Not only is the half-band width a variable, but also the intensity

---

* "Puratronic" grade BaCO$_3$ Johnson Matthey Chemicals Ltd., Orchard Road, Royston, Hertfordshire, 5G8 5HF, England.

E.I.Du Pont de Nemours & Co., Wilmington, DE.

+ "Baker Analysed" Citric Acid, Monohydrate, Crystal, J.T. Baker Chemical Company, 222 Red School Lane, Phillisburg, N.J. 08865.

** 19600 N.W. Walker Road, Beaverton, OR 97006.

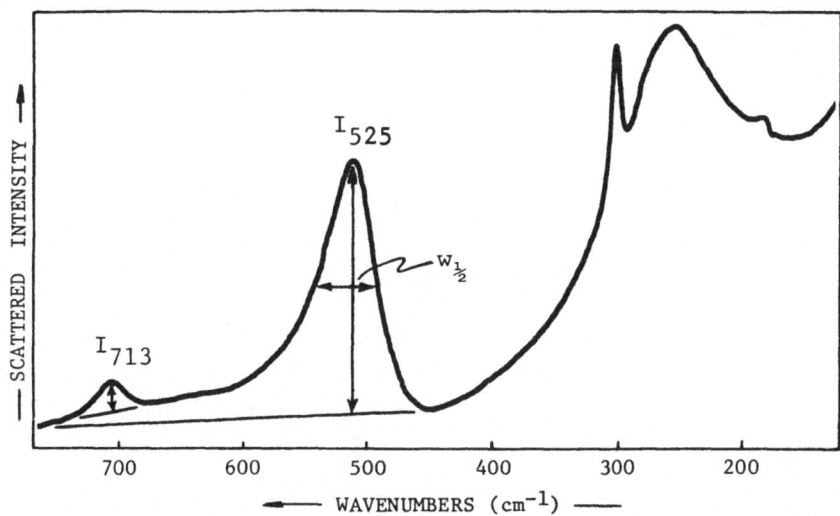

Fig. 1.   A Raman spectrum of $BaTiO_3$ for Ba/Ti = 0.9999 with the
spectral parameters (intensities and half-band width)
defined.   This sample was quenched after 3 hours at 900°C
in 1 atm. oxygen.

ratio of the 713 and 525 cm^{-1} bands, as defined in Figure 1.
This ratio is a minimum for the 0.9999 composition (Figure 2).
This was also the case for the half-band width data of Loehr and
Eror.[1]   A minimum predicted at the one-to-one stoichiometry,
between 0.9999 and 1.0000, is reasonable based on the precision
of the preparation technique.   Very good agreement is seen
between the results for the two sets of samples (MTU and OGC
starting powders).   Our half-width data, $w_{\frac{1}{2}}$, for the OGC samples
agree with their powder results.   Therefore these results confirm
the spectral variation in $w_{\frac{1}{2}}$ as well as the stoichiometric
minimum measured by Eror and Loehr.   Also the second variable
makes it possible to uniquely characterize the Ba/Ti ratio for a
sample quenched from 900°C.

A plot of the half-width of the 525 band as a function of
intensity ratio, I_{525}/I_{713}, is shown in Figure 3.   The two
independent lines, Ba-rich and Ti-rich, have a common origin for
the sample closest to 1/1 stoichiometry, Ba/Ti = 0.9999.   One
can, therefore, define the relative cation composition in any
$BaTiO_3$ sample at room temperature.   Such quantitative information
for a room temperature sample is very difficult or impossible to
obtain via other methods.

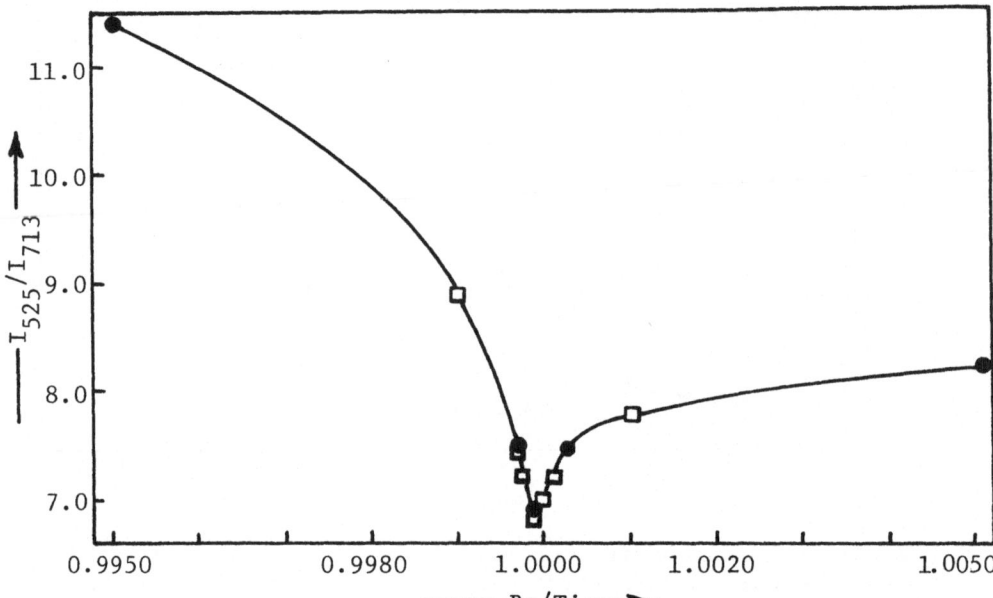

Fig. 2. Dependence of the I_{525}/I_{713} ratio upon the Ba/Ti ratio for OGC samples (▢) and MTU samples (●). The samples were quenched from 900°C in 1 atm. oxygen.

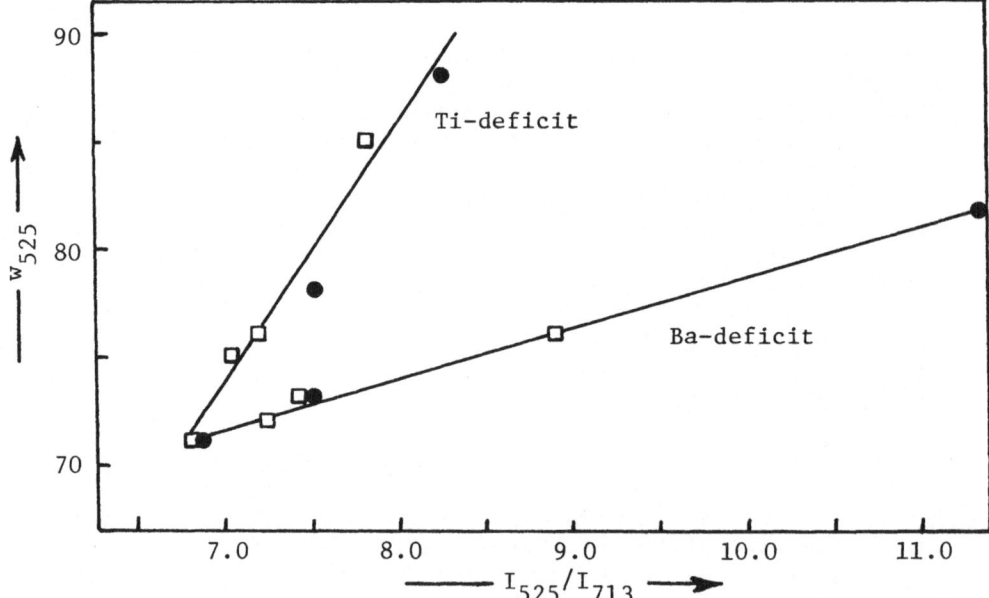

Fig. 3. Correlation of the intensity ratio and half-band width for Ba-deficit and Ti-deficit samples (900°C, 1 atm. O_2).

The dependence of the spectral parameter, I_{713}/I_{525}, upon oxygen partial pressure is nonlinear and the sign of the slope changes. The ratio first decreases and then increases as P_{O_2} is decreased. This suggests that this ratio is sensitive to changes in the predominant type of defects. The latter are known to differ at high P_{O_2} versus low P_{O_2}. Electrical conductivity (σ) data indicate acceptor dopant effects at higher pressure (typically above 10^{-5} atm. at 900°C) and the existence of intrinsic oxygen vacancies at lower pressures of oxygen.[6-8]

The oxygen partial pressure dependence of the Raman intensity ratio ($R=I_{713}/I_{525}$) can be interpreted in a manner similar to that for $\log \sigma - \log P_{O_2}$ plots, where σ is directly proportional to the carrier concentration through equations (1) and (2).

$$\sigma_p \propto [h^\bullet] \propto P_{O_2}^{\frac{1}{4}} \qquad (1)$$

$$\sigma_n \propto [e'] \propto P_{O_2}^{-1/6} \qquad (2)$$

These carriers are compensated by acceptor impurities, A_M' , and doubly ionized oxygen vacancies, $V_O^{\bullet\bullet}$.[6-8] Slopes are significant in the log-log plots because they can be related to the type of defects. Slopes of 1/4 and -1/6 are indicated in equations (1) and (2). Conductivity measurements are sensitive to the concentration of carriers (holes and/or electrons), not the point defects themselves. The Raman can be sensitive to different variables, i.e. the acceptors or vacancies which introduce disorder. Therefore, the results from the conductivity or Raman data are not expected to be exactly the same.

A plot of $\log R$ and $\log P_{O_2}$ is not linear; nor is it expected to be linear. The absolute intensities are not proportional to the defect concentration, rather the change in intensity is proportional to the changing point defect concentration. If the minimum value for R on the $\log R - \log P_{O_2}$ plot, R_o, is subtracted from each R value to give ΔR ($\Delta R = R - R_o$), the $\log \Delta R - \log P_{O_2}$ plot displays two linear portions (Figure 4). The oxygen partial pressure dependence of ΔR represents the "changing" point defect concentrations, and the slopes are now significant in terms of the defect types. This background subtraction method was found successful for V_{Ni}'' analysis in NiO.[3]

Our goal is to understand the origin of the spectral variation in terms of the predominant point defect types. Below 10^{-9} atm., a slope of -1/6.3 is observed (see Figure 4). This P_{O_2} dependence indicates an oxygen vacancy origin for the change of intensity ratio in this region. The doubly ionized oxygen vacancy concentration ($[V_O^{\bullet\bullet}]$) theoretically has a -1/6 slope,[8] which is in very good agreement with our observed slope. At lower oxygen partial pressures, conductivity data also demonstrate a -1/6 slope.[6-8] Electrons from

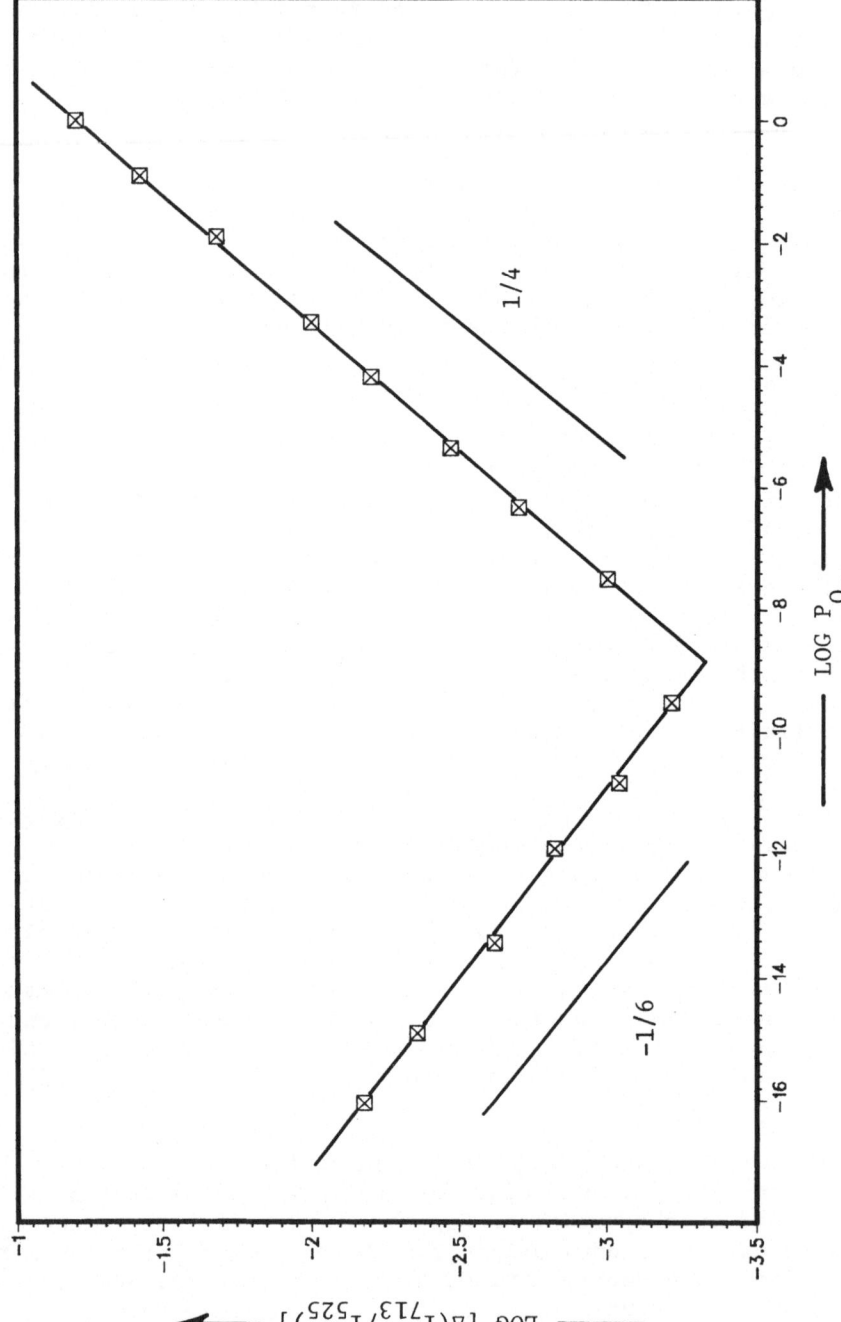

Fig. 4. Oxygen partial pressure dependence of the "change" in Raman band intensity ratio for samples with Ba/Ti = 0.9999 ($R = I_{713}/I_{525}$, $\Delta R = R - R_0$, as described in text).

the doubly ionized vacancies of oxygen predominate in the latter case. Singly ionized vacancies were reported by Daniels et al. although the more recent conductivity results have been interpreted in favor of the doubly ionized oxygen vacancy.[7-9] The Raman data is therefore sensitive to the intrinsic oxygen vacancy disorder in this region, substantiating the presence of $V_O^{\bullet\bullet}$.

A $+1/4.3$ P_{O_2} dependence is observed at higher pressures (refer to Figure 3). Electrical conductivity data have a $-1/4$ and a $+1/4$ dependence at pressures above the intrinsic region.[8] These slopes are controlled by electrons or holes, respectively, which compensate accidental acceptor impurities. It is therefore likely that the Raman dependence also reflects point defects associated with the presence of acceptor dopants. If the phonons couple with electron holes from singly ionized acceptors a $1/4$ dependence is expected in direct analogy with the conductivity data. Further work is in progress to define the origin of this higher pressure spectral variation.

The above assignment to intrinsic oxygen vacancy donors and extrinsic acceptor behavior implies that the minimum at 10^{-9} is an intrinsic/extrinsic minimum, where the acceptor concentration equals the donor concentration. This minimum is at lower P_{O_2} than that observed in the electrical conductivity data (10^{-5}).[8] This Raman technique may provide a means of obtaining true extrinsic/intrinsic minima. In contrast, conductivity minima reflect the mobilities as well as concentrations of carriers. Comparison of the two minima suggests that the mobilities of the electrons are higher than those of the holes; this shifts the σ minimum to higher P_{O2} values. Greater electron mobilities have been suggested by Seuter on the basis of Hall effect measurements.[10]

CONCLUSIONS

The ratio of the intensities of two Raman bands, I_{713} and I_{525}, are defect dependent. These vibrational modes are sensitive to both the cation (Ba/Ti ratio) and the oxygen nonstoichiometry in $BaTiO_3$; therefore, it is possible to monitor this nonstoichiometry in $BaTiO_3$. A plot of intensity ratio versus half-band width allows characterization of the cation ratio for an unknown formed at $900°C$ in oxygen. The stoichiometric compound displays a minimum 525 cm^{-1} half-band width, confirming earlier work, and a minimum I_{525}/I_{713} ratio. It has been found possible to measure the intrinsic oxygen vacancy content, at lower P_{O_2}, and the accidental acceptor dopant content, at higher P_{O_2}. The point defects are believed to introduce disorder which subtly changes the Raman polarizability and therefore band intensities and shapes. The log-log treatment of the Raman spectral parameter defines an intrinsic/extrinsic minimum that is a function of acceptor and donor concentrations, independent of mobilities.

These results demonstrate the potential of Raman spectroscopy for further study of point defect structures in doped $BaTiO_3$ and in other dielectric materials.

ACKNOWLEDGMENTS

We thank Dr. N.G. Eror for making his $BaTiO_3$ powders of varying cation nonstiochiometry available to us. We are also grateful to Dr. N.G. Eror and Dr. D.M. Smyth for helpful discussions.

REFERENCES

1. N. G. Eror and T. M. Loehr, J.Solid State Chem. 12:319 (1975).
2. W. Hayes and R. Loudon, "Scattering of Light by Crystals," John Wiley & Sons, Inc., N.Y. (1978).
3. B. C. Cornilsen, E. F. Funkenbusch, C. P. Clarke, P. Singh and V. Lorprayoon, Characterization of the structure and nonstoi-chiometry of CaO-NiO solid solutions, in: "Advances in Materials Characterization," Vol. 15, D. R. Rossington, R. A. Condrate and R. L. Snyder, eds., Plenum Press, N.Y. (1982).
4. M. P. Pechini, U.S. Patent 3,330,697, July 11, 1967.
5. N. G. Eror, T. M. Loehr, and B. C. Cornilsen, Ferroelec. 28:321 (1980).
6. S. A. Long and R. N. Blumenthal, J.Amer.Cer.Soc. 54:515 (1971).
7. N. G. Eror and D. M. Smyth, J.Solid State Chem. 24:235 (1978).
8. N.-H. Chan, D. M. Smyth, J.Amer.Cer.Soc. 67:285 (1984).
9. J. Daniels, K. D. Härdtl, D. Hennings and R. Wernicke, Philips Res.Reports 31:487 (1976).
10. A. M. J. H. Seuter, Philips Res.Rep., Suppl. No. 3:37 (1974).

RESTSTRAHLEN EFFECTS IN DRIFT EXPERIMENTS

FOR NaCl IN THE FAR-INFRARED REGION*

John R. Ferraro†

Loyola University
Chemistry Department
Chicago, IL. 60626

INTRODUCTION

The alkali metal halides possess the NaCl structure under ambient conditions of temperature and pressure (O_h^5, Fm3m with Z = 1). No activity is observed in the first-order Raman spectrum, but an F_{1u} mode (triply degenerate) is observed in the far infrared region.

For a three-dimensional lattice, motions of atoms can occur in the three dimensions, which involve a transverse motion (ν_{TO}) in two directions and a longitudinal motion (ν_{LO}) in the third direction. The transverse motion is doubly degenerate. The electrical field of the electromagnetic radiation is transverse and can couple with the dipole moment change created during the vibration for the phonon propagation in two of the directions. This involves all of the unit cells in the crystal. However, coupling cannot occur with the longitudinal branch, and since the radiation is transverse, no (ν_{LO}) lattice mode will appear in the infrared under normal or near normal incidence of light. In thin films and where radiation is at oblique angles, Berreman[1] has demonstrated that the ν_{LO} mode can become infrared-active. The ν_{TO} mode occurs at lower frequency than the ν_{LO} mode. The splitting is caused by the interaction of the unit cells even at k = 0 and the degeneracy of the F_{1u} mode is removed giving a doubly degenerate ν_{TO} mode and a non-degenerate

*Taken in part from Kathleen Martin's Ph.D. thesis, Loyola University, 1984.
†Searle Professor of Chemistry

ν_{LO} mode. The intensity of the ν_{TO} mode is much greater than that of the ν_{LO} mode.

RESULTS AND DISCUSSION

We are presently examining several alkali metal halides in the far infrared region (FIR) using DRIFT (diffuse reflectance Fourier Transform interferometry) techniques. When using neat polycrystalline NaCl (particle sizes ranging from 850 μm to <150 μm), we have observed inverse peaks (maxima in transmittance on a %T plot, and minima in A on an Absorbance plot) in experiments conducted by the DRIFT technique. Similar results were obtained when deposits of NaCl (same particle sizes as above) were made on an aluminum foil surface, constituting chiefly a specular reflectance experiment. Neat NaCl pressed pellet, neat NaCl window or neat NaCl single crystal also provided inverted peaks, when light was reflected off the surface.

The inverted peaks appeared to be centered in two regions of the FIR. A strong low frequency peak was found near the transverse phonon mode for NaCl (assigned at 164 cm^{-1} from the film experiments)[2,3]. It was always accompanied by shoulders on both sides of the main band. The main band occurred at higher frequency than 164 cm^{-1}. The second peak was found very close to the longitudinal phonon mode (calculated at 264 cm^{-1} for NaCl by the Lyddane-Sachs-Teller equation)[4]. The relative intensities for these peaks followed the relationship of low frequency > high frequency, similar to results obtained by Berreman[1] and others[5-21] from thin film experiments.

The shoulders appearing on the low frequency peak may be ascribed to the following possible causes:
1) observation of frequencies associated with various particle sizes in the sample.
2) observation of normally silent acoustic overtones from the k^+ and k^- branches of the Brilluoin zone (overall k = 0 and no violation of selection rules occur for cubic salts).
3) combinations of the acoustic modes from k^+ and k^- with the main inverted peak.

DRIFT spectra of materials with higher refractive indices (n) than NaCl (n = 1.54) all produce normal diffuse reflectance spectra with no inverted peaks for both the neat and diluted samples (for example, AgCl, n = 2.21; yellow HgO, n = 2.37; Fe_2O_3, n = 3.01).

Although the literature is replete with Reststrahlen effects occurring for alkali metal halides, the effects for NaCl demonstrated in this paper and obtained by DRIFT experiments in the far infrared region for neat solids are new. In fact DRIFT experiments in this region were only recently initiated.[22] However, Reststrahlen effects in the mid infrared have been cited previously by Fuller and Griffiths,[23,24,25] and very recently by Grim et al.[26] For materials with high absorbances, large particle size and a rapidly changing index-of-refraction play an important role in determining whether one obtains surface effects with specular reflectance predominating, or bulk effects occurring with diffuse reflectance resulting. With strong absorbing solids possessing large particle sizes, and particularly for neat materials or for powders minimally diluted in a matrix, surface effects are dominant and specular reflectance occurs when a DRIFT-type experiment is attempted.[23-26] Inverted peaks are observed as illustrated in Fig. 1 for NaCl in the far infrared region and for kaolin and caffeine in the mid infrared region.[26]

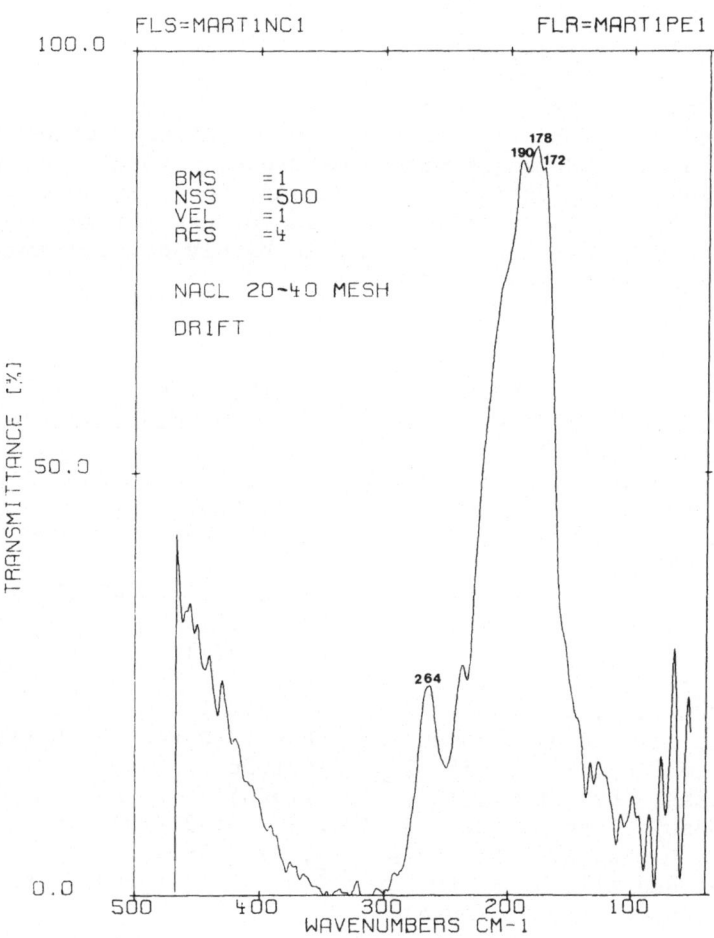

Fig. 1. Drift results for neat NaCl.

Grim et al.[26] showed that the strong specular reflectance inverted peaks in kaolin and caffeine were shifted to higher frequency in the DRIFT experiment from the true peaks, similar to our results obtained for NaCl. They also demonstrated that these solids, when diluted to 1% in KBr and particle size of < 10 μm, gave a true diffuse reflectance spectrum (normal spectra). Since the inverted peaks are shifted toward higher energy from what is obtained in a true DRIFT spectrum, some problems are possible in making assignments from such spectra, as Grim et al. have cited.[26]

Dilutions of NaCl to 3% in polyethylene powder continued to provide inverted peak spectra, and dilutions of less than 3% were too dilute to observe any NaCl spectrum. Our work continues with other alkali metal halides to determine to what extent Reststrahlen effects are common to all members of this family of salts.

EXPERIMENTAL

For these experiments an IBM IR/98 vacuum interferometer interfaced with a Harrick Praying Mantis diffuse reflectance accessory was used. 500 scans were co-added at a resolution of 4 cm^{-1}. A Happ-Genzel upodization was used, and since, in most cases, we were dealing with neat solids, we did not use the Kubelka-Munk correction, which is suitable only in highly diluted samples.

REFERENCES

1. D. W. Berreman, Phys. Rev., 130, 2193 (1963).
2. S. S. Mitra, AFCRL-69-0468, Air Force Cambridge, Bedford, Ma. 01730, (1969).
3. G. O. Jones, D. H. Martin, P. A. Mawer, and C. M. Perry, Proc. Roy. Soc. (London) A261, 10 (1961).
4. R. H. Lyddane, R. G. Sachs and E. Teller, Phys. Rev., 59, 673 (1941).
5. R. B. Barnes and M. Czerny, Z. Physik, 72, 447 (1931).
6. R. B. Barnes, Z. Physik, 75, 723 (1932).
7. H. W. Hohls, Ann. Physik, 29, 433 (1937).
8. M. Klier, Z. Physik, 150, 49 (1958).
9. G. Heilmann, Z. Physik, 152, 368 (1958).
10. F. Abeles and J. Mathieu, Ann. Phys. (Paris) 3, 5 (1958).
11. D. Frohlich, Z. Physik, 169, 114 (1962).
12. M. Hass, Phys. Rev., 117, 1497 (1960).
13. M. Hass, J. Phys. Chem. Solids, 24, 1159 (1963).
14. S. S. Mitra, Sol. St. Phys., 13, 57 (1962).
15. C. M. Randall, R. M. Fuller and D. J. Montgomery, Sol. St. Comm., 2, 273 (1964).
16. H. W. Hohls, Ann. Physik, 29, 433 (1937).
17. J. R. Jasperse, A. Kahan, J. N. Plendl and S. S. Mitra, Phys. Rev., 146, 526 (1966).

18. M. Czerny, Z. Physik, 65, 600 (1940).
19. A. Hadni, J. Claudel, D. Chanel, P. Strimer, and P. Vergnat, Phys. Rev., 163, 836 (1967).
20. A.D.B. Woods, B.N. Brockhouse, R.A. Cowley and W. Cochran, Phys. Rev., 131, 1025 (1963).
21. J.S. Reid, T. Smith and W.J.L. Buyers, Phys. Rev. B, 1, 1833 (1970).
22. J.R. Ferraro and K. Martin, Appl. Spectrosc., 38, 270 (1984).
23. M.P. Fuller and P.R. Griffiths, Am. Lab., 10, 69 (1978).
24. P.R. Griffiths and M.P. Fuller, Chapter 3, Mid-Infrared Spectrometry, V. 9, R.J.H. Clark and R.E. Hester, eds. Heyden, London (1982).
25. R.K. Vincent and G.R. Hunt, App. Opt., 7, 53 (1968).
26. W.M. Grim, J.A. Graham and W.G. Fateley, Transform Times, No. 1, Nov. (1983), pp. 1-2.

NEWER TECHNIQUES IN OPTICAL MICROSCOPY

V. D. Fréchette

New York State College of Ceramics
at Alfred University
Alfred, New York

ABSTRACT

High-contrast optics, especially when used in conjunction with new replication techniques; vicinal illumination for sensitive detection of microcracks; subsurface exploration for three-dimensional details of microstructure; multiple-beam interferometry for astonishingly sensitive description of surface contours; particle size estimation procedures (not new but commonly overlooked); the modern spindle stage with computer programs to yield optical data from a single grain; and a new method for mounting particulates for automated stereology combine to substantiate the opinion that the optical microscope is the most valuable single instrument for characterizing materials.

The optical microscope, one of the earliest of scientific instruments, is still the most useful of all. It hasn't just now been invented, but some new wrinkles have been added - new pieces of accessory equipment, new techniques, and rediscoveries in both of these categories. A few seem to be particularly applicable to some of our present needs.

Replication

Optical examination of ceramic surfaces has been handicapped because over 90% of the incident light is transmitted through the surface, is scattered by subsurface grain boundaries and re-emerges to confuse our image of the surface. The problem can be solved by forming a plastic replica of the specimen; the replica preserves only the surface details. The replica can be studied in transmitted light or in reflected light as convenient. Replicas are also useful as intermediate records of structure in sequential tests. Distinction among several phases by their characteristic colors or reflectivities is lost, of course, but phase recognition is usually still possible through characteristic morphology, etch patterns, and inclusions.

Techniques for replication have been highly developed by transmission electron microscopists. But restrictions which they face, such as opacity of materials to the electron beam and lack of contrast without shadowing, are not met in optical microscopy, so that simpler techniques can be used as alternatives to those which work best for electron microscopy.

For replicas that must be made at room temperature, cellulose acetate tape is immersed for ten to fifteen seconds in acetone and laid on the acetone-flooded surface, against which it is pressed with a smooth-faced PVC or silicone rubber block (Fig. 1). After five minutes, the block is removed and drying is continued for a few minutes longer before stripping off the replica and taping it to a microscope slide. If taped face down, it appears as a positive replica of the surface when viewed by transmitted light from above.

More pliant replicas, which lie tight against the microscope slide without taping, can be made with unpigmented, two-part silicone elastomer (Fig. 2). The desired viscous liquid mixture is

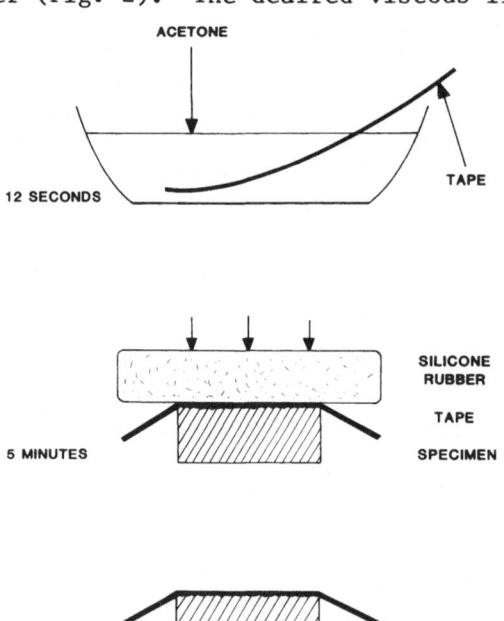

Figure 1. Steps in replication of surfaces using cellulose acetate tape.

104

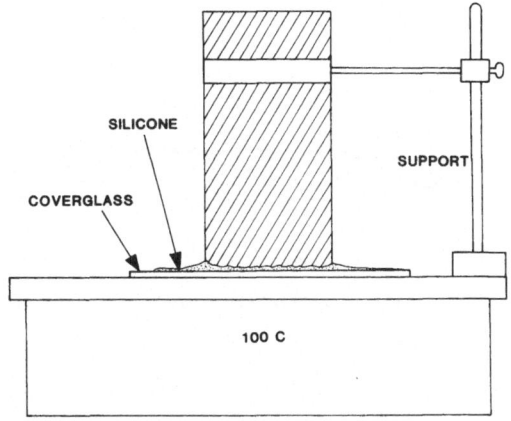

Figure 2. Replication using silicone elastomer.

poured to form a pool on a large microscope cover glass on a hot
plate at about 100 degrees and the prewarmed specimen in a
convenient clamp is lowered into it, leaving a clearance of a few
tenths of a millimeter. When cured (it takes only a few minutes) the
replica is carefully separated from the specimen with the help of a
knife blade. It can be observed in transmitted light from the cover
glass side, appearing as a positive replica.

PVC sheet can be used to replicate specimens whose porous or
rough surface would lead to tearing of the comparatively fragile
silicone (Fig. 3). It is laid on the surface of a specimen
preheated to 180-200 degrees, depending on the particular PVC, and
rolled into contact with it using a Teflon rod. After cooling, but
still warm, it can be easily stripped and is then mounted on a slide
for study. Alternatively, a PVC sheet may be pressed against a cold
coverglass, transferred to a hot plate to melt it and then quickly
placed on a thermally insulating support. The preheated specimen,
suitably clamped to position it, is promptly lowered against it and
allowed to cool. This procedure is especially useful for small
specimens. (Objectionable fumes from hot PVC make it necessary to
work in a hood).

Vicinal Illumination

Elusive three-dimensional microstructural details can often be
seen in polished section by flooding the surface with immersion oil
to suppress surface reflection, illuminating intensely with the
field iris diaphragm restricted to the smallest possible opening (a
pinhole diaphragm may be substituted for the iris), and focusing
down beneath the surface. The region of the specimen in the
vicinity of the bright spot can be scanned in three dimensions to
reveal voids, grain boundaries and inclusions. This is called

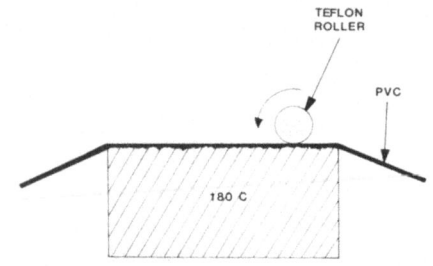

Figure 3. Replication with PVC.

vicinal illumination. Voids seen in this way cannot be confused
with pullouts, inclusions cannot be mistaken for preparational
contaminants and there is no confusion with respect to size, shape
or position (Fig. 4, 5).

Vicinal illumination can also be used on dry specimens to
detect microcracks. The principle is simply that the subsurface
illumination encounters critical reflections at crack walls, so that
a line dividing dark from light areas identifies the path of the
crack. This is most effective with magnification in the range of
200X or higher, putting it within the range of the higher power
stereoscopic microscopes; with them the light source is placed in
the phototube and adjusted to focus on the specimen.

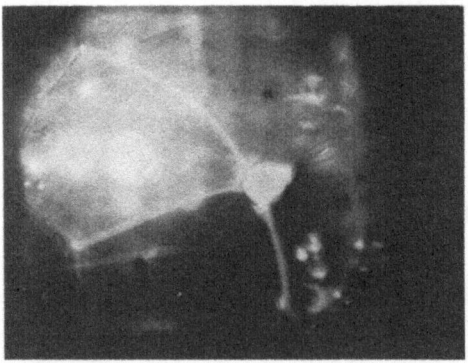

Figure 4. Subsurface grain boundaries and inclusions seen by vicinal
illumination. Photo width 45 mμ.

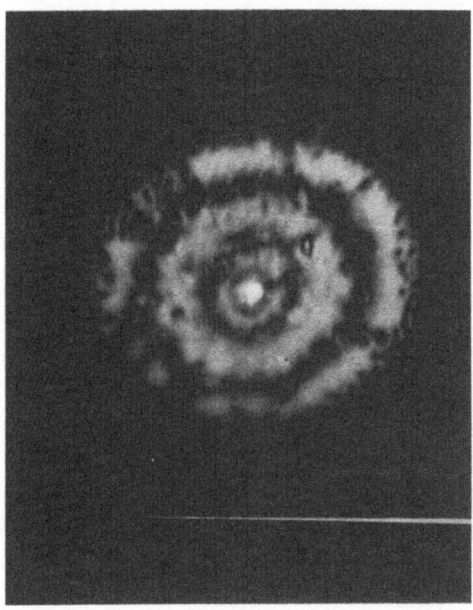

Figure 5. Intergranular void in BeO ceramic. The Newton rings
indicate the amount of separation between the crystals caused
by the presence of an inclusion (bright spot at the center).
Photo width 45 mµ.

Differential Interference Contrast

In any society show of photomicrographs there is a liberal
sprinkling of color prints, rich in yellows, reds and blues, taken
with differential interference optics. The differential interference contrast principle [1,2] of which the Nomarski is the best
known version, consists in splitting the incident light into two
rays which travel through the specimen along paths that are close to
one another (Fig. 6). After passage through the specimen or, in the
case of polished sections, against the specimen, differences in
their optical paths are made visible by interference when they are
later recombined. Differences in the optical experience of the two
rays, i.e., differences in the refractive indices of media through
which they have passed or differences in the level of surfaces
against which they have impinged, are greatly amplified in terms
of the brightness of the image or, when the sensitive tint plate is
introduced, in terms of color changes. Even slight changes are
detectable because the eye is astonishingly sensitive to slight
change from the sensitive tint or "purple of transition"; slight
addition to its retardation shifts it to blue, while slight
subtraction shifts it to yellow.

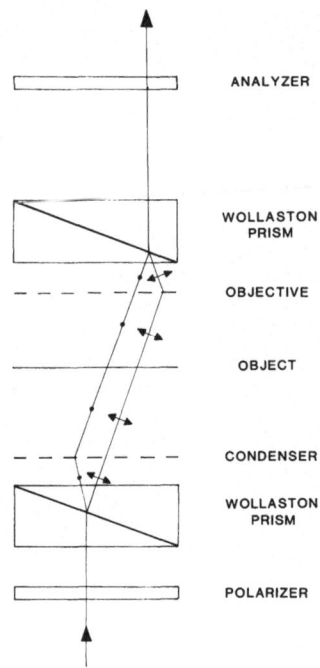

ANALYZER

WOLLASTON
PRISM

OBJECTIVE

OBJECT

CONDENSER

WOLLASTON
PRISM

POLARIZER

Figure 6. Arrangement of optics in differential interference
contrast.

Demonstration that this provides a truly unique capability has
come to us often in the study of markings on crack-generated
surfaces of glass. Some markings that have been impossible to
locate with the scanning electron microscope show up clearly and are
easily measurable under the microscope with differential interference contrast (Fig. 7).

Multiple-beam Interferometry

Description of surface topography on the finest of scales can
be done with the microscope using the method of multiple-beam
interferometry, best known from the work of Tolansky [3,4]. It is
an old method, but it seems not to be known by many who could use it
to advantage. Considering the power of the technique and its
simplicity, that seems a pity. Therefore this capsule summary.

The specimen surface is given a metallic coating (silver is
best but aluminum works well) to raise its reflectivity to maximum.
A reference surface, such as a microscope cover glass coated to
yield a reflectivity of between 55 and 65%, is laid coated side down
on the specimen in the field of a reflected light microscope
illuminated with monochromatic light, as from a filtered mercury or
thallium lamp or from a low-powered laser.

108

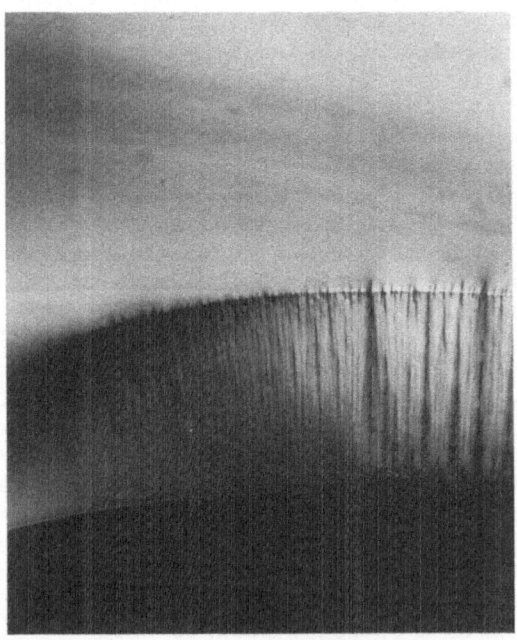

Figure 7. Arrest line (bottom); band of precavitation hackle; and
cavitation scarp (top) on a crack surface generated in glass in
the presence of liquid water. Photo width 300 mμ; reflected
light; differential interference contrast.

Interference fringes form, tracing the contours of the specimen
surface. But, in contrast to the fringes formed in ordinary interferometry, the multiple-beam fringes are narrow and sharp, so that
measurements are easily possible to less than one-twentieth of a
fringe interval and differences in surface elevation of 10 a.u. have
been detected.

Particle Size Estimation

Automated devices for particle size distribution measurement
are prominent on the equipment scene. They certainly qualify for
consideration by laboratories having routine need for large numbers
of determinations, where microscopic methods, although direct and
effective [5], are tedious. For others, who have yet to justify
expensive equipment, abbreviated microscopic methods may be highly
satisfactory.

As is the case with any measurement of particle size, it is
essential to identify which of the many possible size and shape
parameters is relevant to the application. For example, if particle
size is of interest with respect to strength, only the largest
grains are significant, and possibly only the lengths of these,
since they are the ones which will form the origins from which

cracking due to particle size will be generated. Restriction to the relevant parameter simplifies measurement and correspondingly speeds it.

As an example of this, consider the problem of grain growth during the later stages of sintering. Photomicrographs of polished sections of specimens after the several heat treatments can be quantitatively compared simply by adjusting the photomicrograph of the test specimen to a distance where it exactly matches a reference photo held at a fixed distance. The ratio of the distances then indicates the ratio of the grain sizes. For this purpose it is necessary to have both photos in focus at once. The simplest way to do this is to observe them, brightly lighted, through a pinhole, the universal lens. More elaborate and more satisfactory arrangements include use of zoom optics or a comparison eyepiece.

The very fine powders encountered more and more frequently in modern processing present a special problem since these may be below the resolving power of the microscope. Nevertheless, the microscope may still be able to provide the necessary indication of their particle size, even if this is as little as a hundred angstroms. The procedure [6] is to disperse a weighed amount of the powder in a measured volume of a liquid viscous enough to suppress Brownian motion when cold. A drop of the warmed suspension is placed in a hemacytometer cell and after allowing the particles to settle out, cooled. The number of particles per unit volume is counted. A small particle will appear as a point of light when viewed in dark field in the microscope; the presence of a particle far below the resolving power of the microscope can be detected if the lighting is sufficiently intense. From the number of particles per unit volume and the weight of powder per unit volume, the weight per particle is given and other parameters such as specific surface or average particle size can be calculated.

The Spindle Stage

Identification of a crystal phase constituting a single tiny crystal, e.g., from an inclusion, a condensate, or a raw material contaminant, may tax the capacity of even well-instrumented laboratories. The polarizing microscope may be effective in recognizing the mineral, but in some cases the problem of measuring a sufficient number of the optical parameters may be intractable because of small particle size. The four-circle universal stage is in theory capable of making the necessary measurements, but the procedure is so tedious that even in those laboratories where it is available the universal stage is virtually never used.

An inexpensive single-axis stage, or spindle stage, is far less complicated and yields results quickly [7]. In principle it is simply a horizontal needle, rotatable around its own axis in five-degree increments. This is mounted on the stage of the polarizing microscope. The crystal, cemented to its tip, fits into a narrow cell in which an oil of matching refractive index is held by capillarity. Thus the rays from the microscope condenser pass through the crystal without deviation. Measurement of the extinction position of the crystal as a function of the angle of rotation of the spindle can be plotted to indicate whether the crystal is uniaxial or biaxial, to measure its optic angle and sign, and to identify the settings of the spindle stage at which the principal refractive indices can be measured by any of the oil-immersion methods.

It is an advantage that the method is nondestructive, so that the specimen may be recovered for subsequent study by other techniques.

Automatic Stereology

Automated devices for stereological microscopy are being offered by many instrument makers in varying degrees of sophistication. They perform such tasks as quantitative analysis by phases, particle size distributions and averages, shape parameters and preferred orientation information. They work well with polished sections. However, measurements on particulate suffer from the problem of achieving sharp focus, without which the data can be misleading. If the measurements on a particle are to be useful its equatorial plane must be in focus. Yet if the particles are lying on a microscope slide, it is clear that the equatorial plane of a small particle will lie below the level of the equatorial plane of a large one.

The problem can be solved by mounting the powder sample between thin sheets of an elastomer in a vacuum jar. When brought out into the air, the atmospheric pressure forces the particles against the upper and lower sheets equally, and every particle then lies with its equatorial plane at the same level [8].

To accomplish this, a two-part silicone elastomer is cast between microscope cover glasses to a thickness of about a millimeter, and one of the coverglasses is stripped away. The sample powder is scattered on the exposed silicone surface. A second cast, with its cover glass, is laid silicone-side-down on top of this in a vacuum dessicator (the manipulation can be accomplished by a magnet acting through the dessicator wall), and the specimen removed for study.

These are a few of the special ways in which materials investigations can be facilitated by selective application of microscopic techniques. It should be added that the modern polarizing microscope is far more comfortable to use than the older ones. Binocular eyepieces, especially the newer wide-field types, add greatly to visual acuity and to comfort, while the modern halogen lamps run cooler. Photomicrography for record has been enormously simplified by the instant process films and by automated 35-mm systems.

More than ever, when it comes to materials characterization, the polarizing microscope is well worth looking into!

REFERENCES

1. A. E. J. Vickers, "Modern Methods of Microscopy," Butterworths Scientific Publications, London (1956).
2. M. Francon, "Progress in Microscopy," Row, Peterson and Company, Evanston. Ill. (1961).
3. S. Tolansky, "Multiple-Beam Interferometry," Clarendon Press, Oxford (1948).
4. S. Tolansky, "Surface Microtopography," Interscience, New York (1960).
5. V. D. Frechette and H. I. Sephton, A method of particle size determination by means of the microscope, Am. Cer. Soc. Bull. 28:496 (1949).
6. C. R. Amberg, Size determination with the haemacytometer, J. Am. Cer. Soc. 19:207 (1936).
7. F. Donald Bloss, "The Spindle Stage: Principles and Practices," Cambridge University Press, New York (1981).
8. V. D. Frechette, Method to overcome the focus problem in automated stereology of particulates, to be published.

LASER RAMAN MICROPROBE STUDY OF THE IDENTIFICATION AND THERMAL

TRANSFORMATIONS OF SOME CARBONATE AND ALUMINOSILICATE MINERALS

Richard G. Herman, Charles E. Bogdan
and Andre J. Sommer

Center for Surface and Coatings Research
Sinclair Laboratory # 7
Lehigh University
Bethlehem, PA 18015

INTRODUCTION

Raman microprobe spectroscopy is a recently developed technique [1-6] that can give molecular information on a microscopic scale. It covers the complete spectrum of lattice and molecular vibrations and extends the tools of vibrational spectroscopy for chemical and structural studies to the microscopic domain. Information can be derived for both inorganic and organic samples, whether crystalline or amorphous and whether transparent or opaque. The samples can be solids or liquids, e.g. liquid inclusions in minerals.[3] Thus, the Raman microprobe is unique in many of its capabilities and complements the techniques of scanning electron microscopy and electron probe microanalysis that provide sizes and shapes, which the Raman microprobe can also provide, and elemental compositions of single particles in the micrometer size range.

The Raman microprobe utilizes a laser as the monochromatic source of light that is used to irradiate the sample. A microscope objective lens is employed for focusing the laser beam and for viewing the sample. Usually a high numerical aperture (N.A.) objective is employed so that a small diameter laser probe is produced. This ensures high irradiation levels and good spatial resolution, and a high N.A. objective collects light scattered over a large solid angle so that more of the Raman light is detected. However, for a particular image magnification, the maximum useful N.A. is limited by the size of the back focal plane of the objective projected onto the monochromator grating.[7,8] If the grating is overfilled, stray light is introduced into the system and poor spectra are produced.

113

Thus, there is a practical limit to the degree of focusing of the laser beam. For the instrument utilized in this work, the N.A. objective relationship is N.A. (max) = $M/(36(\Pi)^{1/2})$, where M is the magnification factor of the objective.

Utilizing a high N.A. objective, spatial resolutions on the order of 1 micron^2 (μm^2) are achieved. Therefore, particles having a cross-sectional diameter as small as this can be readily analyzed by the laser Raman microprobe. It has been pointed out that for organic particles in the range of 1-25 μm and inorganic particles in the 1-10 μm range, there is no routine alternative to the Raman microprobe for obtaining molecular information.[9] With particles of this size, the spectra determined by this technique can be thought of as single crystal spectra, and one might expect that the normal selection rules for Raman measurements on bulk materials would not strictly apply. However, the high aperture of collection and diffraction effects arising from the particle size generally cause all modes, whose symmetry allows Raman activity, to be observed in a single measurement, and the microprobe spectra are quite analogous to the averaged spectrum observed from a bulk powder sample in regard to band frequency, relative intensity, and multiplet pattern of the bands.[10] Exceptions to the analogous relative intensities are observed in laser Raman spectra, as they are in ordinary Raman spectra,[11] for crystalline materials that exhibit a very high degree of preferred orientation, e.g. graphite and graphitic substances. However, this is useful because the bands broaden as the crystalline domain size decreases (the planar material becomes more disordered and amorphous) and the relative intensities of bands in a spectrum can sometimes be used[11] to determine the crystalline planar domain size in imperfectly crystallized samples. For three-dimensional materials, the spectra are relatively free of artifacts associated with crystal size and shape.

Although the laser Raman microprobe is typically utilized under ambient conditions and no special sample preparation is required, difficulties can still be encountered with this technique. The major limitations of laser Raman microprobe analysis are:
1. fluorescence or luminescence,
2. laser-induced thermal or photo decomposition, and
3. a limited reference data collection.
Fluorescence is due to the absorption of light followed by emission of light at a longer wavelength, and it generally occurs within 1-100 nanoseconds after the absorption of the original radiation. It can sometimes be reduced by using a different exciting laser line or covering the sample with aerated water as a quencher. Laser-induced decomposition can be suppressed by using a defocused beam or by spinning the sample so that the beam is not focused only on one spot. Advantages of laser Raman microprobe spectroscopy include the following:

1. little or no sample preparation of solids,
2. simple spectra with good molecular specificity,
3. spatial resolution down to 1 micron,
4. absolute detection limits down to the nanogram, and sometimes to the picogram, range,
5. favorable signal-to-noise ratio,
6. noninterference of liquid water, and
7. generally nonintrusive and nondestructive of the sample.

It is evident that the laser Raman microprobe is particularly useful for studies of mineralogical samples, mineral inclusions, composition and uniformity of glasses, ceramics, and semiconductors, solid state phase transformations, composite materials, catalysts, and organic contamination. This report describes the application of the laser Raman microprobe to the identification of carbonate minerals and selected aluminosilicates.

EXPERIMENTAL

Raman spectra were recorded on an Instruments S.A. Molecular Optics Laser Examiner (MOLE) Raman microprobe. A schematic diagram of the apparatus is shown in Figure 1. A Spectra Physics Model 164 argon ion laser served as the photon source. The spectra were usually obtained by using the 514.5 nm green excitation line of argon, but the 488.0 nm blue laser line was used in selected cases. Scattered radiation was collected 180° to the incident beam with a 100 X (0.90 N.A.) metallurgical objective from Leitz. Detection was accomplished with a Model 126 wide range photometer from Pacific Precision Instruments and an RCA 31034 photomultiplier tube cooled to -30°C in a Products for Research thermoelectric housing (Model TE-104F).

Each of the presented spectra is produced from two to four spectral scans, which were recorded at the rate of $50 \, cm^{-1}$/min with a time constant of 1 sec. The spectral slit width was 5 cm^{-1} and the wavenumber accuracy was within $\pm$ 2 cm^{-1}. A wavenumber marker was often spectrally applied by using external fluorescent light having a wavenumber of 1123 cm^{-1}, which was near the middle of the 100-2100 cm^{-1} range recorded. With the large inorganic mineral particles, a laser power of 20-30 mW at the sample was used.

Most of the samples were obtained from Ward's Natural Science Establishment, Inc. Some field collected specimens were provided by the Department of Geological Sciences at Lehigh University.

RESULTS

The principal aims of this investigation were to determine the

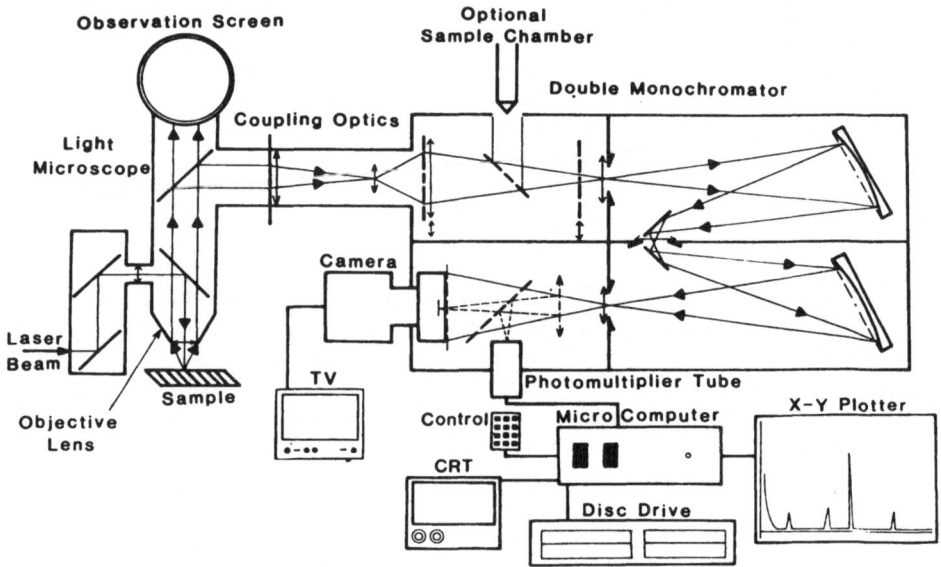

Fig. 1 A schematic diagram of the MOLE laser Raman microprobe system.

laser Raman spectra of selected carbonate and aluminosilicate minerals
and to demonstrate the feasibility of utilizing the laser Raman microprobe for identifying widely occurring and utilized minerals and for
characterizing the impurities contained in these minerals. In addition, it is shown that thermal transformations can be followed by this
analytical technique.

Carbonate Minerals

Carbonate minerals are formed by most of the alkali and alkaline
earth metals and the first row transition metals. Some of these form
extensive rock masses in sedimentary and metamorphosed sedimentary
rocks. Others occur in evaporite deposits. Minerals of interest
include calcite ($CaCO_3$), magnesite ($MgCO_3$), siderite ($FeCO_3$), rhodochrosite ($MnCO_3$), smithsonite ($ZnCO_3$), and dolomite ($CaMg(CO_3)_2$), as
well as sodium carbonate (Na_2CO_3). The laser Raman spectrum obtained
with a specimen of calcite from Cherokee County, Kansas is shown in
Figure 2A. The bands at 1087 and 714 cm^{-1} arise from vibrational
modes of the carbonate ion, while those at 283 and 154 cm^{-1} are due
to lattice vibrations. The vibrational modes can be assigned as A_{1g}
(int), E_g(int), E_g(ex), and E_g(ex), respectively.[12] The band positions are very reproducible and agree very well with those previously
reported for laser Raman spectra obtained with other calcite samples,
[6,13] as well as those observed with ordinary Raman spectroscopy with
bulk samples.[12,14,15]

116

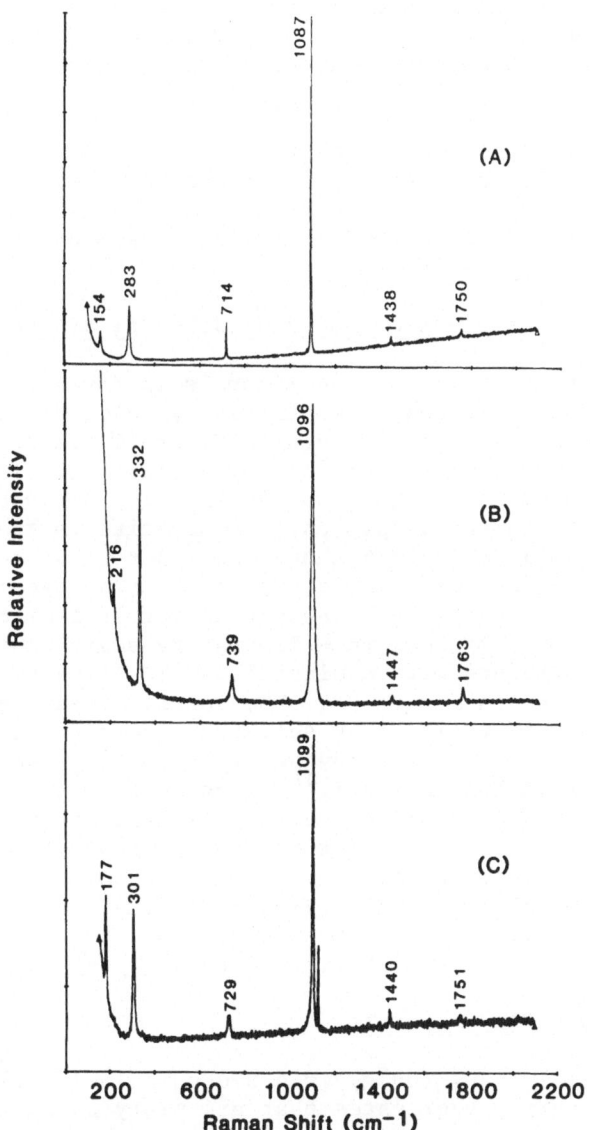

Fig. 2 Raman spectra, obtained with the laser Raman microprobe, of the carbonate minerals: (A) calcite-$CaCO_3$, (B) magnesite-$MgCO_3$, and (C) dolomite-$CaMg(CO_3)_2$.

The laser Raman spectra produced by magnesite from California and dolomite from Essex County, New Jersey are shown in Figures 2B and 2C. The spectrum in Figure 2B was reproduced by magnesite samples from Canada and the U.S.S.R. A comparison of the band positions in Figure 2 demonstrates that a distinctive spectrum is produced by each mineral.

Other minerals that are isostructural with calcite and magnesite include siderite, rhodochrosite, and smithsonite. The laser Raman spectra for these three carbonate minerals are shown in Figure 3. For comparison, the spectrum obtained with sodium carbonate is shown in Figure 4.

Some Thermal Transformations of the Carbonate Minerals

Calcination of the carbonate minerals in dry air produces the metal oxides. This process can be monitored by laser Raman microprobe spectroscopy, and an example is given in Figure 5, where the conversion of siderite to hematite (α-Fe_2O_3) has been complete. In the case of calcite, CaO is produced by calcination in air. However, if a sulfur-containing environment is present, the calcite is transformed into anhydrite ($CaSO_4$). Figure 6 shows the laser Raman spectrum, obtained with a 488 nm defocused beam to prevent thermal decomposition of the coal matrix, of a particle of anhydrite formed from calcite during the mild air oxidation of coal in a laboratory scale fluidized bed reactor. The spectrum obtained with a mineral sample of anhydrite from Balmot, New York is shown in Figure 6B. The calcite-to-anhydrite transformation has also been verified by scanning electron microscopy accompanied by energy dispersive X-ray analysis in the electron microscope with Pennsylvania Montour coal.[16] Thus, the calcite has functioned as a trap for sulfur produced by the oxidation of pyrite (FeS_2) contained in the coal. The latter process proceeds through a series of iron compounds including ferrous sulfate ($FeSO_4$), the laser Raman spectrum for which is shown in Figure 7. It is evident from Figures 6 and 7 that small crystals of anhydrite and ferrous sulfate are readily distinguished by this technique.

Aluminosilicate Clay Minerals

Kaolinite ($Al_4Si_4O_{10}(OH)_8$) is a hydrous aluminosilicate that usually occurs as earthy aggregates that are platy in nature. Kaolinite is formed by weathering or hydrothermal decomposition of other aluminosilicates, especially of feldspars (as found in Czechoslovakia, China, and Cornwall, England) and granite (as eroded and deposited in Georgia and North Carolina, USA). It is typically very fine grained and has industrial importance in the manufacture of ceramics, as a filler in paper products, and as a pigment extender and filler in printing inks. Commercial kaolin deposits often contain small amounts of anatase, rutile, gibbsite, mica, quartz, pyrite, and graphite. Small amounts of iron and titanium can occur in kaolinite by isomorphous substitution of aluminum or silicon.

118

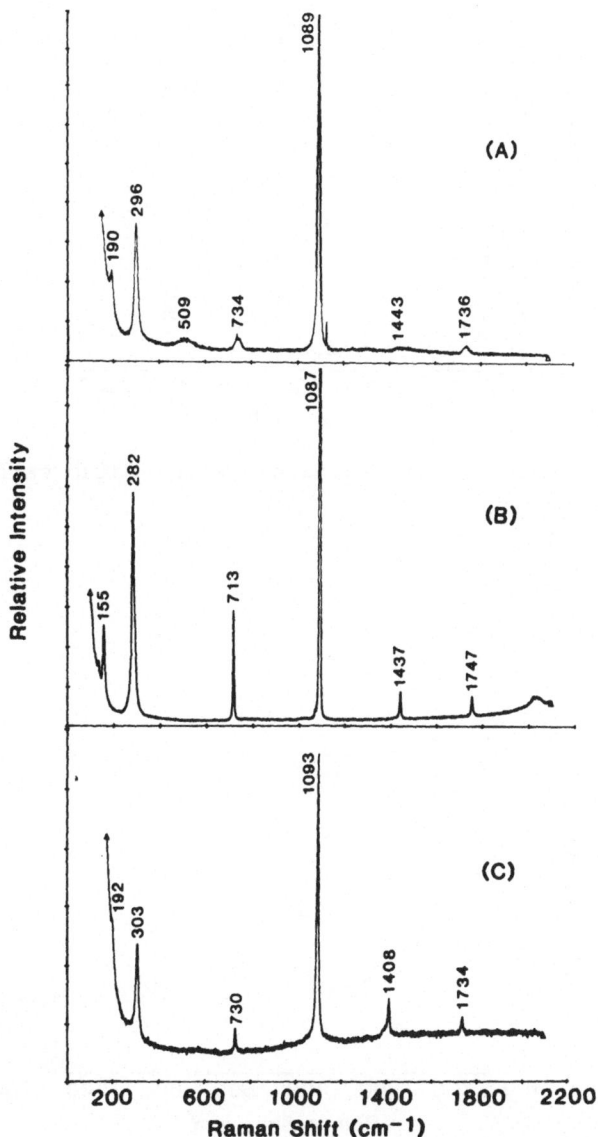

Fig. 3 Raman spectra, obtained with the laser Raman microprobe, of the carbonate minerals: (A) siderite-$FeCO_3$, (B) rhodochrosite-$MnCO_3$, and (C) smithsonite-$ZnCO_3$.

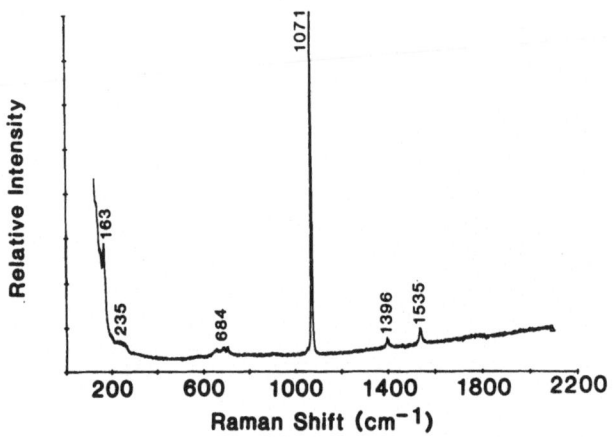

Fig. 4 Laser Raman spectrum obtained with sodium carbonate (Na_2CO_3).

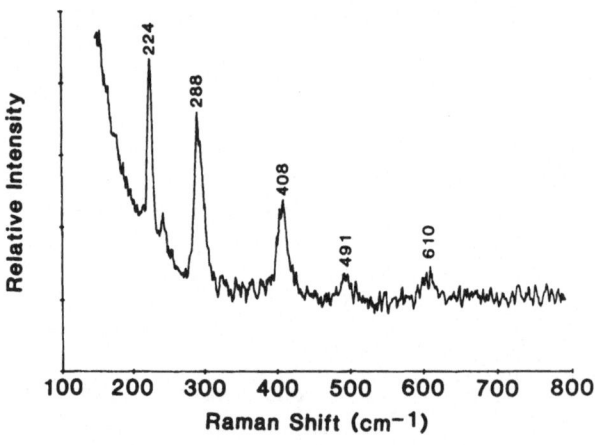

Fig. 5 Laser Raman spectrum of hematite (α-Fe_2O_3).

Other groups of clay minerals are chlorite, mica, and montmoril-
lite. These four clay minerals have many physical features in common,
and it is difficult to distinguish between these soft minerals by

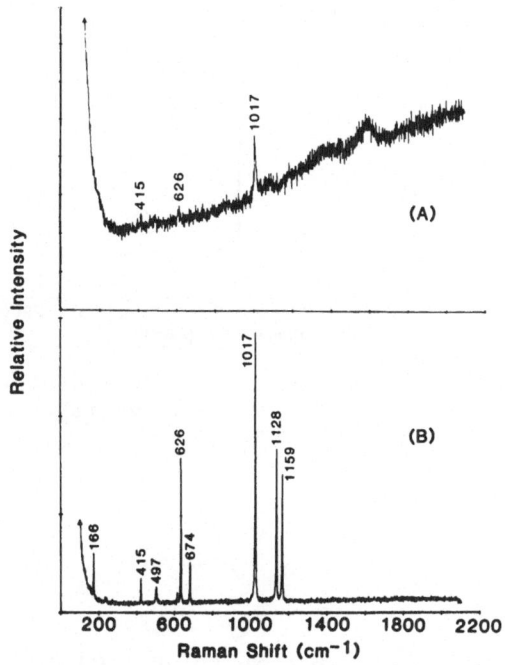

Fig. 6   Laser Raman spectrum (A) of $CaSO_4$ formed from calcite in
Illinois No. 6 coal by treatment in an air-fluidized bed
reactor at 350°C. The spectrum (B) obtained with mineral
anhydrite ($CaSO_4$) is also shown.

their physical properties. Indeed, it has been pointed out that the
positive identification of these minerals in clays is one of the most
exacting problems for a mineralogist.[17] In particular, chlorite in
clays is always mixed with other clay minerals and is difficult to

detect. Although the Raman bands produced by these aluminosilicates have low intensities, the utility of the laser Raman microprobe in detecting and characterizing these minerals will be demonstrated.

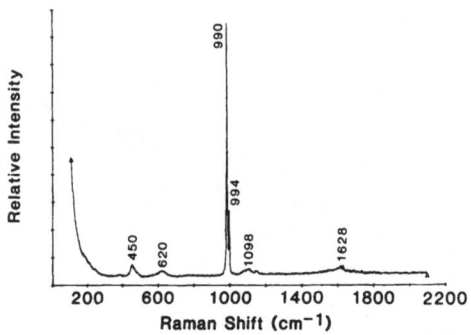

Fig. 7   Laser Raman spectrum of hydrated synthetic ferrous sulfate ($FeSO_4 \cdot 7H_2O$) of at least 99.5% purity.

Soft kaolinite that has been pulverized and air classified has a small and rather uniform particle size. Figure 8 shows the particle size distribution for such a kaolinite from Twiggs County, Georgia. It can be seen that the fraction under 6 microns constitutes 84% of the sample, while the same proportion of the sample falls in the 0.2-10 micron range. This sample has a finer particle size than does a typical Spanish kaolinite.[18] Impurities such as quartz and mica tend to have larger particle sizes and can be separated from the kaolinite at the coarse end of the distribution. However, TiO_2 tends to have a small particle size and to distribute itself throughout the kaolinite. Figure 9A shows the laser Raman spectrum of TiO_2 contained in the white kaolinite, while Figure 9B is the spectrum produced by a pure sample of TiO_2. Other samples of kaolinite that were examined also produced the TiO_2 spectrum, but this impurity was not observed in a sample of halloysite, a polymorph of $Al_4Si_4O_{10}(OH)_8$, from Eureka, Utah.

Montmorillonite ($Al_2Si_4O_{10}(OH)_2 \cdot xH_2O$) is an earthy layered mineral that always deviates from the ideal formula given. The laser Raman spectrum of a montmorillonite sample is given in Figure 10A. Benton-

122

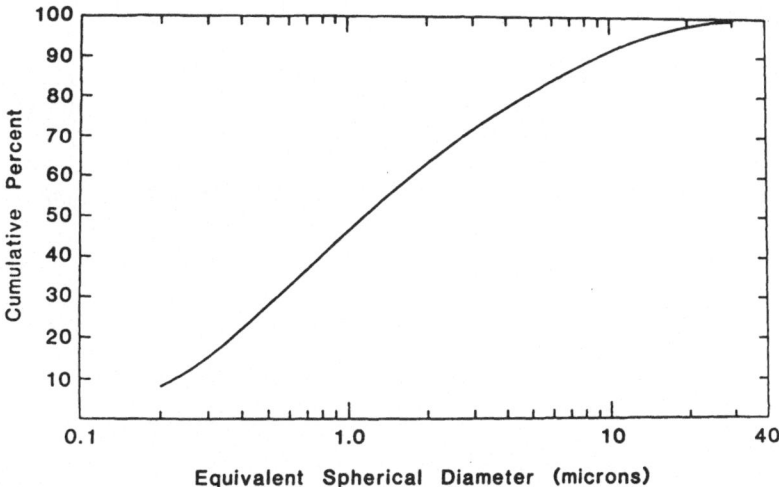

Fig. 8   Particle size distribution of air-classified white kaolinite
from Georgia.   The centrifugal Casagrande hydrometer method
was utilized for the analysis.

ite is a commercially important rock that is composed principally of
montmorillonite. Because of its rheological characteristics, the
primary consumer of bentonite is the drilling mud industry. However,
bentonite is also utilized as a thickening agent in aqueous organics
such as adhesives, caulks, inks, and paints. The laser Raman spectrum
of a gray bentonite from Bellefonte, Pennsylvania is shown in Figure
10B, where the spectrum was produced from approximately one-half the
number of counts that were collected for the generation of the spectrum in Figure 10A. Broad bands at approximately 1345 and 1610 cm^{-1}
are clearly evident, and these are due to degraded organic compounds.
These bands can be compared to those in Figure 6A that were produced
by the coal matrix. Therefore, it is shown that this bentonite contained carbonaceous material. In addition, trace quantities of TiO_2
(with bands at about 640, 513, and 400 cm^{-1}) and $CaCO_3$ (with bands at
approximately 1085, 713, and 278 cm^{-1}) were present. The 1008 cm^{-1}
band might be due to a pyroxene mineral, e.g. $MgSiO_3$, although a
weaker band at 650-680 cm^{-1} would also be expected to be present.[19]

The layered structure of the aluminosilicate clays is most evident
in the micas. The laser Raman spectrum of muscovite mica ($KAl_2(AlSi_3
O_{10})(OH)_2$) from Maine is shown in Figure 11A. This mica was very
transparent, and the spectrum shown was produced from three scans of
the spectrum. For comparison, the spectrum produced by a sample of
biotite mica from Bancroft, Ontario is exhibited in Figure 11B. In
this mica, the two aluminum atoms in the six-coordination sites are
replaced by iron and magnesium. For comparison, the laser Raman
spectrum of a green chlorite ($(Mg,Fe)_5Al(AlSi_3O_{10})(OH)_8$) sample from
Calaveras County, California is shown in Figure 12.

DISCUSSION

The results presented in the previous section clearly demonstrate the utility of using the laser Raman microprobe in identifying carbonate minerals and aluminosilicate clays. In addition, the resolution and sensitivity of this analytical technique provides for the detection and identification of very small quantities of impurities in

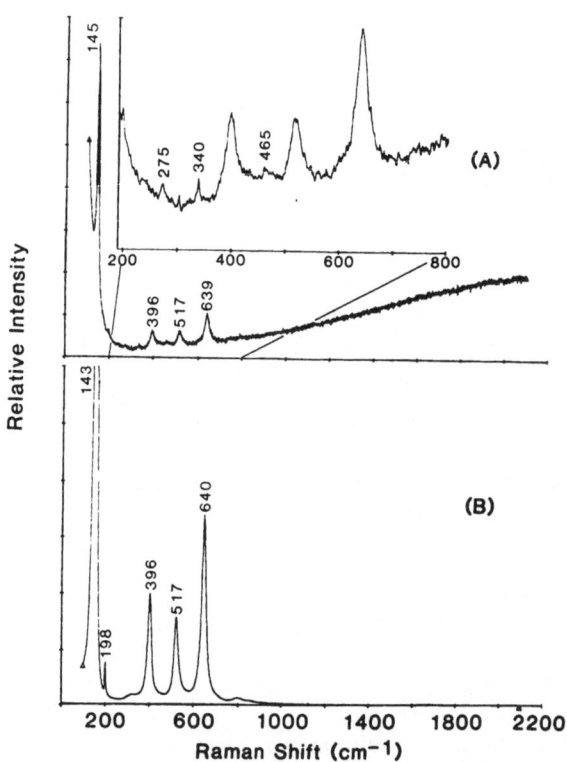

Fig. 9 Laser Raman spectrum (A) obtained with the Georgia kaolinite, for which the dominant bands are due to a TiO_2 impurity. The 200-800 cm^{-1} expansion indicates three bands attributable to the kaolinite. The spectrum (B) of an authentic TiO_2 sample is shown for comparison.

materials. The wing of the elastic scattering peak, which begins at about 300 cm^{-1} does not interfere with the analyses. Fluorescence that occurs at higher wavenumbers with some of the samples is usually not severe enough to interfere with observation of the Raman spectra produced by the laser Raman microprobe.

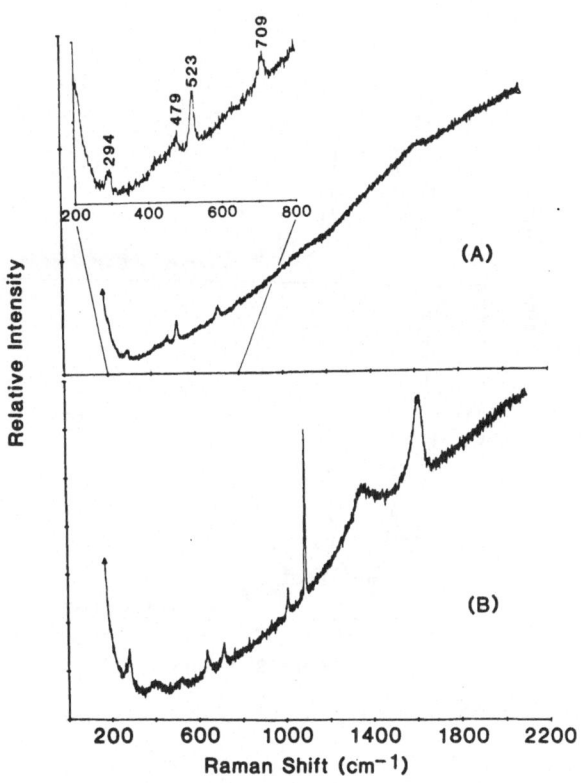

Fig. 10 Laser Raman spectra of (A) montmorillonite and (B) bentonite.

The Raman active modes of the planar triangular carbonate anion in aqueous solutions of the alkali carbonates are observed at wavenumbers that are similar to those produced by the solid phase alkali carbonates. For example, the A_{1g} (ν_1), E_g (ν_3), and E_g (ν_4) vibrational modes of the aqueous alkali carbonates are observed at 1064–1069, 1415, and 680 cm^{-1}, respectively,[15] and they can be compared with the values given in Figure 4 for solid Na_2CO_3, which were obtained with the laser Raman microprobe. The set of five bands produced by normal Raman spectroscopy with a bulk sample of $CaCO_3$ is

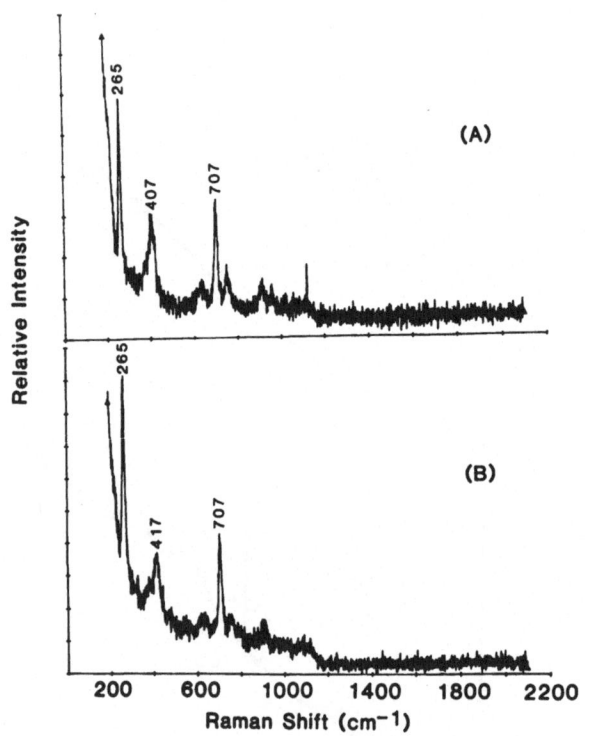

Fig. 11 Laser Raman spectra of (A) muscovite mica and (B) biotite mica

found at 1441, 1086, 718, 283, 156 cm^{-1}.[15] The analogous data obtained from a large single crystal of calcite using a He/Ne laser source are 1432, 1088, 713, 283, 156 cm^{-1}.[20] The spectrum for calcite obtained by the laser Raman microprobe (Figure 2) yield the corresponding wavenumbers of 1438, 1087, 714, 283, and 154 cm^{-1}. Thus, the Raman spectra of carbonate samples with different crystallite sizes are essentially identical. In addition, the internal vibrational modes of CO_3^{2-} are the same whether the particular carbonate is in the solid state or is in solution.

126

The five simple carbonate minerals included in Figures 2 and 3, excluding dolomite, are isostructural and have the calcite crystal structure. However, the divalent cation size is different in each case and decreases in the series Ca/Mn/Fe/Zn/Mg = 0.99/0.80/0.76/0.74/0.65 Å. Considering the two lower wavenumber bands in the Raman

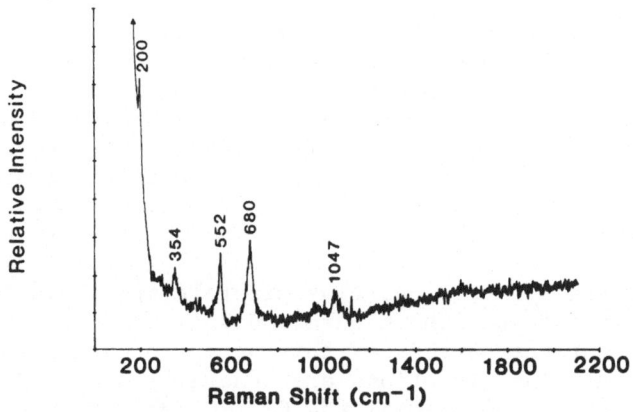

Fig. 12   Laser Raman spectrum of green chlorite.

spectrum of each mineral, that is the E_g(ex) lattice vibrational modes, the band positions increase in wavenumber in the series Ca<Mn<Fe<Zn <Mg. Thus, the frequencies of the two vibrational bands are directly related to the ionic size of the cation. The trend in increasing wavenumbers in the divalent cation series then also follows the trend in increasing charge/radius ratio along the series, which correlates to the degree of compression of the carbonate anions around the metal cation. The internal carbonate vibrational modes do not follow this systematic sequence.

The seven metal carbonates that were examined by the laser Raman microprobe have distinctive spectra (Figures 2-4). The four or five Raman bands at lower wavenumbers form a unique set for each carbonate mineral. Thus, magnesite can easily be distinguished from calcite, dolomite, and the other carbonates that have been characterized. Likewise, siderite can be readily identified (Figure 3) and distinguished from rhodochrosite and smithsonite, as well as from the alkali and alkaline earth carbonates. These analyses can be achieved on a microscopic scale and utilize very small quantities of samples on a non-destructive basis. In addition, the transformation of these materials into other compounds such as oxides (Figure 5) and sulfates (Figures 6 and 7) can be monitored by this technique.

The clay minerals have layer structures that are based on $(Si_2O_5)_n^{2n-}$ units, where the layers are made up of tetrahedra with three shared corners. Kaolinite has a two-layer dioctahedral structure with Si-O tetrahedrons on the bottom half of the layer and Al-O,OH octahedrons on the top half of the layer. Montmorillonite and mica have three-layer dioctahedral structures with layers of Al-O,OH or Al,Mg-O,OH octahedrons sandwiched between connected layers of Si,Al-O tetrahedrons. On the other hand, chlorite has a three-layer triochtahedral structure consisting of montmorillonite units spaced by separate Al, Mg-OH octahedrons. The space occupied by the latter units in chlorite are filled with water in montmorillonite and with potassium ions in muscovite mica. Thus, these minerals have distinct structures that should each produce a unique Raman spectrum, although the spectra of montmorillonite and mica should have some similarities.

The spectra for the four clay minerals are shown in Figures 9-12, and it is evident that the minerals display distinct Raman bands. While it is not the purpose of this paper to make band assignments, the Si-O stretching vibrations are generally observed in the 700-1200 cm^{-1} range, while the Si-O-Si and O-Si-O bending modes produce bands in the 150-800 cm^{-1} region. For example, three-dimensional α-quartz produced a very strong band at 464 cm^{-1} that corresponds to the O-Si-O bending vibration. Librational modes of the OH groups are found in the 600-700 cm^{-1} range. The spectral bands for these materials are typically weak, but repetitive scanning enhances the resultant spectrum.

The weak laser Raman spectra of the aluminosilicate clays can be used to advantage during the detection and identification of impurities in these materials. An example is given in Figure 9, where the TiO_2 was not detected in the kaolinite by X-ray powder diffraction. The TiO_2 produces a very strong spectrum, and it is evident that it can be detected at low impurities concentrations. In this case, the TiO_2 is present as anatase and is not a substitutional impurity in the kaolinite. It is of interest to note that isomorphically substituted titanium is a luminescent activator in kaolinite.[21] However, when TiO_2 is present as an extrinsic oxide impurity, it is a quencher of luminescent phenomena in inorganic solids. Another example of the sensitivity of the laser Raman microprobe is given in Figure 10, where it is evident that carbon, calcite, and anatase are present as impurities in the bentonite sample.

CONCLUSIONS

It has been shown that laser Raman microprobe spectroscopy is a very sensitive technique for providing molecular information on a microscopic scale. Thus, only small quantities of samples are required and small particles and crystallites can be vibrationally characterized. This provides for ready molecular identification, and it has been shown that the common carbonate minerals can be distinguished from one another by this method of analysis. The same is true for the aluminosilicate clay minerals. The detection and identification of trace impurities in natural minerals has also been demonstrated, and it has been indicated that chemical transformations of materials can be monitored and followed by utilizing this method of laser Raman microanalysis.

ACKNOWLEDGEMENTS

This research was partially supported by grants from the U.S. Department of Energy (Contract No. DE-FG22-81PC40802) and the Lehigh University Unsponsored Research Fund. C.E.B. is pleased to acknowledge the receipt of a Lehigh University Buch Fellowship. The authors thank Dr. Dale R. Simpson of the Department of Geological Sciences at Lehigh University for supplying some of the mineral samples that were utilized in this study.

REFERENCES

1. M. Delhaye and P. Dhamelincourt, Raman Microprobe and Microscope with Laser Excitation, J. Raman Spectrosc., 3:33 (1975).
2. G. J. Rosasco, E. S. Etz, and W. A. Cassatt, The Analysis of Discrete Fine Particles by Raman Spectroscopy, Appl. Spectrosc., 29:396 (1975).
3. G. J. Rosasco, E. Roedder, and J. H. Simmons, Laser-excited Raman Spectroscopy for Nondestructive Partial Analysis of Individual Phases in Fluid Inclusions in Minerals, Science, 190:557 (1975).
4. M. Delhaye, Y. J. M. Moschetto, and P. Dhamelincourt, Apparatus for the Non-destructive Examination of Heterogeneous Samples, U.S. Patent 4,030,827 (June 21, 1977); assigned to Institut National de la Sante et de la Recherche Medicale.
5. G. J. Rosasco and E. S. Etz, The Raman Microprobe: A New Analytical Tool, Res. Dev., 28:20 (1977).
6. P. Dhamelincourt, F. Wallart, M. Leclercq, A. T. N'Guyen, and D. O. Landon, Laser Raman Molecular Microprobe (MOLE), Anal. Chem., 51:414A (1979).
7. P. Dhamelincourt, Ph.D. Dissertation, Lille University, France (1979).
8. M. E. Andersen and R. Z. Muggli, Microscopical Techniques with the Molecular Optics Laser Examiner Raman Microprobe, Anal. Chem., 53:1772 (1981).

9. M. E. Andersen, Integrating a Raman Microprobe into a General Microanalytical Problem-solving Scheme, Microbeam Anal., 1984: 115.

10. J. J. Blaha and G. J. Rosasco, Raman Microprobe Spectra of Individual Microcrystals and Fibers of Talc, Tremolite, and Related Silicate Minerals, Anal. Chem., 50:892 (1978).

11. F. Tuinstra and J. L. Koenig, Raman Spectrum of Graphite, J. Chem. Phys., 53:1126 (1970).

12. H. N. Rutt and J. H. Nicola, Raman Spectra of Carbonates of Calcite Structure, J. Phys. C., 7:4522 (1974).

13. H. Buiteveld, F. F. M. deMul, J. Mud, and J. Greve, Identification of Inclusions in Lung Tissue with a Raman Microprobe, Appl. Spectrosc., 38:304 (1984).

14. W. P. Griffith, Raman Spectroscopy of Terrestrial Minerals, in "Infrared and Raman Spectroscopy of Lunar and Terrestrial Minerals," ed. by C. Karr, Jr., Academic Press, New York, 299 (1975).

15. W. B. White, The Carbonate Minerals, in "The Infrared Spectra of Minerals," ed. by V. C. Farmer, Mineralogical Society, London, 227 (1974).

16. R. G. Herman, G. W. Simmons, D. A. Cole, V. Kuzmicz, and K. Klier, Catalytic Action of Minerals in the Low Temperature Oxidation of Coal, Fuel, 63:673 (1984).

17. L. G. Berry and B. Mason, "Mineralogy," W. H. Freeman and Co., San Francisco, 505 (1959).

18. F. Sandoval del Río, M. Lacaba Velasco, and J. M. González Pena, Quantitative Electron Microscope Study of Ceramic Kaolins, J. Mat. Sci. Letters, 2:781 (1983).

19. W. B. White, Structural Interpretation of Lunar and Terrestrial Minerals by Raman Spectroscopy, in "Infrared and Raman Spectroscopy of Lunar and Terrestrial Minerals," ed. by C. Karr, Jr., Academic Press, New York, 325 (1975).

20. S. P. S. Porto, J. A. Giordmaine, and T. C. Damen, Depolarization of Raman Scattering in Calcite, Phys. Rev., 147:608 (1966).

21. L. Coyne, M. Sweeney, and W. Hovatter, Luminescence Induced by Dehydration of Kaolin--Association with Electron Spin Active Centers and with Surface Activity for Dehydration-Polymerization of Glycine, J. Lumin., 28:395 (1983).

SAMPLE CHARACTERISTICS AFFECTING QUANTITATIVE ANALYSIS BY X-RAY POWDER DIFFRACTION

James P. Cline and Robert L. Snyder

New York State College of Ceramics

Alfred University Alfred, New York 14802

Introduction

In the field of materials science, an experimenter often wishes to perform a quantitative phase analysis on a given product. While wet chemistry can give an elemental analysis without difficulty, a phase analysis by methods other than diffraction is difficult if not impossible to perform. With the advent of the powder diffractometer, quantitative phase analysis by X-ray diffraction has become widely known and accepted. Unfortunately this acceptance of has been clouded, as results are often fraught with anomalous inconsistencies.

Recently, two factors have allowed for the isolation of the principal sources of error. The first is the use of spray drying as a sample preparation method [1-3] which eliminates preferred orientation, long the scourge of accurate intensity measurements. The second is the emergence of the automated diffractometer and associated software [4-9], allowing for routine intensity measurements within a specified precision and accuracy. Recent work utilizing these procedures has allowed for the association of sample characteristics with X-ray powder diffraction measurements [1].

Quantitative X-ray Powder Diffraction

The use of X-ray powder diffraction for quantitative analysis is based on the fact that the intensity of a diffraction pattern from a given phase is related to the concentration of that phase in a sample. As the line intensity is dependent on the absorptivity of the mixture as a whole, there is generally not a linear relationship between intensity and concentration. Quantitative analysis is generally performed by the comparison of intensity ratios from the unknown sample to those of a standard. While the analysis can be done by camera techniques, the powder diffractometer is preferred because of the relative ease and accuracy of line intensity measurements. Also, the absorption effects associated with a diffractometer are independent of diffraction angle, allowing them to be omitted in data analysis.

The intensity of a diffraction line i of phase α from a flat powder specimen is given by [9]:

$$I_{i\alpha} = K_{i\alpha} X_\alpha / (\mu/\rho)_m \tag{1}$$

Where K_i is a constant concerning sample, structure and instrument geometry, X_α and ρ_α are the weight fraction and density of phase respectively, and $(\mu/\rho)_m$ is the mass absorption coefficient of the mixture. In some cases, assumptions concerning $(\mu/\rho)_m$ can be made that will allow for simplified procedures to be followed. The most general case is that of a mixture of several components, with one or more being analysed, and $(\mu/\rho)_m$ being unknown. Such an analysis calls for the elimination of matrix effects by the use of the internal standard method.

The Internal Standard Method

The division of equation (1) for line i of phase α by itself for line j of a second phase S, yields

$$I_{i\alpha} / I_{jS} = K' X_\alpha' / X_S \tag{2}$$

Note that $(\mu/\rho)_m$ has canceled out. In this equation, X_α' is the weight fraction of phase α after the internal standard, phase S, has been added, this is related to the desired quantity X_α by

$$X_\alpha = X_\alpha' / (1 - X_S) \tag{3}$$

Solving equation (2) for X_α' and substituting in equation (3)

$$X_\alpha = K'' I_{i\alpha} / I_{jS}$$

Thus K'' is the slope of the linear plot of $I_{i\alpha}/I_{jS}$ vs. X_α.

A more general interpretation of equation (2) is

$$(I_{i\alpha} / I_{jS} (I_{reljS}/Irel_{i\alpha}) (X_S/X_\alpha) = K = RIR_\alpha \tag{4}$$

This equation defines the Reference Intensity Ratio (RIR) [11-12] which is a measure of the ratio of the inherent intensities of the two phases in question, relative intensity and compositional

132

parameters being scaled out. Its value corresponds to K, the slope of the calibration curve previously discussed. Because the variables of the individual measurement are removed, the RIR is theoretically a constant relating the overall scattering power of the two phases. The $I/I_{corundum}$ (or I/I_c) value reported on recent JCPDS cards is the RIR of the phase reported ratioed to corundum, for the 100% lines in a 50:50 weight mixture of the phases α and S. However it should be noted that, for reasons that follow, this value is seldom accurate and should be used only for rough estimates.

A general procedure is to prepare one or more samples each with a known concentration of α and S. A plot of ($I_{i\alpha} / I_{iS}$) vs. X_α, the slope of which is K'', allows the computation of the RIR. Subsequent measurements of $I_{i\alpha} / I_{iS}$ from the unknown, with knowledge of X_S, allows for comparison with the calibration curve or equation (4) to yield X_α of the unknown.

Preferred Orientation

In the operation of a powder diffractometer, the crystallites are reduced to a size range such that, within the volume illuminated by the incident beam, we can assume their orientations to constitute continuous random distribution. Thus with the powder diffractometer at a certain two-theta position, the form of the sample insures that any corresponding d spacing, regardless of its crystallographic orientation, will indeed diffract X-rays. The randomization of the crystallites is critical to any theoretical treatment of measured diffraction line intensities.

The term from equation (1) concerning the concentration of the diffracting phase can be considered in terms of the number of crystallites so oriented as to diffract X-rays, Nf. The assumption of a continuous random distribution of orientations within the particulate sample holds that Nf will remain constant regardless of what orientation the powder block has to the incident beam. But as samples are loaded by some mechanical means, and due to the anisometry of most crystalline powders, some systematic orientation will take place. This systematic variation in Nf manisfests itself in terms of a yet to be modeled variable added to line intersity considerations. Certain sample loading procedures attempt to obtain uniformity in orientations, with limited success. Orientation effects are enhanced by the degree of crystallite anisometry, though it has been found that even corundum will orient to five percent [10].

This phenomenon, commonly known as preferred orientation, has been one the most celebrated errors in the use of X-ray powder diffraction. Its effects can easily cause distortions in measured intensities in excess of 100%. Weak lines can be enhanced while the strong ones can be nearly obscured by background noise. Preferred orientation has long been the limiting factor in routine, accurate, intensity measurements that are paramount to an accurate quantitative analysis.

Obviously if the crystallites themselves were spherical there would be no orientation effects, thus formation of the powder into spherical agglomerates would serve the same function. One of the first quantitative X-ray diffraction studies utilizing spherical agglomeration concerned the analysis of a large number of shale samples [2-3]. Samples were spray dried into spherical agglomerates of about 50 microns in diameter. This size insures that enough crystallites are contained in any agglomerate so as to guarantee a smooth spherical form yet small enough such that a large number are illuminated by the incident beam. Two mechanisms, or dimensional scales, can be considered whereby these agglomerates serve to rid us of orientation effects. One can consider the dynamics of agglomerate drying to be such that the crystallites are deposited in a random orientation within each sphere. Second, the spheres themselves will be randomly oriented in the sample bed. Figure 1 is an SEM micrograph of a typical spray dried powder.

A spray dryer is essentially a heated chamber into which the suspension, or slip, is atomized forming small droplets. Surface tension insures the spherical form while the heat drives off the

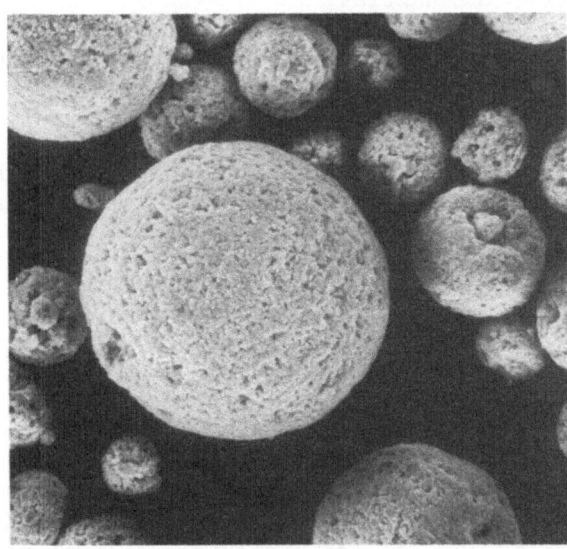

Figure 1. SEM Micrograph of a Typical Spray Dried Material, a 50 - 50 Mix of Si and TiC, 400X.

suspending medium, generally water, before the agglomerates settle
onto the base of the dryer. The slip is a preparation of the powder
dispersed in water, or another non-dissolving volatile liquid, with
a defloculant and binder added. Defloculants are surface active
agents that impart a charge on the particles in suspension, leading
to a progressive repulsive force as interparticle distance falls.
This insures that the particles do not adhere together into flocs,
and also allows for the high solids content desired. Darvan C is
one of the more popular commercial defloculants. Binders are long
chain polymers that simply act as a water soluble glue that holds
the agglomerates together when dry, Evanol is a common PVA compound
used for this purpose.

Spray Drying is used extensively by the many industries
involved in the processing of materials in the powder form. Dryers
of several stories in hight which process hundreds of pounds an
hour are not uncommon. Not so common, however, is one that will
process the one gram samples that the diffractionist may have to
contend with. Early dryers developed at Alfred University for
spray drying of X-ray samples were too large for use in many
laboratories. The authors have recently developed a new type of
dryer that utilizes an ultrasonic atomizer which allows for a
small, table top, size. It should be put into production in the
near future.

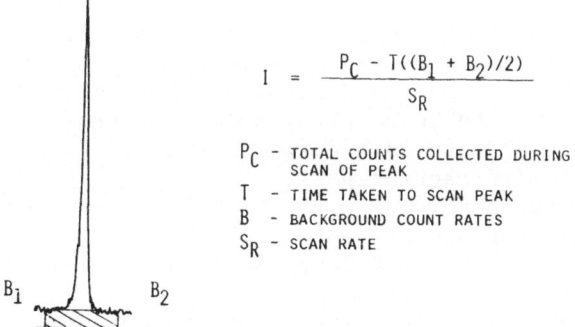

$$I = \frac{P_C - T((B_1 + B_2)/2)}{S_R}$$

P_C - TOTAL COUNTS COLLECTED DURING SCAN OF PEAK

T - TIME TAKEN TO SCAN PEAK

B - BACKGROUND COUNT RATES

S_R - SCAN RATE

Figure 2.   Integrated Intensity Measurement.

Sample and Equipment Parameters

As the intensity of a diffraction line is the parameter that is measured in the use of X-ray powder diffraction for a quantitative analysis, variables affecting this measurement must be evaluated. The measured intensity from a diffractometer may be considered in terms of three variables. The first is that of the elemental composition and structure of the crystalline sample itself. These parameters, addressed in terms of the atomic scattering and structure factors, are constant in this work. The second concerns that of sample parameters relating to its particulate form and history. These would include stoichiometry [13], lattice perfection, crystallite size and shape. The last factor is that of the optical, electrical, and instrumental parameters of the diffractometer.

Two characteristics of any particulate material, crystallite size and strain, affect peak profiles and peak hight, but not their total area. This area, or integrated intensity, is an innate property of the material and should be measured for an accurate quantitative analysis. The integrated intensity is defined as the area under the peak with the background level subtracted out as illustrated in Figure 2.

Of critical interest in this measurement is the statistical accuracy associated with it, which defines a limitation on the ultimate result. The random error, associated with any X-ray intensity measurement is the square root of the number of counts collected. This consideration affects the amount of time one must spend on a given peak, with the time spent being inversely proportional to peak intensity, to obtain a given statistical error. The calculations involved in the propagation of statistical errors through data collection and reduction are very tedious. This problem can be overcome by the use of an automated diffractometer and the implementation of the AUTO and NBS*QUANT84 software packages [4-7].

The crystallite size of the sample has a multitude of most serious effects on intensity measurements. The first to be considered is that of primary and secondary extinction, both of which can be addressed only in terms of dynamical diffraction theory. The net result of both effects is a dramatic increase in the absorptivity of the crystal in diffracting position. For larger and more perfect crystallites, stronger reflections will be measured as less intense relative to weaker ones. The usual solution is to grind the sample to <5um or thermally shock it to destroy crystal perfection.

A second crystallite size effect is the statistical accuracy of intensity measurements. This can be thought of in terms of Nf, which will increase as the crystallite size is reduced. But as Nf is increased, variation in it with respect to a set of sample loadings will decrease. This effect, along with that of extinction was strikingly demonstrated in a study by Klug and Alexander [9]. Results indicate that the sample should be ground to 5um, though <15um is shown to adequate for accuracies to 2%.

Experimental

A study was undertaken to evaluate the use of spray drying as a sample preparation method for X-ray powder diffraction. The evaporation of the suspending medium initiates a flow pattern within the droplet that carries the more mobile species to its surface. This phenomenon may be considered a driving force that may result in an agglomerate with the coarse phase segregated to the interior while the fines coat the surface. Should these fines be highly absorbing, the interior phase may be shielded from the incident beam and thus measured intensities would be distorted.

Filtration theory was studied and applied to the dynamics of microstructure development during the dewatering of droplets. Sample and spray drying parameters identified by the review of filtration mechanisms were varied to produce a "worst case" situation, wherein the tendency for particle migration was maximized. Standard procedures were followed to produce a "homogeneous" series of samples. A third series was prepared in which the slip was not milled prior to spray drying, designated "no grind".

In order to test for the effects of the contrast in X-ray absorptivity in multiphase systems, two sets of samples were prepared varying concentration. The first set of three series utilized silicon ground from single crystal boules and Linde C Al_2O_3, with linear absorption coefficients of $141.2cm^{-1}$ and $126.1cm^{-1}$ respectively. The second set, with two series, was prepared from the same silicon mixed with TiC, $\mu = 814.1cm^{-1}$.

Of the parameters that effect particle migration under the influence of a fluid flow, the relative particle size of the "stationary" vs. "mobile" phases was found to be the most important. To this end, the particle size of the starting materials was varied, with the two phases used in the "homogeneous" series having relatively similar distributions. While the size ranges of those in the "worst case" were prepared to be in sharp contrast.

The particle size of all starting materials was measured using a light scattering sedigraph. Figure 3 indicates the particle size distributions of the material in the $Si-Al_2O_3$ "homogeneous" and "no grind" series. Slip preparation of the "homogeneous" series called for a five minute mill time of the previously ground and sieved silicon. The data presented as Si "milled" is that of the "prepared" silicon milled to simulate the slip preparation procedure.

137

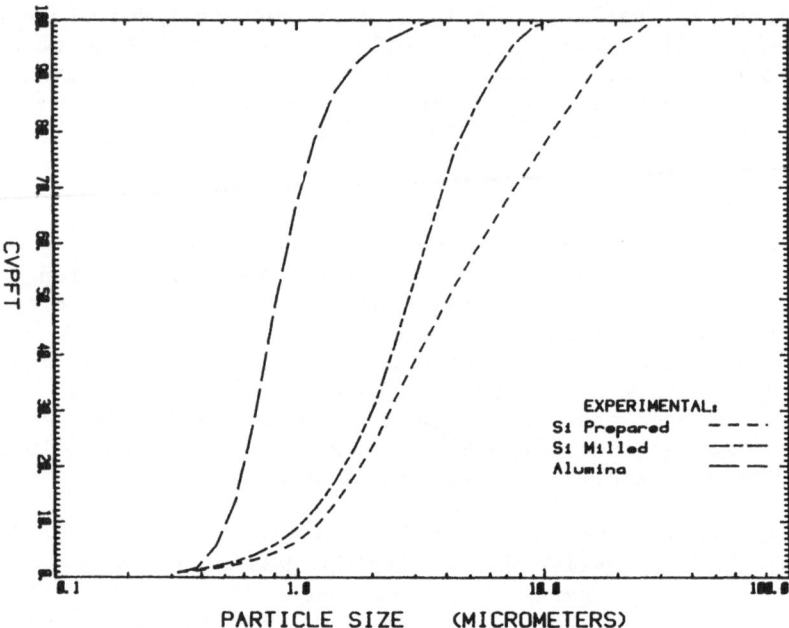

Figure 3. Particle size distributions of the Si and Al_2O_3 contained in the Si-Al_2O_3 "homogeneous" and "no grind" series.

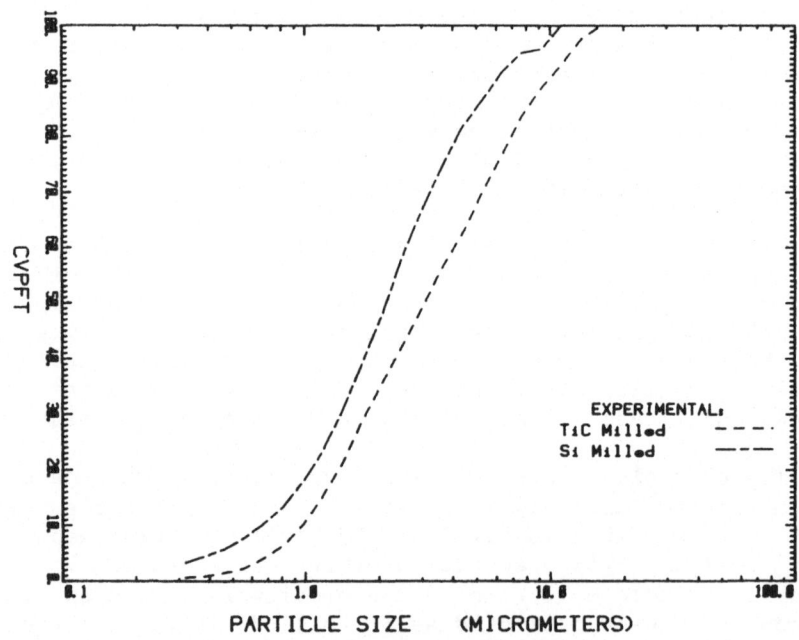

Figure 4. Particle size distributions of the Si and TiC contained in the Si-TiC "homogeneous" series.

138

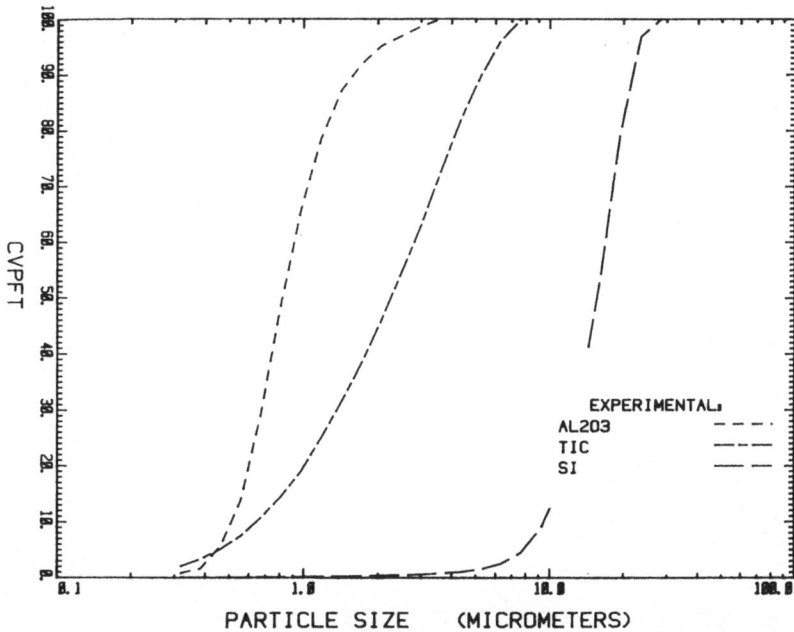

Figure 5. Particle size distributions of the Si, TiC, and
Al_2O_3 used in the Si-Al2O3 and Si-TiC "worst case"
series.

The 1 micron Linde C was de-agglomerated by milling prior to all
measurements and preparations. Figure 4 indicates the size
distributions of the Si and TiC contained in the Si-TiC "homogeneous" series. Figure 5 is that of the material contained in the
"worst case" preparations.

After spray drying, selected samples were vacuum cast in epoxy
and polished to reveal a sectional view of their interior microstructure. Examination on an optical, DICM, microscope indicated
the desired inhomogeneous microstructures had been obtained. These
inhomogeneties were found to have an immeasurable effect on X-ray
data, however, the particle size was found to have a most dramatic
effect on X-ray intensities measured.

All X-ray data was collected on an automated Philips diffractometer using the Auto control algorithm [5]. Integrated
intensities were measured with a precision of up to 1% with statistical errors being tracked throughout data collection and reduction.
Six Al_2O_3 lines were measured, while three lines from the Si and TiC
phases were scanned. Data were reduced via the NBS*QUANT84 [6] to
yield a Reference Intensity Ratio for all combinations of
reflections.

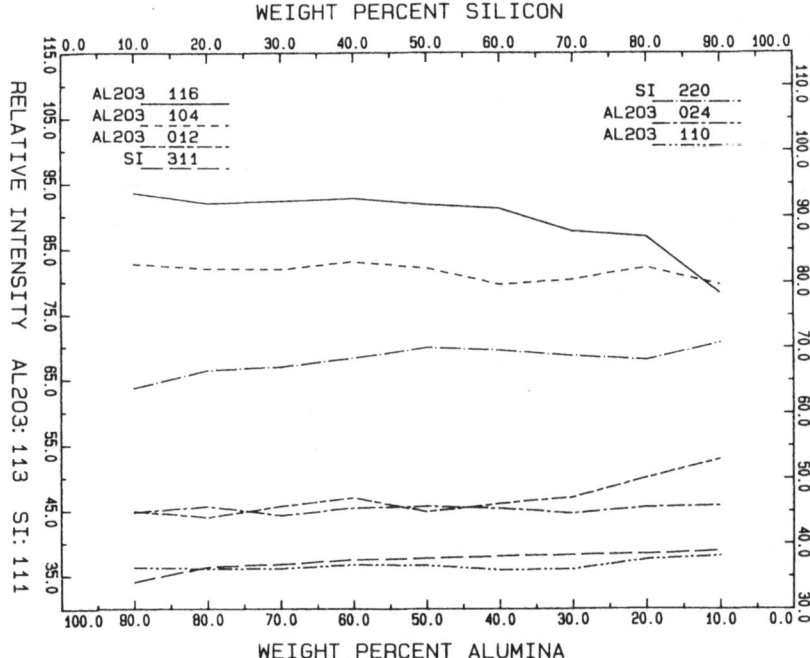

Figure 6.   Relative intensity data from the $Si-Al_2O_3$. "homogeneous" series.

Figure 6 is a plot of the relative intensities of the 3 Si lines and 6 Al_2O_3 lines used for the RIR determination in the $Si-Al_2O_3$ "homogeneous" series. Random counting errors ranged from 2 to 10%. The lack of variation in the relative intensity with each sample is indicative of the consistent elimination of preferred orientation by the spray drying process. Of the two Al_2O_3 lines that do deviate, one was caused by a peak overlap that induced errors in background determination. The second was caused by a diffuse peak of an anomalous Si hydration phase beneath the Al_2O_3 peak. Both lines were eliminated from subsequent RIR calculations.

Shown in Figure 7 is a plot of RIR (Al_2O_3/Si) vs. composition for the three $Si-Al_2O_3$ series. A difference in the relative positions of the lines is noted; RIR values vary from .21 to .32, a 34% change. If a shielding effect had been operative, reflections of the Al_2O_3 would have been enhanced and the "worst case" values would have been greater, not less than the "homogeneous" values. As the 1 micron particle size of the Linde C is considered to be constant, the effects shown are believed to be caused by the variation in the particle size of the silicon. Comparison of the particle size and RIR data indicates that with the size of the silicon particles a corresponding increase in the reflected intensity occurs. As the particle size of the silicon is well

140

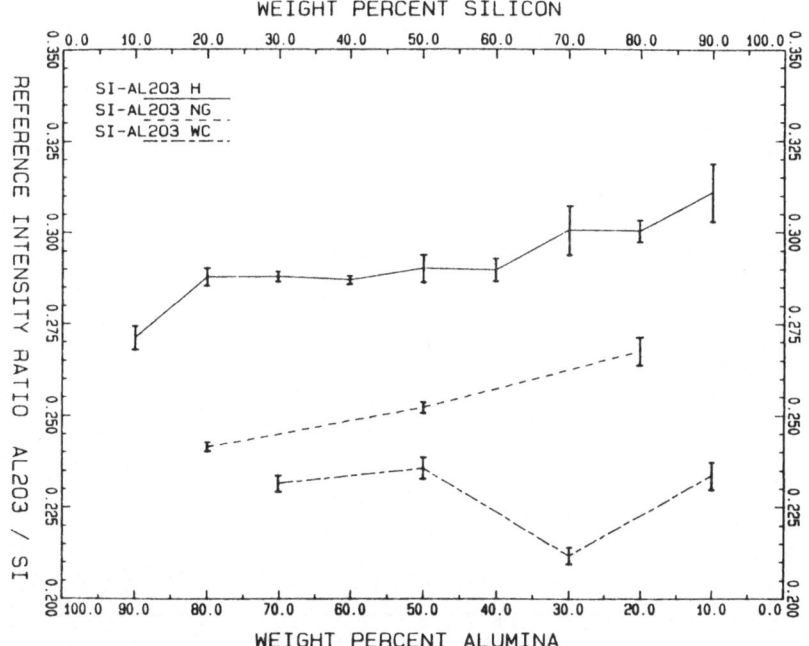

Figure 7. Reference intensity ratio data from the three Si-Al$_2$O$_3$ series.

above that of the alumina in all preparations, the true RIR value, one determined in the absence of particle size effects, would be well above any determined in this study.

A particle size related enhancement of reflected intensities is also noted in the Si-TiC data shown in Figure 8. Inspection of Figure 4 reveals that the size distribution of the TiC is somewhat greater than that of the Si in the "homogeneous" series. This condition is reversed in the "worst case" series with the Si being much coarser than the TiC. Thus the true RIR value would be between these two data sets, consideration of the particle size data would indicate it to be just below the "homogeneous" values.

The positive slope noted in three of the five RIR vs. composition data sets is indicative of a systematic effect related to the change in each mixture's average absorption coefficient. The RIR values of the "homogeneous" and "no grind" series in the system of Si-Al$_2$O$_3$ are both seen to approach the true RIR as the absorption coefficient of the mixture approaches that of the larger particles. The nearly indicernable curvature possessed by these data sets and the nearly parallel nature of them indicate that the true RIR will never be reached by either particle size condition regardless of composition. Thus the effects noted are directly related to the particle size and absorption coefficients of the phases present and

141

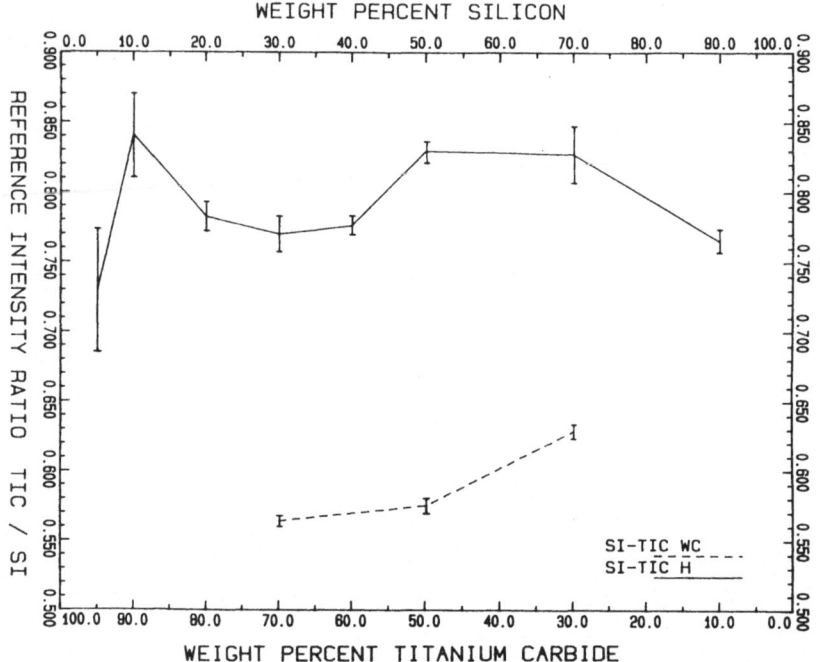

Figure 8. Reference intensity ratio data from the two Si-TiC
series.

are influenced by the absorption coefficient of the mixture. The
trend of the "worst case" Si-TiC series also approaches the true RIR
value in the same manner as did the two Si-Al$_2$O$_3$ series. While the
variation in the data for the "worst case" Si-Al$_2$O$_3$ and Si-TiC
"homogeneous" series cannot be explained at this time, it is noteworthy that the two points containing the highest fraction of Si for
both series indicate an approach to the true RIR value.

G. W. Brindley considered the effect of particle size on
intensity of X-ray reflections in his 1945 analysis [14]. His
theory considers multidomain particles of varying sizes. When a
given domain, within a larger particle, is in the reflecting
position, both the incident and reflected beam will be unduly
influenced by the absorptivity of that particle as opposed to the
mixture as a whole. The absorption of the beam within the particle
was termed "microabsorption" as opposed to the absorption of the
sample bed, "macroabsorption". The term microabsorption has since
been adopted by the X-ray community to explain any anomalous
intensity data. Brindley's theory fails to predict the trends noted
in this work, which isn't very surprising as his theoretical bases
does not apply to the single crystallite materials used in it.

Conclusions and Recommendations

Recent studies relating the character of a particulate sample to subsequent diffraction intensities affords the researcher insights to increase the reliability of X-ray results. Though the errors caused by neglecting these effects have been shown to be most severe, the trends isolated allow for a grounded evaluation of measurements. The following recommendations should be noted:

1) The elimination of preferred orientation by spray drying the sample is essential for an accurate quantitative analysis.

2) Owing to the severity and range of crystallite size effects, samples must be ground to a sub five micron size range.

3) The character of materials used in the determination of the standard curves should be matched to that of the unknown with respect to particle size distribution, crystalline perfection, and stoichiometry.

4) The composition of the samples prepared for calibration purposes should approximate those in the unknown.

5) Particular attention should be paid to particle size and compositional considerations in systems where a large degree of absorption contrast is noted.

5) Use of automated equipment and associated software will allow for a reduction in time expended and an increase both precision and accuracy of results.

Acknowledgement

The authors gratefully acknowledge the support of the The National Science Foundation (Grant No. DMR-8115242) and of the NYS College of Ceramics.

References

1. J. P. Cline and R. L. Snyder, Adv. X-ray Anal., 26, 111-118 (1983) .
2. S. T. Smith, R. L. Snyder and W. E. Brownell, Adv. X-ray Anal., 22, 181-192 (1979).
3. S. T. Smith, R. L. Snyder and W. E. Brownell, Adv. X-ray Anal., 22, 77-88 (1979).
4. R. L. Snyder, C. R. Hubbard and N. C. Panagiotopoulos, Adv. X-ray Anal. 25, 245-260 (1982).
5. R. L. Snyder, C. R. Hubbard and N. C. Panagiotopoulos, National Bureau of Standards Publication NBSIR 81-2229, 102 pages (1981).

6. R. L. Snyder and C. R. Hubbard, NBS Special Publication, (in press 1985).

7. C. R. Hubbard, C. R. Robbins and R. L. Snyder, Adv. X-ray Anal. 26, 149-157 (1983).

8. R. L. Snyder , p.449-464 in "Advances in Material Characterization", D. R. Rossington, R. A. Condrate, and R. L. Snyder, editors, Plenum Press, New York, 1983.

9. H. P. Klug, L. E. Alexander, X-Ray Diffraction Procedures, 2nd ed., John Wiley and Sons, New York, New York, 1974.

10. R. L. Snyder and W. L. Carr, "Surfaces and Interfaces of Glass and Ceramics", ed., V. D. Frechette p. 85-99 (1974).

11. F. H. Chung, J. Appl. Cryst. 8, 17 (1975).

12. C. R. Hubbard, E. H. Evans and D. K. Smith, J. Appl. Cryst., 9, 169 (1976).

13. R. C. Gehringer, G. J. McCarthy, and R. G. Garvey, Adv. X-ray Anal., 26, 119-128, (1983).

14. G. W. Brindley "The Effect of Grain or Particle Size on X-ray Reflections from Mixed Powders or Alloys," Phil. Mag. (7) 36, 347 (1945).

144

LASER RAMAN MICROPROBE CHARACTERIZATION

OF FLY ASH

Barry E. Scheetz[1], William B. White[1], and F. Adar[2]

[1]Materials Research Laboratory, The Pennsylvania State University, University Park, PA 16802
[2]Instruments SA, 173 Essex Avenue, Metuchen, NJ 08840

INTRODUCTION

Fly ash is a byproduct of the combustion of coal and is currently being produced at the rate of approximately 80 million tonnes annually in the United States. The annual usage of fly ash is only about 10% of the production with the remaining 90% requiring disposal. Fly ash has been utilized as both an extender and as a reactive admixture to concretes. In fact, some compositions of fly ash exhibit hydraulic properties when mixed with water.

The chemical reactivity of fly ash in the presence of lime (CaO) is a pozzolanic reaction. This reaction in high-lime fly ashes is in part attributable to the presence of many of the crystalline phases that are found in cement, CaO, $CaSi_2O_5$, brownmillerite and gypsum. However, the majority of the pozzolanicity is attributable to an x-ray amorphous, glassy component in both the high- and low-lime fly ashes. Diamond et al. (1981) measured the amorphous contribution to the x-ray diffraction patterns of twelve fly ashes. Their study indicated that the broad maximum in the diffraction pattern can be linearly correlated with the CaO content of the fly ash up to about 20 weight %. At concentrations of CaO above this value the maximum in the amorphous component of the diffraction pattern was centered at $\approx32°2\theta$. These data suggest the presence of at least two distinct types of glasses in the bulk fly ash.

The disadvantage of bulk characterization techniques for fly ashes is the heterogeneity from particle to particle. Fly ash is an intimate mixture of multiphase materials which are derived from

145

the thermal alteration of discrete minerals during the combustion of the coal. During burning in a modern coal fired power plant, the individual particles of coal and mineral matter move through the burner at high velocity so that particle-particle agglomeration does not take place. Because of these fluid dynamic constraints, the mineralogy of each fly ash grain is dependent only upon the makeup of the coal particle that was burnt. Bulk characterization techniques tell little about the individual particles.

The laser Raman microprobe offers the possibility of structural information from individual fly ash particles. This study examines in detail two fly ashes, with the objective of describing the structure of the glasses which give rise to their pozzolanic reactivity.

COMPOSITION AND MORPHOLOGY OF FLY ASH

Two fly ashes were selected to investigate the effect of the concentrations of CaO. A low-lime fly ash from the Pennsylvania Power and Light Company's Montoursville, PA plant, designated B25, represents the equivalent of the ASTM class F fly ash. The second fly ash was a commercial product sold under the

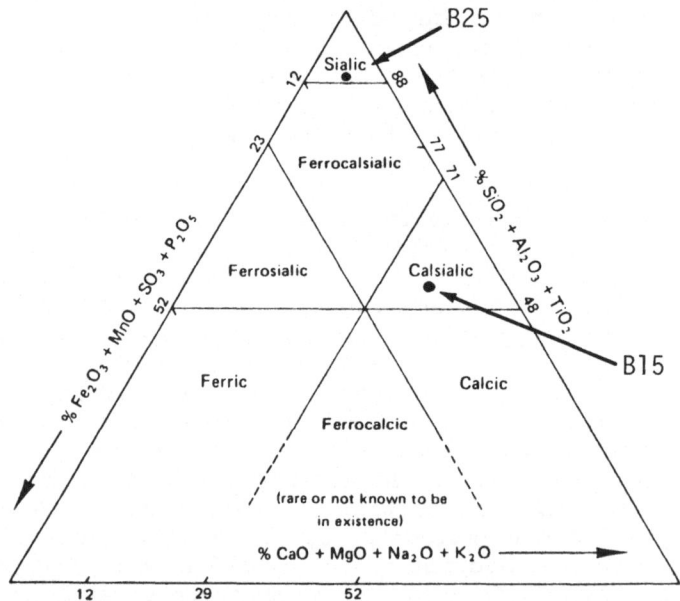

Fig. 1. Compositional classification of fly ashes showing bulk composition of B15 and B25 fly ash samples. Adapted from Roy (1981).

Table 1. Bulk Chemical Compositions in the Fly Ashes Examined.

	B25	B15
SiO_2	58.2	33.2
Al_2O_3	29.8	17.6
TiO_2	1.2	1.55
Fe_2O_3	3.6	6.03
MgO	1.1	5.07
CaO	1.3	29.91
MnO	---	0.07
SrO	---	0.37
Na_2O	0.3	1.03
K_2O	3.2	0.41
P_2O_5	---	0.95
SO_3	0.3	2.77
CO_2*	---	(0.22)
LOI	1.3	0.44
Total	100.3	100.30

*Not included in total.

name of LITPOZ and supplied by Dowell from a power plant in Harrington, Texas. This sample is the equivalent of the ASTM class C fly ash and was designated B15. Following the nomenclature of Roy (1981), the sample B25 would be SIALIC fly ash and the B15 would be classified as a CALSIALIC (Fig. 1). The bulk chemical analyses of these fly ashes are summarized in Table 1.

Although fly ash shows little in the way of diagnostic morphology in the optical microscope, the scanning electron microscope reveals much detail in the individual particles. Melting of the mineral particles as they are swept through the flame produces small spheres such as those illustrated in Fig. 2. Many of the spheres are noncrystalline but some can be demonstrated by acid-etch techniques to be polyphasic, usually with a crystalline phase embedded in glass.

The bulk fly ash contains a mixture of crystalline and noncrystalline particles. Scheetz et al. (1981) characterized the crystalline components by a combination of selective etching and x-ray powder diffraction. Table 2 lists the observed crystalline phases and gives their JCPDF reference numbers. Crystalline phases are also readily identified by Raman spectroscopy. The crystals

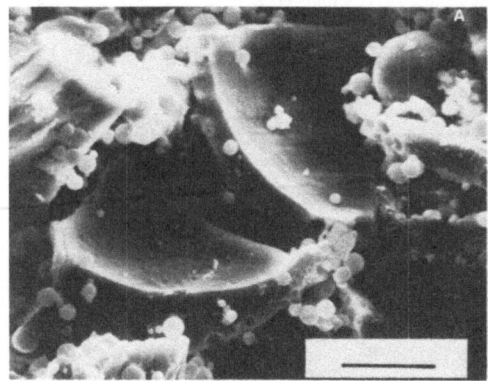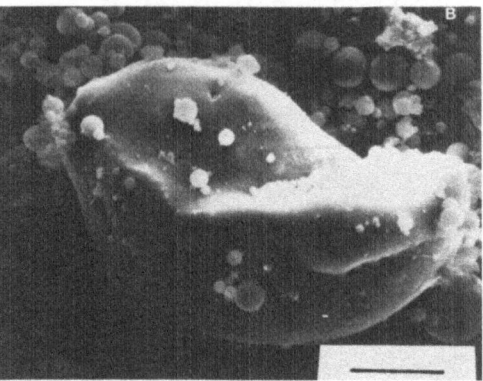
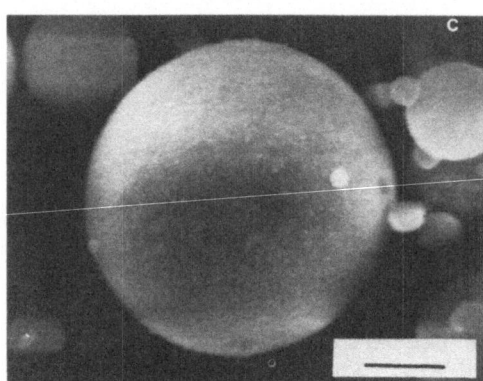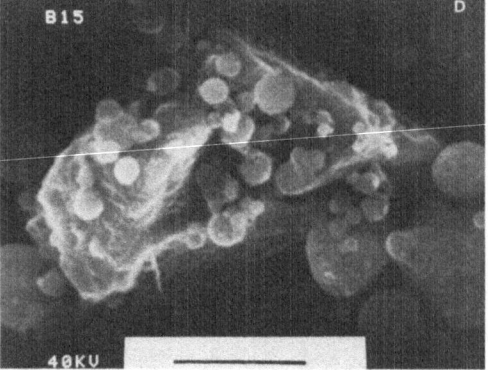

Fig. 2.   SEM images of fly ash particles.   Three frames are from
          sample B25; lower right figure from B15.   Scale bars in
          upper photos are 20 μm; in lower photos the scale bars
          are 5 μm.

with few exceptions produce sharp, well-defined vibrational bands
that are easily recognized above the spectra of the glasses.
Raman spectra of the crystalline phases are discussed elsewnere
(Scheetz and White, in press).

RAMAN SPECTRA OF FLY ASH

     Raman spectra were measured on individual fly ash particles
by spreading a sample of the powder on a microscope slide and
focusing the 514.5 nm line of an argon ion laser onto the
particle.   Spectra were measured on an Instruments SA Ramanor 1000

Table 2. Summary of Crystalline Phases Identified
in Fly Ash Samples.

fly ash	observed phase	JCPDF #	comments
B-15	quartz	5-490	major
	periclase	4-829	minor
	anhydrite	6-226	trace
	magnetite	19-629	minor
	mullite	15-776	minor
	brownmillerite	11-128	minor
	alite	11-593	major
	gehlenite	27-81	trace
	CO_3-apatite	19-272	possibility
	calcium oxide	4-771	possibility
B-25	quartz	5-490	major
	mullite	15-776	major
	α-hematite	13-534	minor
	periclase	4-829	trace
	spinel	16-367	trace
	magnetite	19-629	possibility

microfocus Raman spectrometer. Individual scans of the spectra
were stored in a computer and printed on a Hewlett-Packard
plotter. The spectra shown in Figs. 3-5 are direct traces of the
computer output. Multiple scans and spectral averaging would
remove some of the background noise but do little to improve the
spectral features. The spectra are characteristic of highly
disordered glasses. Distinct bands are clearly defined but are
very weak and very broad.

Figure 3 shows both crystalline and glassy particles from
sample B-15. Figure 3-A is the spectrum of quartz, the most
common crystalline component in the fly ash. As is typical of
framework silicates, the 465 cm^{-1} band, associated with the
Si-O-Si bridging bond is strong but the high wavenumber Si-O
stretching bands are very weak. There is little evidence for
noncrystalline material in this particle. In comparison, the
particle whose spectrum is shown in Fig. 3-B is mostly glassy but
with a small amount of quartz that produces a weak 463 cm^{-1} band.
The glassy phase produces a broad and weak scattering near 1100
cm^{-1}. From comparisons with alkali silicate glasses (Brawer and
White, 1975) and with a selection of natural glasses (White and
Minser, 1984) it is argued that the 1100 cm^{-1} band scattering
feature is due to SiO_4 tetrahedra with one non-bridging oxygen.
The 1050-1100 cm^{-1} scattering is better displayed in Fig. 3-C.

149

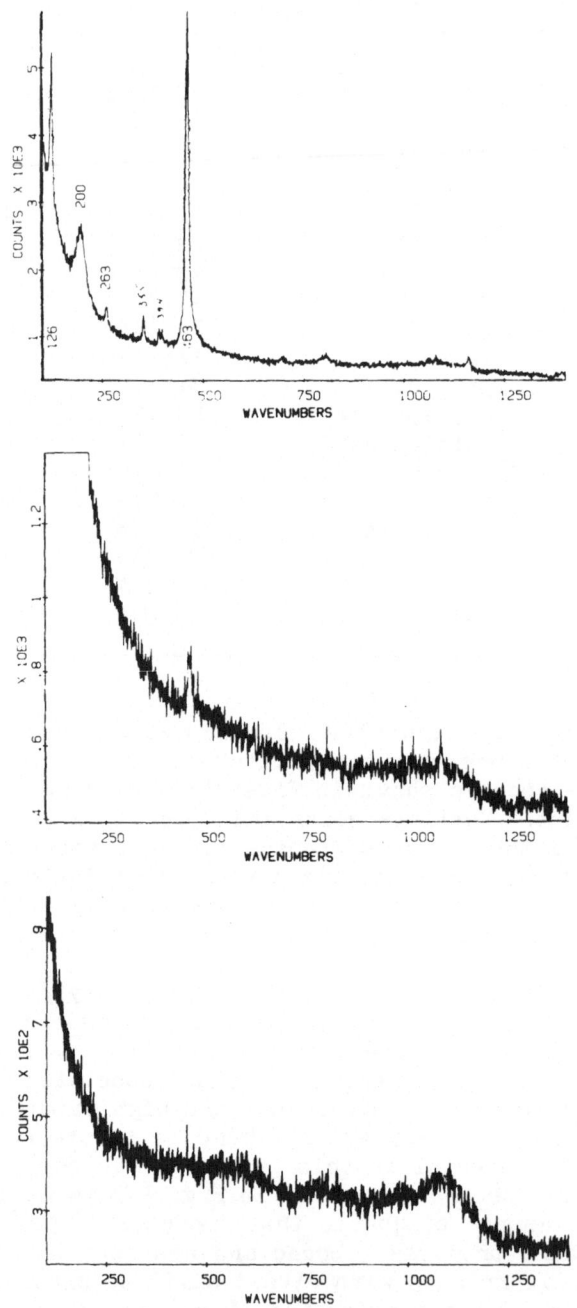

Fig. 3.   Raman spectra from three fly ash particles from sample
          B15.

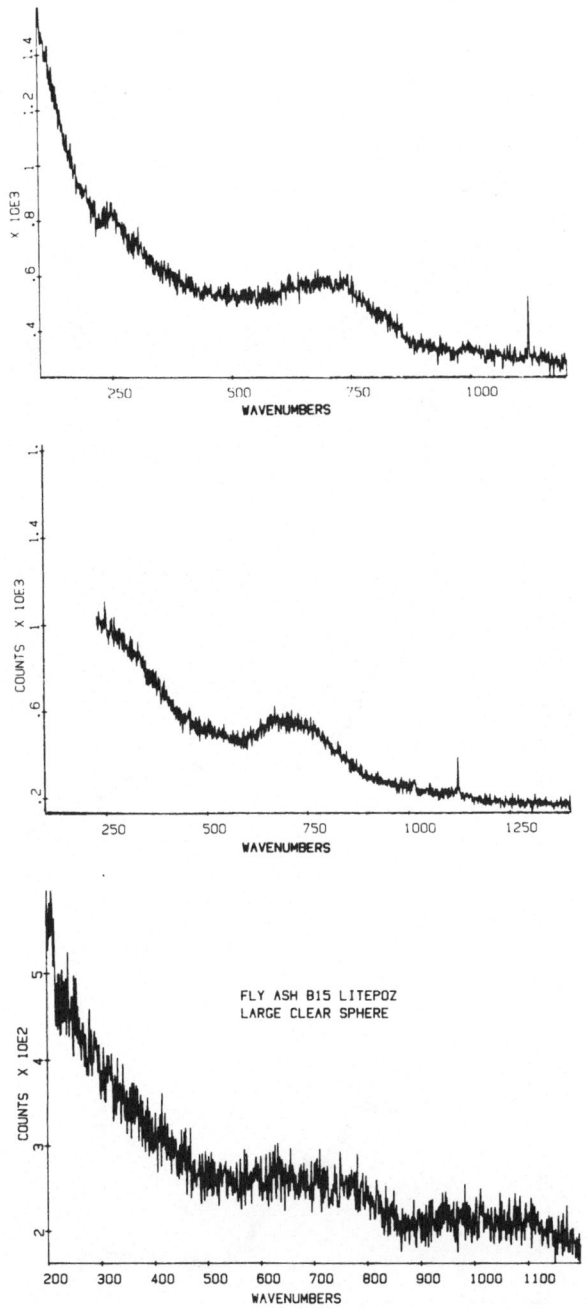

Fig. 4.   Raman spectra of three fly ash particles from fly ash
          B15.

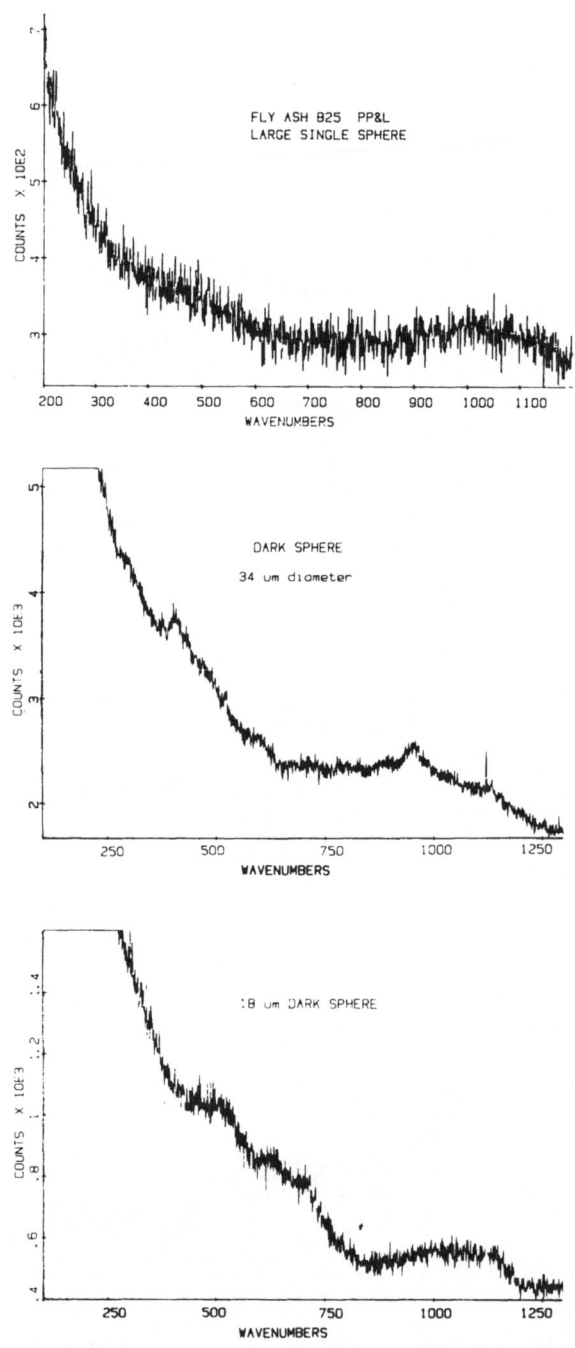

Fig. 5. Raman spectra of spherical particles from fly ash B25.

These particles are silicate glasses partially depolymerized by network modifier cations, probably Ca^{++}. Scattering from a completely polymerized SiO_4 (and AlO_4) framework is very weak, as was illustrated in the spectrum of the quartz particle. Thus one cannot estimate the total number of non-bridging oxygens in these glassy particles.

The spectra shown in Fig. 4 suggest a rather different kind of glass. The strongest scattering (Fig. 4-A,B) occurs at 700-750 cm^{-1} and there is little or no scattering in the 1000-1100 cm^{-1} region. It is instructive to compare these spectra with those of calcium aluminate glasses (McMillen and Piriou, 1983). There is good agreement in the position of the 750 cm^{-1} band with the spectrum of the calcium aluminate glasses. Considering the high calcium and aluminum concentrations in the B-15 fly ash, it is reasonable to attribute the spectra shown in Fig. 4-A,B to calcium-aluminate glasses. It is possible that these glasses are stabilized by some silica in the glass structure. Small quantities would not modify the spectra.

Figure 4-C is the spectrum of a large clear sphere. The two broad bands of low scattering intensity suggest a heterogeneous glass with both silicate and aluminate components.

Figure 5 presents the Raman spectra of three glassy spheres from fly ash B25. The large clear sphere gives the spectrum shown in Fig. 5-A. It appears to be a silicate glass but with very broad and weak scattering. In contrast the spectra of the dark spheres exhibit some narrower bands that are of intermediate line width between the other glass spectra and the spectra of crystals. These bands suggest partial crystallization in these glasses.

The B25 fly ash is more silica-rich than the B15 fly ash and this is reflected in the higher percentage of silicate glass particles. However, calcium aluminate glass particles are also found in the B25 fly ash. It appears that the bulk composition of the fly ash only controls the population of the two types of glass.

CONCLUSIONS

Many fly ash particles are completely glassy. Examination of a selection of these particles by microfocus Raman spectroscopy reveals two populations of glasses. One is a high silica glass with spectra similar to the spectra of obsidians and other natural glasses. The other population of particles is characterized by a broad Raman band near 750 cm^{-1} that is a good match for calcium aluminate glasses, possibly with some silica stabilization. Both kinds of glass particles are present in high silica and high alumina fly ashes. It is suggested that the pozzolanic behavior of fly ash is related to the presence of the aluminate glasses.

REFERENCES

Brawer, S.A., and White, W.B., 1975, Raman spectroscopic investigation of the structure of silicate glasses. I. The binary alkali silicates, J. Chem. Phys., 63:2421-2432.

Diamond, S., 1983, On the glass present in low-calcium and high-calcium flyashes, Cem. Concr. Res. 13:459-464.

McMillan, P., and Piriou, B., 1983, Raman spectroscopy of calcium aluminte glasses and crystals, J. Non-cryst. Solids 55:221-242.

Roy, W.R., Thiery, R.G., Schuller, R.M., and Suloway, J.J., 1981, Coal flyash: A review of the literature and proposed classification system with emphasis on environmental impacts, Illinois State Geol. Survey, ENG 96, Champaign, IL 61820.

Scheetz, B.E., Strickler, D.W., Grutzeck, M.G., and Roy, D.M., 1981, Physical and chemical behavior of selectively etched flyashes, Mat. Res. Soc. Meeting, Symposium N, S. Diamond, Ed., 24-34.

Scheetz, B.E., and White, W.B., in press, Characterization of crystalline phases in fly ash by microfocus Raman spectroscopy, Proc. Materials Res. Soc.

White, W.B., and Minser, D.G., 1984, Raman spectra and structure of natural glasses, J. Non-cryst. Solids 67:45-59.

RAMAN AND LUMINESCENCE SPECTROSCOPY OF ZIRCONIUM OXIDE WITH THE USE OF THE MOLE MICROPROBE

T. E. Doyle and J. L. Alvarez

EG&G Idaho, Inc.
P. O. Box 1625
Idaho Falls, ID   83415

ABSTRACT

Raman and luminescence spectroscopy with the use of the MOLE microprobe has been used to characterize ZrO_2 originating from oxidized fuel-rod cladding in nuclear accidents. The spectra provide information about temperatures, hydrogen production, and eutectic formation inside nuclear reactor cores during accident conditions.

INTRODUCTION

The zirconium-water reaction ($Zr + 2H_2O \rightarrow ZrO_2 + 2H_2$) produced essentially all of the hydrogen in the Three Mile Island accident.[1] The hydrothermal oxidation of the fuel-rod cladding, which consists of a zirconium alloy, not only forms hydrogen but also influences fission-product release and severe fuel damage during a nuclear accident. To explore these processes, we have examined the solid end-product of the zirconium-water reaction, ZrO_2, discovered in samples from Three Mile Island Unit 2 (TMI-2) and the Power Burst Facility (PBF - a nuclear test reactor at the Idaho National Engineering Laboratory). Micro-Raman spectroscopy is uniquely suited to investigating the crystal structures of microscopic ZrO_2 particles, and the spectra provide information about processes that occurred in the TMI-2 core. We have also used luminescence spectroscopy to differentiate oxidized fuel-rod cladding from ZrO_2 ceramics.

Raman spectra of the monoclinic, tetragonal, and cubic phases of ZrO_2 have been acquired by several researchers, and the spectra are characteristic for each of the three phases.[2-4] As a consequence,

Supported by the U. S. Department of Energy under DOE Contract DE-AC07-76IDo1570.

Raman spectroscopy can be used to identify single and mixed phases of ZrO_2.[5,6] Raman spectroscopy has also been used to study the mono-clinic-tetragonal phase transformation,[3,4,7] to determine the crystal structure of ZrO_2-HfO_2 solid solutions,[8] and to elucidate structural disorder in cubic ZrO_2.[5,7]

Investigators have also observed luminescence in zirconia as a blue-white broadband emission[9] and sharp-line emissions in the green region of the spectrum.[10] Both the broad band and sharp-line luminescence can be ascribed to trace impurities of titanium in solid solution with the ZrO_2.[9,10] Asher et. al.[10] proposed that the sharp-line luminescence arises from phonon-mediated de-excitation of excited states rather than direct electronic transitions. The excited states, consisting of three closely-spaced energy levels, decay by producing both a photon and a single long-wavelength optical phonon. Therefore, the frequencies of the luminescence peaks correspond directly to discrete phonon energies.

The present study examined Raman and luminescence spectra of ZrO_2 from TMI-2 and PBF debris samples. The information obtained was compared to possible reaction mechanisms to produce the samples and the environment of a failing nuclear reactor core. Temperature, hydrogen production, and eutectic formation were considered.

EXPERIMENTAL

The TMI-2 samples were obtained from Filter MU-F-5B, one of two make-up and purification demineralizer filters in the TMI-2 coolant purification system. The samples consisted of debris transported by water from the TMI-2 core during the accident. Possible sources of ZrO_2 in the TMI-2 samples were oxidized fuel-rod cladding and frag-ments of ZrO_2 ceramic spacers.

The PBF samples consisted of debris retrieved from a filter in the PBF coolant system. The debris was transported by water from the fuel assembly during a severe-fuel-damage test. Possible sources of ZrO_2 in the PBF samples were oxidized fuel-rod cladding, fragments of ZrO_2 fiberboard (used to insulate the fuel assembly), fragments of ZrO_2 spacers, and ceramic-coated thermocouples.

Debris particles were mounted on glass slides and placed under the microscope of a MOLE Raman microprobe* for analysis. Spectra were taken of the particles by monochannel scanning and photon counting. The excitation source was an Ar^+ laser at wavelengths of 488.0 nm (blue) and 514.5 nm (green). The luminescence peaks were differentiated from the Raman peaks by changing the excitation wavelength.

---

* Instruments SA MOLE 77 Raman Microprobe

Raman and luminescence spectra of a tetragonal ZrO_2 ceramic, a cubic ZrO_2 synthetic gemstone, and ZrO_2 ceramic fiberboard particles were also obtained. X-ray fluorescence verified that the ceramics and the gemstone were stabilized with yttria. These samples are solid solutions in the $ZrO_2-Y_2O_3$ system, and display luminescence due to titanium impurities.

RESULTS AND DISCUSSION

Analysis of TMI-2 Debris

Particles were discovered in the MU-F-5B filter debris sample which produced Raman spectra identifying ZrO_2 in different phases. The expected form was monoclinic but several tetragonal particles were also found.

A spectrum taken of a white, rectangular, 6 X 8 μm particle (Figure 1, top) displayed weak, broad lines at 256 cm^{-1}, 465 cm^{-1}, and 628 cm^{-1}, characteristic of the tetragonal crystal structure.[4,7] The 180 cm^{-1} and 192 cm^{-1} lines of the monoclinic phase were missing. Subsequent spectra (Figure 1, middle and bottom) showed the particle undergoing a phase transformation. The monoclinic phase became more dominant with successive spectra. Heating from the laser induced this phase transformation. The final spectrum (Figure 1, bottom) was not entirely monoclinic but showed a disparity in the relative heights of the 337 and 476 cm^{-1} lines, and a weak line at 216 cm^{-1} was not found in the standard spectrum. In addition, the major monoclinic lines of the intermediate spectrum (Figure 1, middle) were shifted to higher frequencies (lower wave numbers). The 180 cm^{-1} line was shifted to 173 cm^{-1}, the 192 cm^{-1} line to 186 cm^{-1}, and the 476 cm^{-1} line to 465 cm^{-1}. The intermediate spectrum also showed a strong peak at 255 cm^{-1}, not a monoclinic line.

The origin of these particles is of considerable interest. Since ZrO_2 is insoluble in either cold or hot water, the particles most likely originated in the TMI-2 core and were transported by suspension. The existence of monoclinic ZrO_2 in the purification system is not a surprise, since a large portion of the zircaloy fuel-rod cladding is believed to have oxidized, and the most stable form of ZrO_2 below 1170 C is monoclinic.[11] Tetragonal ZrO_2, which was also found, is unstable at low temperatures and cannot be quenched.[12] The conventional method of maintaining the tetragonal phase at room temperature is to stabilize the phase with an impurity (e.g., Y_2O_3 or CaO). Four possible explanations were considered to account for the results:

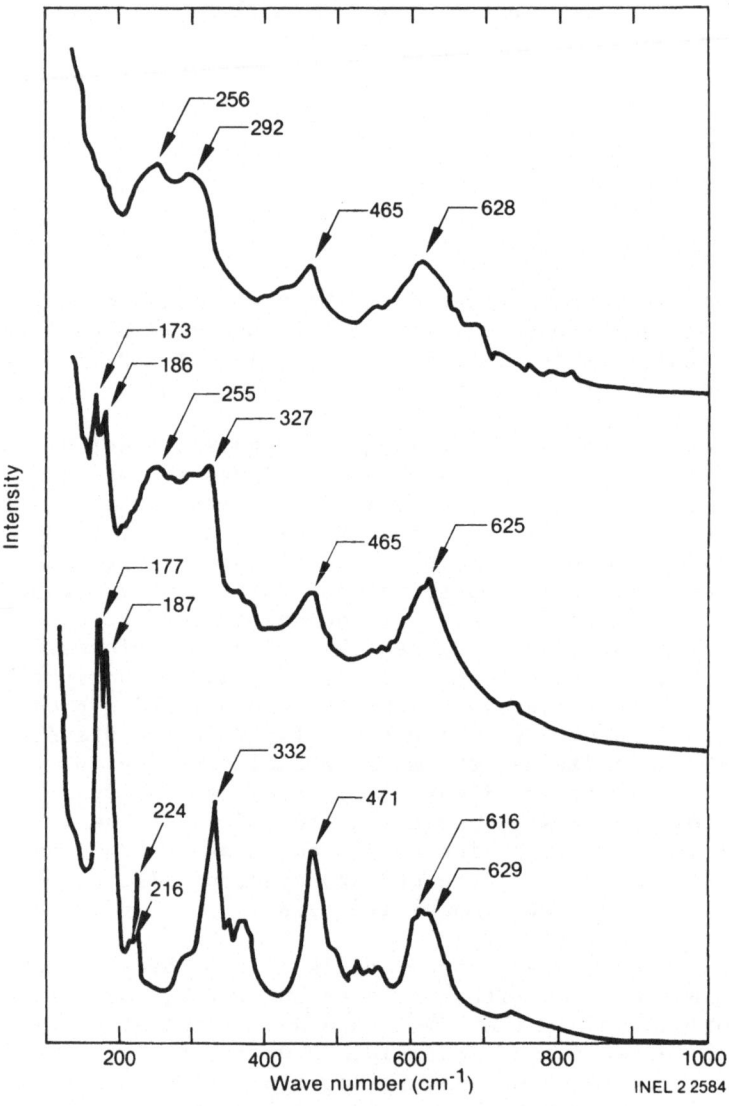

Fig. 1 — Raman spectra of TMI-2 debris particle.

INEL 2 2584

1. The particles are fragments of the cubic ZrO_2 ceramic spacers in the core. The fragments were subjected to stresses (i.e., heat and pressure) that have caused the phase changes to tetragonal and monoclinic.

2. The ZrO_2 in the particles is in a solid solution. Eutectics may have formed from zirconium and constituents from the fuel and control rods, subsequently oxidizing or forming in the oxidized state, and thereby stabilizing the tetragonal phase.

3. The tetragonal phase was stabilized by a nonstoichiometric oxide. The differences in lattice spacing may produce stresses sufficient to stabilize the tetragonal phase.

4. Particle size influences the phase transformation of tetragonal ZrO_2 to monoclinic ZrO_2. The observed tetragonal phase may have been metastable because the particles were smaller than a critical size.

We tested some of these explanations by analyzing the ZrO_2 particles on the scanning electron microscope (SEM). The ceramic spacers in the TMI-2 core contained 15% (mol.) calcia to stabilize the ZrO_2 in the cubic phase. The presence of calcium, as revealed by SEM analysis, would determine whether the ZrO_2 particles were fragments of ceramic spacers or oxidized fuel-rod cladding. Calcium was not detected by the SEM on any of the ZrO_2 particles.

SEM analysis also did not indicate the presence of eutectics. The ZrO_2 in the particles is either pure or, most likely, contains trace amounts of impurities. The SEM is not capable of detecting these trace quantities, and we do not believe that they would have stabilized the ZrO_2 in the tetragonal phase. Heating from laser irradiance transformed the particles to the monoclinic phase; this phenomenon has not been reported for free particles or ceramic grains stabilized in the tetragonal phase by impurities. Also, it has not been the experience of zirconia researchers that trace impurities stabilize the tetragonal phase.

Krebs and Condrate[8] studied solid solutions of ZrO_2 and HfO_2, and reported peak shifts in the Raman spectra due to composition. These peak shifts, however, do not fully correspond to the peak shifts we observed. Krebs and Condrate also observed peak broadening and small peak shifts caused by induced stress in particles containing ZrO_2 and 10% (mol.) HfO_2. Induced stress may have created the effects we saw in the nonstandard monoclinic spectra produced by some TMI-2 particles. Residual strain may have been left in the particles after the phase transformation.

Zircaloy samples that were oxidized at different rates were studied. These samples produced two types of ZrO_2 spectra. Rapidly oxidized samples produced the standard monoclinic spectrum. Raman spectra of slowly oxidized zircaloy displayed similar line shifts and peak intensity differences as observed on spectra of some ZrO_2 particles (from the TMI-2 filter debris sample). The lines were shifted to lower wavenumbers with some differences in peak intensities.

The gradual formation of monoclinic ZrO_2 on the zircaloy creates stress in the crystal structure. The monclinic lattice tries to conform to the atomic spacings of the underlying zircaloy. Similarly, nonstoichiometries in the zirconium oxide of some TMI-2 particles would also create stress or decrease the lattice dimensions in the monoclinic crystal structure.[13] The nonstoichiometries would result from incomplete oxidation of the zircaloy. Instead of being characterized by vacant anion sites, evidence suggests that incompletely oxidized zirconium, known as "black" ZrO_2 because of its color, is comprised of additional Zr atoms residing interstitially in a monoclinic ZrO_2 lattice.[14] Because "black" ZrO_2 retains the monoclinic crystal structure, we concluded that the tetragonal phase we observed was not stabilized by a nonstoichiometric oxide.

We conducted experiments with various sized particles of monoclinic ZrO_2 precipitated by a method described by Garvie.[15] The experiments duplicated the conditions present during the acquisition of the nonstandard monoclinic spectra. These experiments demonstrated that laser heating by itself could not have caused the peak broadening and frequency shifts observed in the nonstandard monoclinic spectra.

Garvie[15,16] reports that tetragonal ZrO_2 is metastable below a critical crystallite size (about 30 nm). He also reports the transformation of the tetragonal phase to the monoclinic on heating of the crystallites above 500 C. The particles we observed were likewise not fully-stabilized in the tetragonal phase. Because of the high velocity of the transformation front, microscopic crystals or grains of ZrO_2 transform instantaneously from the tetragonal to monoclinic phase; there is no gradual transformation or intermediate phase for small, single crystals.[17] The ZrO_2 particles from the MU-F-5B sample gradually transformed, evidencing that they are a conglomerate of smaller particles or crystals. SEM micrographs showed the particles to be conglomerates. In addition to crystallite size, the results of Mitsuhashi et. al.[17] suggest that strain induced by adjacent crystallites in polydomain particles may contribute to the stabilization of tetragonal ZrO_2.

Furuta and Kawasaki[18] found that the presence of a metastable tetragonal phase in oxidized fuel-rod cladding is closely associated with both a critical level of hydrogen concentration and hydrogen absorption. At high temperatures (950-1100 C) in a steam-hydrogen gas mixture, tetragonal ZrO_2 forms when the hydrogen fraction reaches a level where it is appreciably absorbed by the zircaloy. The absorption of hydrogen contributes substantially to the embrittlement of the cladding and may have been a process in cladding failure during the TMI-2 accident. The tetragonal phase is thought to be formed and stabilized by hydrogen.

Bradhurst and Heuer[19] observed similar processes during the high-temperature (700-1300 C) steam oxidation of zircaloy-2. However, they suggest that the formation and stabilization of the tetragonal ZrO_2 may be linked to a local concentration of tin, an alloy element in zircaloy-2 and in zircaloy-4. The experiments of Furuta and Kawasaki, and of Bradhurst and Heuer, were not at the higher pressures and temperatures that were present in the TMI-2 core.

Our studies suggest that a crystallite size effect is involved in the formation and stabilization of the metastable tetragonal ZrO_2. We do not believe the tetragonal ZrO_2 was stabilized by impurities. Elemental analysis of the particles revealed that impurities were not present in greater than trace quantities. Additionally, the tetragonal phase cannot be quenched.

We can conclude from the gradual transformation and SEM observations that the observed particles are polydomain particles (conglomerates of crystallites). The tetragonal phase was most likely stabilized by a crystallite size effect and by strain induced by adjacent crystallites. The nonstandard monoclinic spectra result from residual strain in the monoclinic crystal lattice.

The tetragonal ZrO_2 we observed may be related to the hydrogen content of the ambient gas inside the TMI-2 core during the accident. If this is so, it would be fruitful to find a mechanism which explains the formation of metastable tetragonal ZrO_2 crystallites in the presence of a critical level of hydrogen.

Analysis of PBF Debris

The PBF experiment contained sources of ZrO_2 other than oxidized fuel-rod cladding. These were investigated along with the filter debris. Micro-Raman analysis of particles from the zirconia fiberboard which surrounded the PBF fuel bundle produced spectra of both monoclinic and tetragonal ZrO_2. Cubic ZrO_2 was not detected. Additionally, spectra of all fiberboard particles analyzed displayed strong, unique peaks at 963, 1026, and 1079 cm^{-1} with the

514.5 nm green line of the argon ion laser as the excitation source (see Figure 2). Changing the lasing frequency to the 488.0 nm blue line verified that these peaks were due to luminescence. X-ray fluorescence revealed that the zirconia fiberboard contained more than trace quantities of yttrium, and evidently yttrium oxide (yttria) was used as a stabilizing agent for the ZrO_2.

These spectra were compared to a tetragonal ZrO_2 ceramic and a cubic ZrO_2 gemstone that were yttria-stabilized. Both produced weaker luminescence peaks than the zirconia fiberboard. The luminescence peaks can be attributed to trace impurities of titanium in the ZrO_2. The presence of yttria in the crystal lattice shifts the luminescence peaks to different wavelengths as compared to yttria-less ZrO_2 with titanium impurities. Subsequent research with ZrO_2-Y_2O_3 solid solutions supports these observations. We used these luminescence peaks as markers in the PBF filter debris examinations to differentiate between the yttria-stabilized zirconia fiberboard and oxidized fuel-rod cladding.

Monoclinic, tetragonal, and cubic ZrO_2 were identified in the PBF filter debris samples. One of the particles, identified as tetragonal ZrO_2, produced the strong, unique luminescence peaks indicating that it originated from the zirconia fiberboard. Spectra of other ZrO_2 particles, however, did not show luminescence peaks. It is very probable that these particles came from oxidized fuel-rod cladding. Interestingly, oxidized fuel-rod cladding does not luminesce although zircaloy contains titanium impurities to 50 ppm.[20] Iron impurities also in the zircaloy may quench the luminescence.[21] One particle was monoclinic while two other particles were cubic. Figure 3 is the spectrum of a cubic ZrO_2 particle.

Still other ZrO_2 particles, identified as tetragonal and cubic, produced spectra with luminescence peaks similar to those of the ceramic and gemstone -- weak but existing. These particles most likely originated from a third source of ZrO_2 in the PBF fuel assembly -- the ZrO_2 spacers and ceramic-coated thermocouples. It seems highly unlikely that the titanium would diffuse or drift out of the ZrO_2 from the zirconia fiberboard. Zirconia fiberboard particles that were fragmented (crushed) and heated to 1000 C showed no reduction in luminescence.

The presence of oxidized fuel-rod cladding as cubic ZrO_2 is interesting because cubic ZrO_2 is a high temperature crystalline form of ZrO_2. ZrO_2 transforms to the cubic phase at 2370 C.[12] The phase transformation is reversible so the ZrO_2 must be stabilized with an impurity to remain in the cubic form below 2370 C. An explanation which accounts for the cubic ZrO_2 is that oxidized cladding combined with an impurity, probably an oxide, to

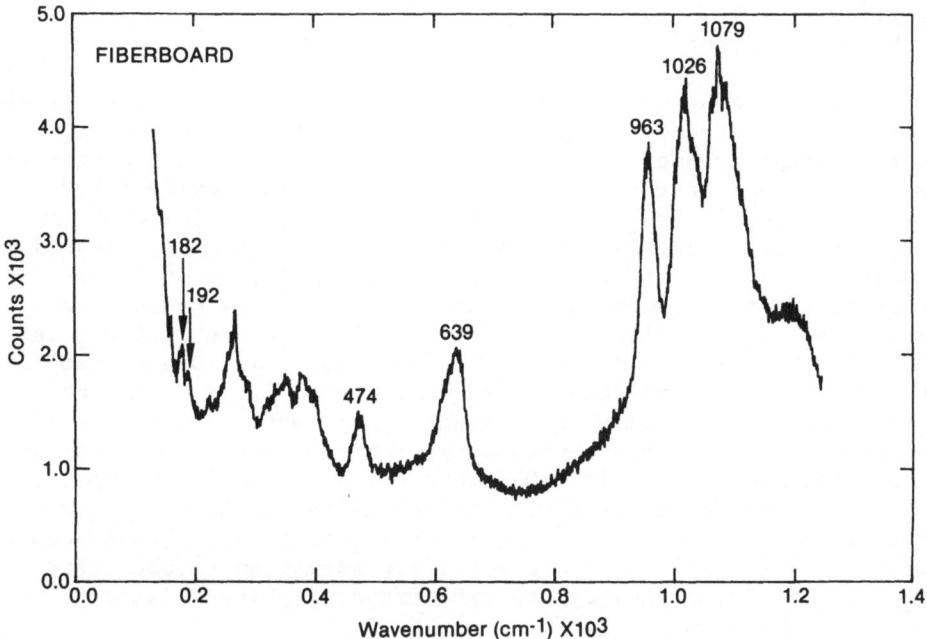

Fig. 2 — Spectrum of ZrO$_2$ fiberboard particle.

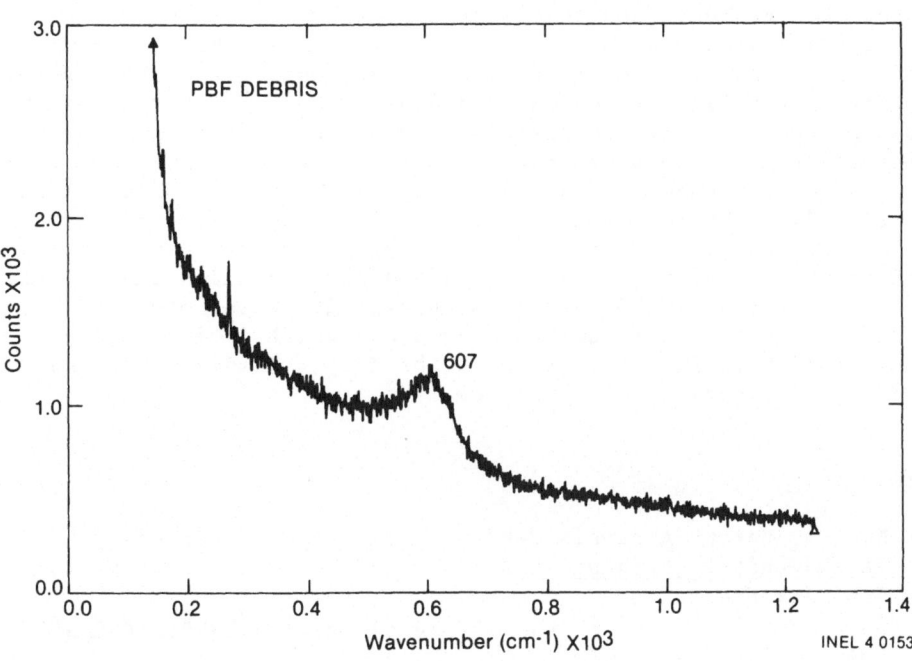

INEL 4 0153

Fig. 3 — Spectrum of PBF debris particle.

stabilize the cubic phase. We were unable to analyze these particles on the SEM to identify the impurity oxide.

Cubic ZrO_2 grown on zirconium was detected by Ploc with the use of selected area electron diffraction.[22,23] The formation of cubic and tetragonal ZrO_2 on zirconium and zircaloy-2 at low temperatures (500 C) has been attributed to stress at the metal-oxide interface.[24] The cubic ZrO_2 we observed were free particles not associated with a metal-oxide interface. The particles appeared to be single, translucent crystals, pentagonal and 5 μm in diameter. If the particles had formed as an oxide film on the zirconium, the compressive stresses induced by the underlying atoms of the metal would have been relieved when the oxide separated. Consequently, the cubic phase would have transformed to the monoclinic. The cubic phase, however, was fully stabilized and not metastable since all attempts to induce a phase transformation by laser heating failed. Because the cubic phase in ZrO_2 cannot be quenched,[13] an impurity must be present in the particles to stabilize the ZrO_2. The most probable form of the stabilizing impurity would be an oxide such as CaO or UO_2. The minimum temperature for formation of the cubic ZrO_2 would be determined by the ZrO_2-impurity phase relations.

Summary

Raman and luminescence spectroscopy with the use of the MOLE microprobe has been used to characterize ZrO_2 originating from oxidized fuel-rod cladding in nuclear accidents. Micro-Raman analysis of samples from Three Mile Island Unit 2 and the Power Burst Facility identified tetragonal and cubic ZrO_2. The tetragonal and cubic phases are high-temperature polymorphs of ZrO_2 and provide information about temperatures and hydrogen formation in the TMI-2 core. The data suggest that the tetragonal ZrO_2 in TMI-2 samples was stabilized by a crystallite size effect, whereas cubic ZrO_2 in PBF debris samples was stabilized by impurities. Luminescence was used to differentiate yttria-stabilized ZrO_2 ceramics and oxidized fuel-rod cladding in PBF debris samples. The ZrO_2 ceramics produced strong, sharp luminescence peaks which indicated the presence of titanium and yttria in the ZrO_2. Oxidized fuel-rod cladding displayed no luminescence.

References

1. Nuclear Safety Analysis Center, _Analysis of Three Mile Island-Unit 2 Accident_, NSAC-80-1, NSAC-1 Revised March 1980, Appendix HYD, p. 5.
2. C. M. Phillippi and K. S. Mazdiyasni, _J. Am. Ceram. Soc., 54_, 254-258 (1971).

3. V. G. Keramidas and W. B. White, J. Am. Ceram Soc., 57, 22-24 (1974).
4. M. Ishigame and T. Sakurai, J. Am. Ceram. Soc., 60, 367-369 (1977).
5. V. G. Keramidas and W. B. White, J. Phys. Chem. Solids, 34, 1873-1878 (1973)
6. D. R. Clarke and F. Adar, J. Am. Ceram. Soc. 65, 284-288 (1982).
7. A. Feinberg and C. H. Perry, J. Phys. Chem. Solids, 42, 513-518 (1981).
8. M. A. Krebs and R. A. Condrate, J. Am. Ceram. Soc., C144-145 (1982).
9. J. F. Sarver, J. Electrochem. Soc., 113, 124-128 (1966).
10. I. M. Asher, B. Papanicolaou, and E. Anastassakis, J. Phys. Chem. Solids, 37, 221-225 (1976).
11. E. C. Subbarao, "Zirconia - An Overview," Science and Technology of Zirconia, edited by A. H. Heuer and L. W. Hobbs, Columbus, Ohio: The American Ceramic Society, Inc., 1981, pp. 1-24.
12. W. A. Lambertson and M. H. Mueller, J. Am. Ceram. Soc., 36, 365-368 (1953).
13. S. C. Carniglia, S. D. Brown, and T. F. Schroeder, J. Am. Ceram. Soc., 54, 13-17 (1971).
14. K. M. Nair, Unfamiliar Oxidation State of Zirconium and the Effect of Oxygen Activity on the Structure of ZrO_2, Ph.D. Dissertation, University of Washington, 1969, 68 pp.
15. R. C. Garvie, J. Phys. Chem., 69, 1238-1243 (1965).
16. R. C. Garvie, J. Phys. Chem., 82, 218-224 (1978).
17. T. Mitsuhashi, M. Ichihara, and U. Tatsuke, J. Am. Ceram. Soc. 57, 97-101 (1974).
18. T. Furuta and S. Kawasaki, J. Nucl. Mater., 105, 119-131 (1984).
19. D. H. Bradhurst and P. M. Heuer, J. Nucl. Mater., 55, 311-326 (1975).
20. B. Lustman and J. G. Goodwin, "Zirconium and Its Alloys," Reactor Handbook - Volume I, Materials, 2nd edition, New York: Interscience Publishers, Inc., 1960, pp. 708-738.
21. J. F. Sarver, Am. Ceram. Soc. Bull., 46, 837-840 (1967).
22. R. A. Ploc, J. Nucl. Mater., 110, 59-64 (1982).
23. R. A. Ploc, J. Nucl. Mater., 99, 124-128 (1981).
24. C. Roy and G. David, J. Nucl. Mater., 37, 71-81 (1970).

HIGH RESOLUTION ELECTRON MICROSCOPIC

CHARACTERIZATION OF INTERFACES IN CERAMICS

Mehmet Sarikaya, Ilhan A. Aksay, and Gareth Thomas*

Department of Materials Science and Engineering
University of Washington, Seattle, WA 98195
*Department of Materials Science and Mineral Engineering
University of California, Berkeley, and
National Center for Electron Microscopy
Lawrence Berkeley Laboratory, Berkeley, CA 94720

ABSTRACT

High resolution electron microscopy (HREM) is useful in bringing out the microstructural variations, such as structural details at grain boundaries, at very high resolutions, even at atomic levels. An important aspect is the examination of the surface irregularities at the boundary regions which provide information on the transformation characteristics of the phases. In this paper, we present data on the microstructural characteristics of some high temperature ceramics (silicon nitride, mullite, and aluminum nitride) with particular emphasis on the detection of grain boundary amorphous phases. Furthermore, the HREM technique is reevaluated with reference to other techniques.

INTRODUCTION

Interfaces are of great fundamental and technological importance in polycrystalline ceramics since their properties, such as electrical and mechanical, depend largely on the interfacial characteristics. In the characterization of interfaces in ceramic systems, electron microscopy provides a unique opportunity. In the conventional mode, bright-field (BF) imaging does not have sufficient resolution for interfacial details, especially in very thin regions of second phases which may be present at grain boundaries. However, high resolution electron microscopy (HREM) is an advanced imaging technique which can provide invaluable information on the details of the interfaces, e.g., interfacial steps, and lattice arrangements at or near the interfaces at atomic levels.[1,2] Especially important is the detection of thin film amorphous grain boundary second phases which usually occur in sintered ceramics.[3,4]

The presence of amorphous grain boundary phases in ceramics is generally associated with additives that are used as processing aids. For instance, in the case of silicon nitride ceramics, as a result of reactions between the surface silica present on the matrix powder and the processing aids (such as alumina, yttria, and magnesia) a liquid phase, which wets the grains and acts as a densifying agent, is formed during sintering.[3-7] A similar situation also arises during the sintering of aluminum nitride with (e.g., silica or calcia) additives.[8] In the case of mullite, a metastable liquid phase may be present during the course of densification.[9] In all cases, this amorphous grain boundary phase, with a low softening temperature, becomes responsible for the loss of strength at elevated temperatures.

The goal of this paper is two-fold. First, the technique of high resolution electron microscopy is discussed. Second, the use of HREM in the characterization of grain boundaries in ceramics is illustrated with case studies on silicon nitride, mullite, and aluminum nitride.

TECHNIQUE

An ideal condition for imaging an interface between two grains exists when the interface is parallel to the incident electron beam direction.[3,4,10] In this way, the details of both surfaces on either side of the boundary will be clearly revealed. For example, such a configuration of the grain boundary with a thin amorphous film is schematically illustrated in Fig. 1. Here, the boundary is seen edge-on with respect to the incoming electron beam. In order to image the boundary,

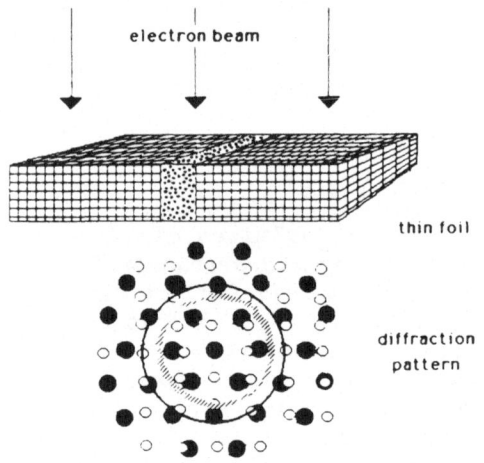

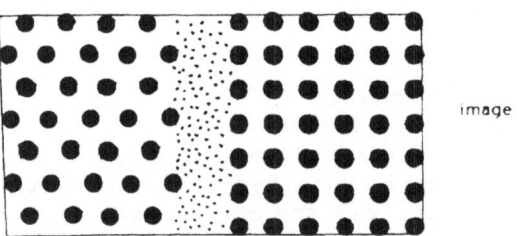

Figure 1. Schematic illustration of an edge-on configuration in imaging of an interface where the grains are in z.a. orientation.

168

an objective aperture is used which encompasses the diffraction spots due to both grains, as well as the diffuse ring (hatched) due to the inelastically scattered electrons from the amorphous region. Under favorable conditions consistent with the specimen-electron beam orientation discussed here and proper performance of all electron-optical alignments, a lattice image of the boundary region can be resolved in a sufficiently thin region of the foil. If both of the grains are in the zone axis (z.a.) orientation, then the atomic periodicities are resolved as schematically shown in Fig. 1, provided also that the correct defocus value has been set with an objective lens with sufficiently high resolving power (~0.2-0.3 nm). The presence of an amorphous layer can then be discerned from the lack of any atomic periodicity right at the boundary, even if the layer is 1.0 nm thick. The sensitivity of this image also depends on the boundary tilt and the defocus value of the image.

Compared to other imaging techniques, such as diffuse dark-field imaging (DDF)[10] in which the objective aperture is placed on the diffuse ring produced in the diffraction pattern due to the presence of an amorphous layer at the boundary, higher spatial resolution is achieved by the lattice imaging technique. In addition, an amorphous layer might be caused due to external effects, such as excessive carbon accumulation at the boundary groove which occurs during the carbon evaporation to the surface of the foil, a process used to prevent decharging during electron microscopy observation of insulators.

INTERFACE CHARACTERIZATION IN CERAMICS

Case I: Silicon Nitride

The silicon nitride compact used in this study was prepared with high purity powders of β-Si_3N_4, α-Al_2O_3, and Y_2O_3 with the use of an aqueous colloidal filtration route.[12] Densification was achieved by liquid phase sintering at 1750°C and 725 KPa nitrogen pressure.[12] The fully dense microstructures produced after this liquid phase sintering process displayed uniform distribution of β-Si_3N_4 grains (Fig. 2(c)). Two different second phases were determined within the microstructure: one crystalline and the other amorphous. Since there is no morphological difference between these two phases, geometrical effects in the crystallization behavior may be disregarded. However, a compositional analysis by energy dispersive x-ray spectroscopy (EDS) indicated a higher yttria content in the crystalline second phase than in the glassy regions, while electron energy loss spectroscopy (EELS) showed higher oxygen content in the glassy regions.[13]

The distribution of amorphous phase is homogeneous as revealed in the DDF micrograph of Fig. 2(b). Although the regions

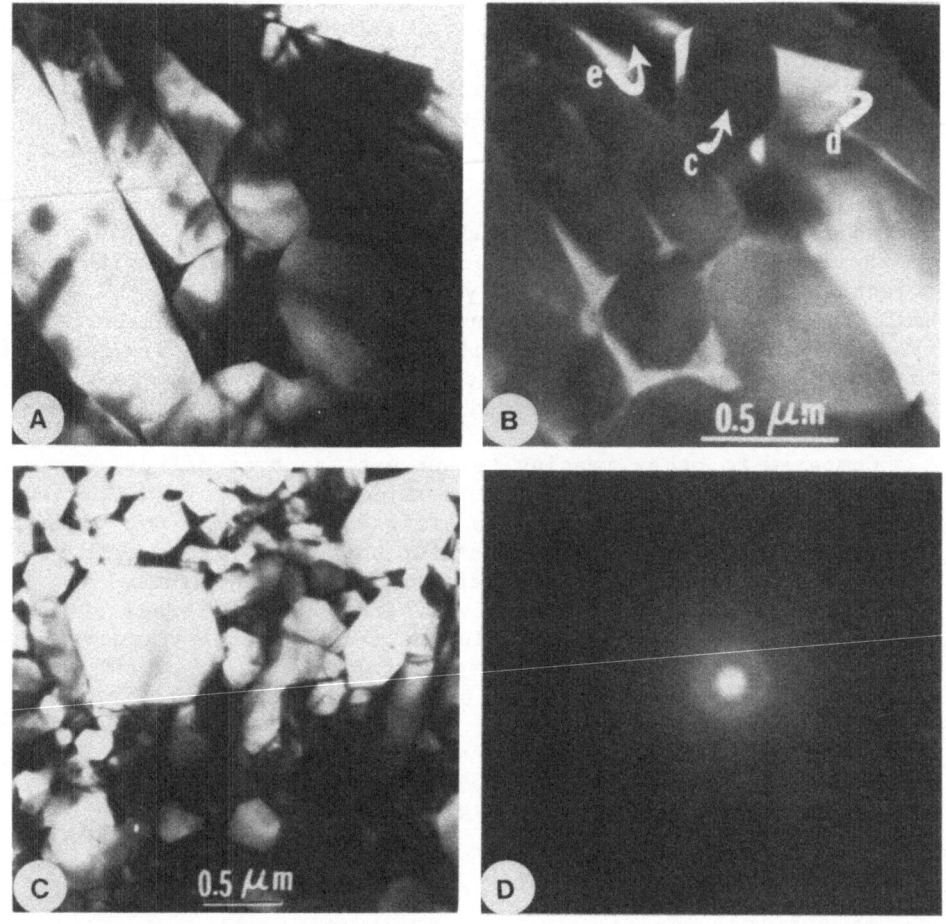

Figure 2. (a) BF and (b) DF images from a thin foil showing
the distribution of glassy phase; (c) low magnification BF
image showing the general microstructure; (d) microdiffraction
pattern from glassy region revealing only diffuse scattering.

indicated as e and d produce similar contrast in the BF image of
Fig. 2(a), only the region d changes contrast in the DDF image. In
fact, microdiffraction from region e produces elastic scattering
while diffuse scattering is produced from region d (Fig. 2(d)).

At low resolutions and magnifications (Fig. 2(c)), it is not
possible to reveal whether or not grain boundaries are also covered
with a thin (usually 2-10 nm thick) amorphous phase, which is
usually the case in all liquid phase sintered silicon nitrides.[5,6]
Therefore, it becomes necessary to perform HREM. The example given

170

in Fig. 3 is of interest for several reasons: First, one of the grains is Si-Al-Y-oxynitride and the other is Si_3N_4. Second, both of the grains are in the z.a. orientations. An amorphous phase is clearly seen in the lower portion as indicated by Am. However, as one follows this amorphous region towards the boundary, it is seen that several nanometers inside the boundary the amorphous phase ends. At this portion of the grain boundary, atomic periodicities in both grains continue right to the boundary and stop when they reach each other. The absence of any discontinuity at the boundary then indicates the absence of an amorphous phase. Fig. 4

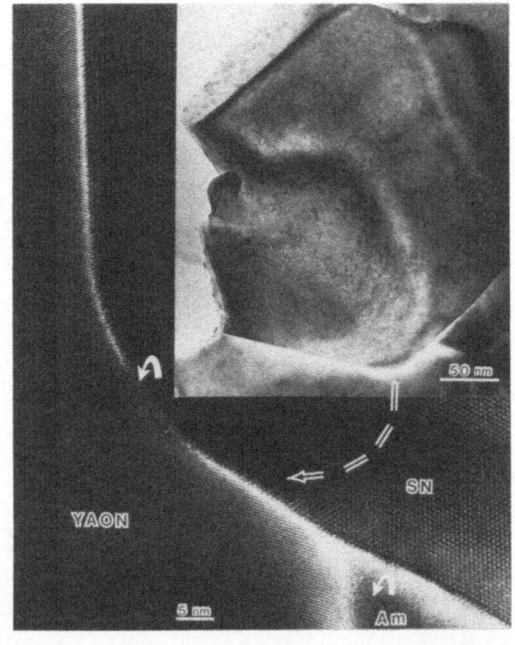

Figure 3. HREM image of an interface between a Y-Si-Al-oxynitride and silicon nitride grains.

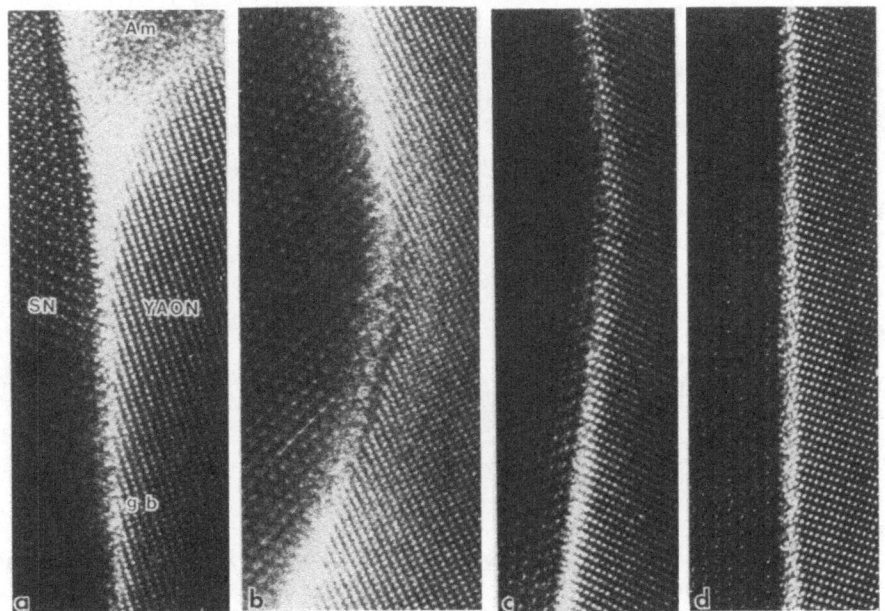

Figure 4. Details of the interface shown in Figure 3.

171

presents a series of images along the boundary revealing the details at very high magnifications. Again, the absence of any amorphous phase film is made obvious by the presence of rows of atoms in both grains at the points of contact.

One of the objectives in reducing the detrimental effects of the amorphous phase is to crystallize it during cooling. This crystallization can be achieved when the liquid phase, which is present at the sintering temperature, reaches a composition of one of the complex yttrium-aluminum-silicon oxynitride phases. Many factors can affect this process, such as compositional variations due to inhomogeneities in mixing, and the temperature and pressure of sintering. An example of such a case is shown in Fig. 5. At the triple junction of Fig. 5, where normally an amorphous phase would be present, lattice fringes are revealed which indicate that the amorphous phase has crystallized. However, upon a close examination of the boundary between this crystalline phase and the silicon nitride grains, one can still see a thin layer of amorphous film. Micrographs taken along the boundary region between the silicon nitride grains indicate that this film is discontinuous and the rest of the boundary is free from any amorphous layer.

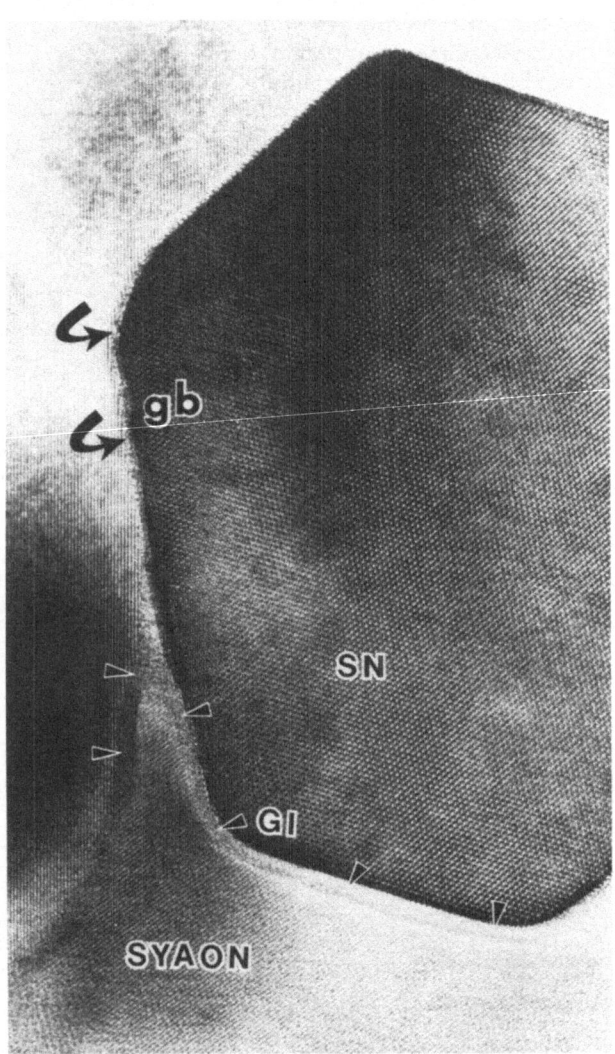

Figure 5. Low magnification HREM image revealing a crystalline phase, formed from the amorphous phase, at a three-grain junction.

172

Case II: Mullite

Mullite ($3Al_2O_3 \cdot 2SiO_2$) is the only stable intermediate compound in the silica-alumina binary system.[14] Mullite's main importance as a structural material comes from the fact that it has favorable thermal shock resistance, low specific gravity, good corrosion resistance, and especially high creep resistance.[15]

In the processing of mullite, recent trends have favored the use of submicron size particles or molecularly mixed systems in order to enhance the mullitization process during densification.[9,16] The mullite compact that was examined in this current study was similarly prepared by a colloidal filtration process of submicron size alumina and kaolinite ($Al_2O_3 \cdot 2SiO_2 \cdot 2H_2O$).[17] The premise of the work was to show that if diffusion distances were shortened, diffusion controlled reactions could be completed at relatively low temperatures ($<1550\,°C$) to form liquid free mullite. The results of the electron microscopy observations are as follows.

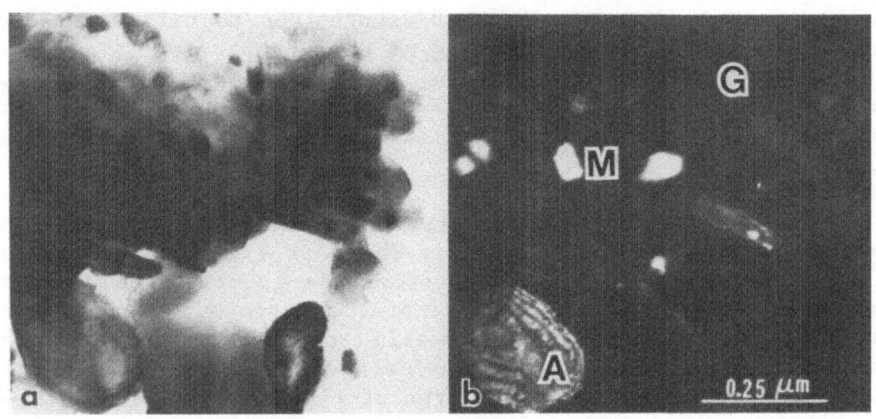

Figure 6. (a) BF and (b) DF micrographs from 1300°C sample. The amorphous phase, G, surrounding the crystals, M, are also revealed. Alumina crystals are still present as indicated by A.

In Fig. 6, the pair of TEM micrographs displays a characteristic region of a sample that was heat treated at 1300°C for 1 hr to achieve partial densification. In the BF image, small mullite crystals with well defined rectangular shapes are seen in various orientations. This is more clear in the dark-field (DF) micrograph (b), which was taken by using a superimposed reflection formed by the diffraction from those crystals which change contrast. The size of these crystals ranges from 10 to 100 nm. Mullite particles are embedded in an amorphous matrix which is mostly SiO_2. (It should be noted here that because of experimental difficulties and uncertainties, quantitative chemical analysis by EDS method was not performed.) The distribution of this glassy phase is seen as a

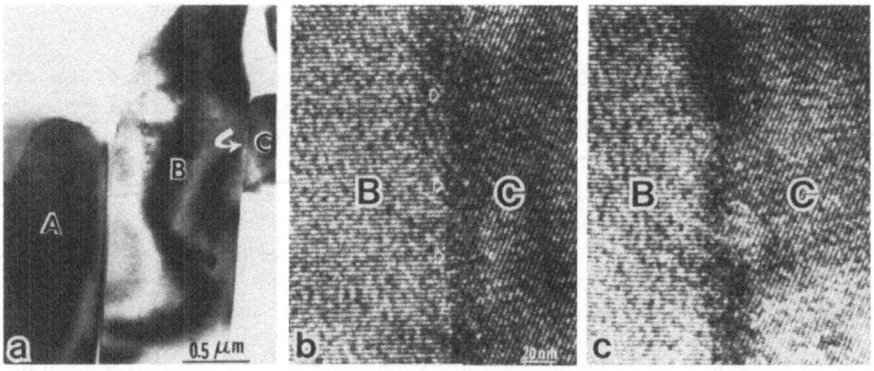

Figure 7. (a) BF image from 1500°C sample. (b) and (c) are are HREM images taken along the boundary shown by arrow in (a) revealing the absence of any amorphous layer.

faint bright background in the DF image (due to diffuse scattering). The duration of sintering at this temperature was not sufficiently long enough to result in complete dissolution of the alumina grains within the amorphous matrix. Therefore, some alumina still remained in the compact (as indicated by letter A in the micrograph).

A portion of the microstructure evolved after a densification treatment at 1500°C, for 1 hr, is shown in Fig. 7(a). In comparison to the 1300°C treatment, considerable growth of mullite crystals has occurred at the expense of the surrounding matrix. Although an amorphous phase could still be detected in certain portions, the amount is reduced considerably and many mullite-mullite grain boundaries appear to be free of the amorphous phase. HREM is utilized to examine the mullite grain boundaries. A favorable configuration of imaging is presented in Fig. 7, where the boundary between grains B and C, each exhibiting one set of fringes, was examined. The fringes in grain C are continuous up to the interface where the fringes in grain B are also revealed. It was ascertained from this and other micrographs that many grain boundaries were free from an amorphous phase.

Case-III: Aluminum Nitride

AlN is a member of the SiAlON system and is used for high temperature applications.[18] Only limited work has been reported on the Al rich corner of the $SiO_2-Al_2O_3-Si_3N_4-AlN$ system.[18-20] Similar to the silicon nitride ceramics, AlN can be sintered with the aid of sintering additives such as silica and calcia. Small amounts of silica (5-10 %) produce alloys near the 2H phase field.[8] The samples in the present investigation (with 10% SiO_2 addition) were prepared by a two stage sintering process: First at 2000 °C and then at 2100°C under 300 kg/cm^2 pressure in nitrogen atmosphere.[8] In this research, our intent was to study the 2H to polytypoid transformation. HREM was utilized to examine the details of the 2H/ polytypoid interface.

Conventional characterization of the microstructure revealed high amounts of different types of polytypoids, in addition to the original 2H structure.[20,21] Polytypoids form because of the change in the stacking sequence. The reason for the local change in the hcp stacking sequence (ABABAB..) to cubic (ABCABC..) in AlN is because of the deviation of Al/N ratio from 1. As the Al/N ratio decreases, more N is taken into the structure. Together with N, Si and probably O are taken into the structure to keep the charge balance. In fact, a recent study with EELS indicated that oxygen is indeed incorporated into the polytypoid structure.[21] There is also 5-10% Si in polytypoids as measured by EDS and no Si in the original grains.[21]

An example of the AlN microstructure examined in this study is shown in Fig. 8. In the low magnification BF micrograph (b), one can see a triangular projection of a three grain junction, which may be interpreted as a "glassy pocket" at this low magnification. However, the HREM micrograph in (a) reveals that this region is indeed crystalline and, interestingly enough, has an original 2H structure rather than

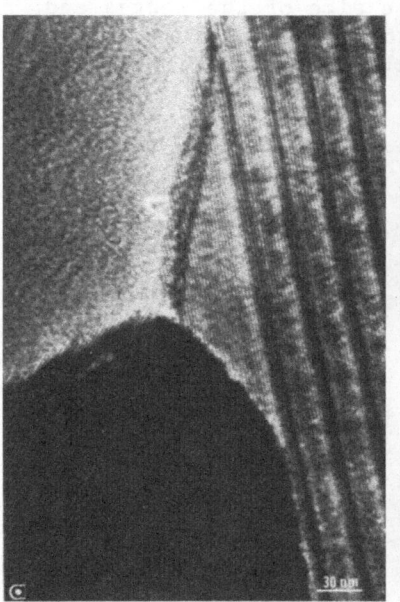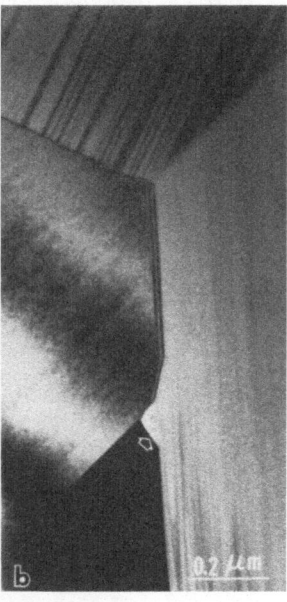

Figure 8. (a) Lattice fringe image and (b) BF image taken from the AlN sample showing the details of a boundary region (arrow).

the polytypoid seen on the right of the micrograph, which repeats itself through the whole grain. Another interesting feature in micrograph (b) is the presence of several different polytypoids in a grain seen at the top of the picture, which are frequently observed in these samples. Therefore, in this grain, faulting is not a regular single type but is a combination of different types which repeat regularly throughout the structure.

It is often very useful to find grains which are half-transformed. Examination of the interfaces between the transformed and the untransformed regions reveals valuable information about the nature of the interfaces which could be related to the mechanism of transformation. Both straight boundaries and boundaries containing ledges were observed in an earlier study.[21] Fig. 9 shows an example of the latter type. In the BF image of Fig. 9(a), the end part of the long grain is seen where, in either side of the grain, two small pockets of the original grain (shown by arrows) were left untransformed. The high resolution image in (b) reveals the details of the interface between the 2H and the polytypoid where the steps, or ledges, at the interface are clearly seen. It was hypothesized earlier[21] that the growth of polytypoids occurs when the concentration of nitrogen reaches a critical value in the original structure. Under a positive nitrogen atmosphere and high temperature, nitrogen is taken into the structure at the boundary by forming ledges. The size of the ledges and, therefore, the type of the polytypoid is then determined by the local concentration of nitrogen in the structure which dictates the Al/N ratio.

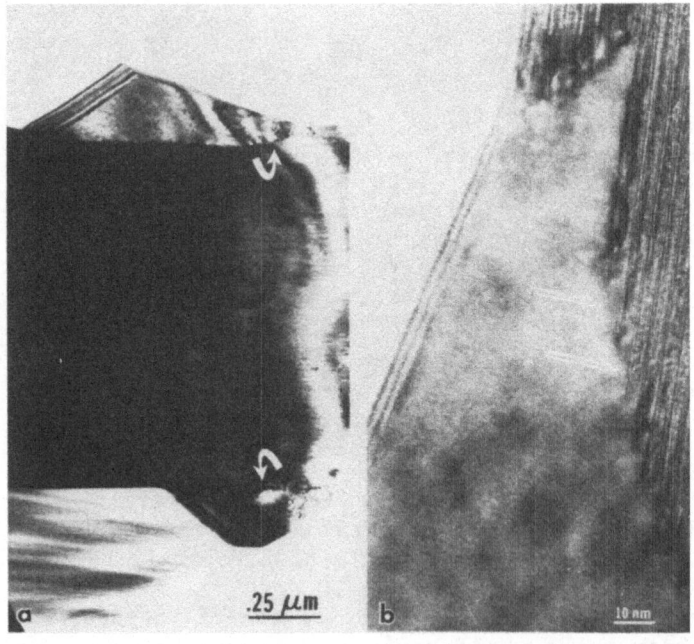

Figure 9. (a) BF image showing 2H/Polytypoid boundaries containing ledges, and (b) HREM image revealing the details of the boundary indicated by B in (a).

176

CONCLUSIONS

High resolution electron microscopy technique was used to reveal microstructural details in some ceramic systems. Lattice image analysis resulted in the following conclusions:

(1) At some grain boundaries of silicon nitride ceramics (fluxed with alumina and yttria) a glassy phase is absent and in some cases even a crystalline second phase is observed at three grain junctions.

(2) In mullite ceramics, mullite-mullite grain boundaries are shown to be glass-free provided that sufficient time is allowed during the densification stage to achieve an equilibrium phase structure.

(3) AlN doped with 10% silica does not contain a glassy phase but has polytypoids in which the Al/N ratio deviates from the value in the original, 2H, structure. Stepped, or ledged, interfaces as well as straight boundaries can be formed between polytypoids/2H grains depending on the transformation stage.

ACKNOWLEDGEMENTS

Financial support from NSF (DMR-83-1317239) for the work on silicon nitride and aluminum nitride and from AFOSR/DARPA (83-0375) for mullite work is greatly acknowledged. We thank K. Komeya for providing the AlN samples.

REFERENCES

1. J. G. Allpress and J. V. Sanders, J. Appl. Crystallogr., 6, [pt.3] 165 (1972).
2. R. Gronsky, "High Resolution Electron Microscopy" in: "Experimental Methods in Materials Science", H. Herman, ed., Academic Press, New York (1982).
3. D. R. Clarke and G. Thomas, J. Am. Ceram. Soc., 60, [11-12] 491 (1977).
4. L. K. Lou, T. E. Mitchell, and A. H. Heuer, J. Am. Ceram. Soc., 56, [9-10] 392 (1978); and ibid., p. 462.
5. F. F. Lange, Final Tech. Rep. No. 74-9D4-POWDR-R2, Westinghouse Corp., (1976).
6. G. E. Gazza, J. Am. Ceram. Soc., 56, [12] 662 (1973).
7. O. L. Krivanek, T. M. Shaw, and G. Thomas, J. Am. Ceram. Soc., 62, [11-12] 585 (1979).
8. K. Komeya and A. Tsuge, Yogyo-Kyokai-Shi, 89, 615 (1983).
9. M. Sacks, Ph. D. Thesis, University of California, Berkeley, California, LBL Report # 10372, 1979.

10. O. L. Krivanek, T. M. Shaw, and G. Thomas, $\underline{J. \; Appl. \; Phys.}$, $\underline{50}$, [6] 4223 (1979).

11. D. R. Clarke and G. Thomas, $\underline{J. \; Am. \; Ceram. \; Soc.}$, $\underline{61}$, [3-4] 114 (1978).

12. I. A. Aksay and C. A. Ambarian, $\underline{in:}$ "Development of Homogeneity in Si_3N_4 Ceramics by Colloidal Filtration," Report submitted to Garrett Turbine Engine Co., Phoenix, AZ, 85010, 1984.

13. M. Sarikaya, P. Rez, and G. Thomas, paper to be submitted to $\underline{J. \; Am. \; Ceram. \; Soc.}$ (1985).

14. I. A. Aksay and J. A. Pask, $\underline{J. \; Am. \; Ceram. \; Soc.}$, $\underline{58}$, [11-12] 507 (1975).

15. R. F. Davis and J. A. Pask, $\underline{in:}$ "High Temperature Oxides", Part. IV, A. M. Alper, ed., Academic Press, New York (1972).

16. S. Prochazka and F. J. Klug, $\underline{J. \; Am. \; Ceram. \; Soc.}$, $\underline{66}$, [12] 874 (1983).

17. I. A. Aksay and M. Sarikaya, to be published.

18. K. H. Jack, $\underline{J. \; Mat. \; Sci.}$, $\underline{11}$, [10] 1611 (1983).

19. G. Van Tandeloo and G. Thomas, $\underline{Acta. \; Met.}$, $\underline{31}$, [10] 1611 (1983).

20. G. Van Tandeloo, K. T. Faber, and G. Thomas, $\underline{J. \; Mater. \; Sci.}$, $\underline{18}$, 525 (1983).

21. M. Sarikaya and G. Thomas, $\underline{in:}$ "Electron Microscopy", p.227, A. Csanady, P. Rohlich, and D. Szabo, eds., Proc. 8th EUREM, Budapest, Hungary, Aug. 13-18 (1984).

CHARACTERIZATION OF GRAIN BOUNDARIES IN ALUMINA

K. J. Morrissey and C. B. Carter

Department of Materials Science
Cornell University
Ithaca, NY 14853

INTRODUCTION

Alpha alumina is used in such applications as substrates in electronic packaging or envelopes for Na vapor lamps. Thus, a large amount of research has been devoted to improving the sintering of alpha alumina. MgO is routinely used as an additive in alumina to enhance the densification of alpha alumina compacts (1). This additive not only enhances densification but also inhibits grain growth: exaggerated grain growth is detrimental to product performance. However, it is still not fully understood how this additive affects the grain boundaries of the compacts (2,3). It has previously been reported that the combination of this additive and impurities which may be present in concentrations exceeding the solid solubility limit, causes the formation of second-phase particles (2). The purpose of the present paper is to review some recent studies on alpha alumina grain boundaries and to report some new results. Reports on TEM observations on grain boundaries in alumina in most cases have not related the structure at the boundaries, for example (3,4). Of the studies that did report boundary structure, the most frequently discussed boundaries were twins, specifically basal and rhombohedral twins (8). For example Heuer (5) has reported the characterization of basal and rhombohedral twins. In that report deformation twinning was discussed for single crystal alumina specimens which were fractured in four-point bending tests at different temperatures (5), twinning was noted to occur on 0111 planes, morphological indices were used in the paper. Rhombohedral twinning in Al_2O_3 has also been discussed by Scott (6), who presented models for the structures of this boundary, and by Bhandari et al. who discussed energy aspects of the twin (7). Basal twinning is the more common type of twinning reported in most cases and the structure of this type boundary was originally discussed by Kronberg (8). Hockey used TEM to study twins produced during sample deformation (9).

Studies of grain boundaries in metals, semiconductors and some ceramics have been interpreted using the terminology of CSL, DSCL, and the O-lattice (10,11). This terminology is used to give a common basis for understanding the crystallography of grain boundaries but it has not been used as extensively to discuss grain boundaries in ceramics. The structure of twins and other boundaries in alumina can be complex as different arrangements of the ions may occur at the boundaries (12).

When two identical interpenetrating crystal lattices are rotated with respect to one another through certain angles a large fraction of lattice sites in one crystal may be coincident with the same large fraction of lattice sites in the other crystal. This lattice of coincident sites is referred to as the coincident site lattice (CSL). The inverse of the fraction of coincident sites is defined to be Σ. A difficulty with the CSL model is that the two adjoining grains are often translated relative to one another so that there are no coincident atoms. Wagner, Tan, Balluffi (13), proposed the generalized CSL in which the coincident lattice sites need neither be identical nor be occupied by atoms. It is in this sense that the model is used in this study.

The structure of α alumina may be characterized by a sublattice of hexagonally close-packed oxygen ions with aluminum ions in 2/3 of the octahedral interstices (8), the actual crystal structure is rhombohedral. Basal twins were first discussed in terms of the CSL theory by the present authors (14) who showed that Σ = 3 for this type of interface. It was also shown that the favored planes for the boundary would be the basal plane, the {11$\bar{2}$0}, {10$\bar{1}$0} or {10$\bar{1}$2} type planes: this analysis was confirmed by Hansen and Phillips (17). It was also pointed out that the oxygen sublattice is, to a first approximation, not affected by the twin boundary for either the mirror or the glide configuration. A discussion of dislocations which can occur in these twins has been given previously (22). The rhombohedral twin was illustrated in (12) and the structure was discussed in terms of near CSL with Σ = 8. The structure and an exact Σ = 13 grain boundary was also discussed for which the principal plane is the basal plane. This grain boundary is produced by a 27.8° rotation about the 0001.

The presence of these special grain boundaries was often related to the morphology of the grains in the matrix (16); for example basal twins were often associated with lath-like grains in the matrix. Hanson and Phillips (17) reported that there is a film of amorphous material along all grain boundaries other than the basal twin and that this film causes or is associated with faceting of grain boundaries. Faceting can be due to the presence of an amorphous film, or to the forces acting on a boundary causing migration. It has been shown that it is misleading to say that all boundaries have a glassy phase, since many faceted special grain boundaries have been observed including the exact Σ = 13 interface and other twin boundaries (15). The structure of special grain boundaries has also been discussed by Fortunee (18), who gave a theoretical analysis of boundaries that might be expected to occur in alpha alumina compacts. Rhombohedral twins have also been studied in the mineral hematite (19) which is isostructural with α Al_2O_3. Shiue and Phillips (20) have recently extended the analysis of Fortunee (18) and Morrissey and Carter (12) for the rhombohedral twin giving an O-lattice formalism and confirming that faceted grain boundaries other than the basal twin do in fact exist in alumina.

EXPERIMENTAL

The observations in this paper were made on two types of polycrystalline alpha alumina. The first was produced by sintering and contained MgO added as a densification aid. This material was prepared for observation in the transmission electron microscope by mechanically polishing the samples to a thickness of 50 μm. 3 mm diameter discs were made by ultrasonically cutting the material and were thinned to perforation in an ion thinner operating at 4 kV. The disadvantage of having to prepare the samples in this manner is that both mechanical polishing and ion thinning can produce artifacts at the surface of the samples.

The second type of sample was produced by the thermal oxidation of aluminum substrates. The oxidation of aluminum in air at 863 K for three hours produces a thin film of γ-alumina (21), which can then be chemically detached from the metal using a solution of mercuric chloride. The films were caught on Mo, Pt and Ni TEM, heated in vacuum at 1473 K for 5 min. The resulting films contained delta and α-alumina with small amounts of γ- and θ-alumina. The presence of these different modifications of alumina indicates that the different forms do not always occur at the 'accepted' temperatures of formation.

OBSERVATIONS

When studying the characteristics of grain boundaries in the alumina compacts it is important not only to understand the boundary structure but also to relate the geometry of the grain boundary to the local grain morphologies (16). The examples

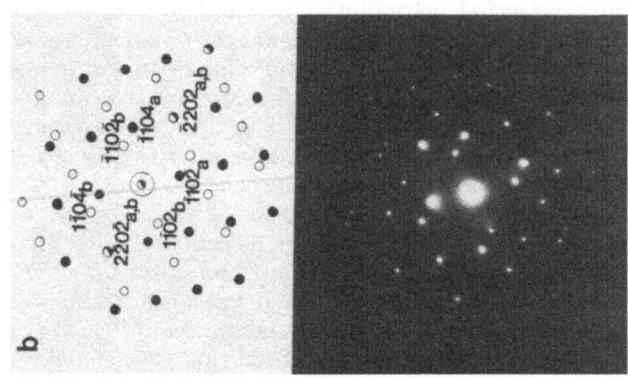

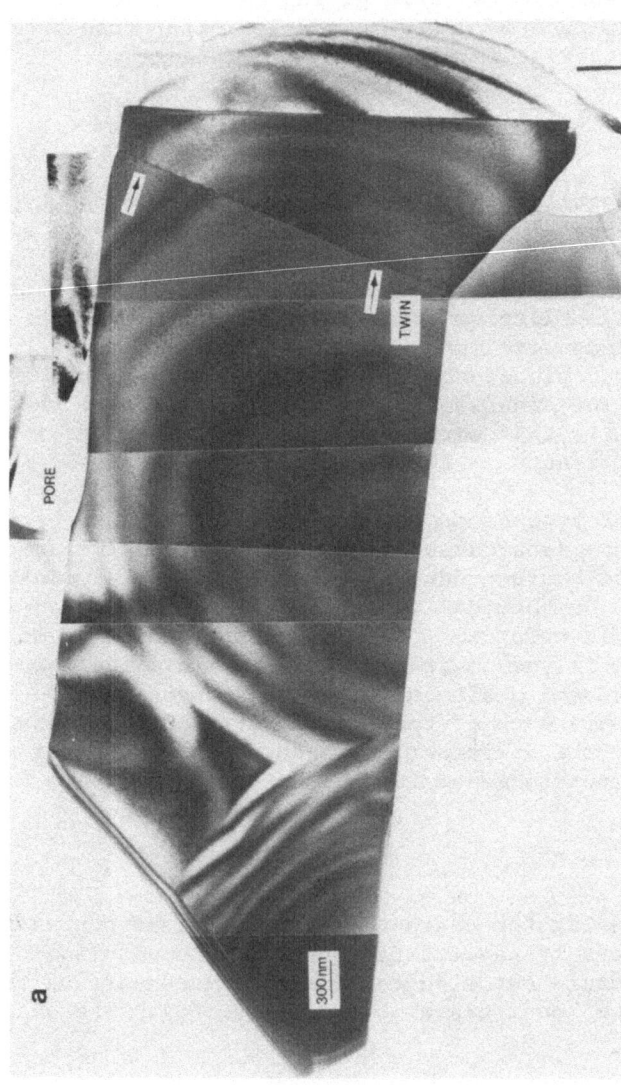

Fig. 1. A bright-field image of a 10$\bar{1}$1 type twin, arrowed. The corresponding SAD is shown in Fig. b.

which will now be discussed have been chosen to illustrate this relationship: due to page limitations the reader is referred to references for a more detailed discussion of the structure of the grain boundaries. The grain boundary shown in Fig. 1 is a new type of twin interface for which $\Sigma = 11$. The plane of the interface shown in the figure is the symmetric 1101 plane. The selected area diffraction pattern shown in Fig. 1 gives the crystallographic information on the boundary, the two grains are related to one another by a rotation of approximately 81° about the common <1120> pole. The morphology of the long part of the grain is similar to that reported previously (16) for a 'grain' containing a basal twin boundary: i.e. this segment is lath-like with its sides parallel to the basal plane.

The grain morphology may be more complicated as illustrated by Fig. 2 where the grain again is much larger than the surrounding matrix. In this example the grain is considerably larger than is the previous example and contains several basal twins which are identified by arrows in the figure. The other boundaries in the grain are low angle boundaries. There is also a large number of pores present within this grain: this suggests that this grain grew to be much larger than the surrounding matrix during the sintering process before complete densification, i.e. pore removal, could take place. The lower boundary in the figure contains both $\Sigma = 13$ and $\Sigma = 39$ segments. The $\Sigma = 13$ boundary changes into the $\Sigma = 39$ boundary when intersected by a basal twin. The interpretation of this image is that the lower boundary was migrating and as it moved past the pores a growth accident occurred which resulted in the formation of the basal micro twin labelled 'twin' in Fig. 2. This mechanism for the formation of a twin boundary is thus fundamentally different to that which occurred in the area shown in Fig. 1. The boundary in Fig. 2 has its origin in a growth accident while that in Fig. 1 originated when two grains sintered together: it separates the original two grains.

The final illustration shows a basal twin boundary in alumina prepared by the oxidation method. The films of a α-alumina contained regions of transition aluminas and the areas of α-alumina were often surrounded by porosity (21). In the example shown the area of the film is mostly α-alumina and there is little porosity either in or surrounding the grains. The basal twin boundary in this case is highly curved and most certainly moving during the heating of the film. The films of α-alumina actually contain a large number of basal twins, due presumably, in part, to the same reason that the alumina compacts contain a large number of such twins, i.e. there is a unique structure at the boundary for which the oxygen ions are not disturbed. When basal twinning occurs the growth of the oxygen sublattice is essentially undisturbed. These films present a unique sample for

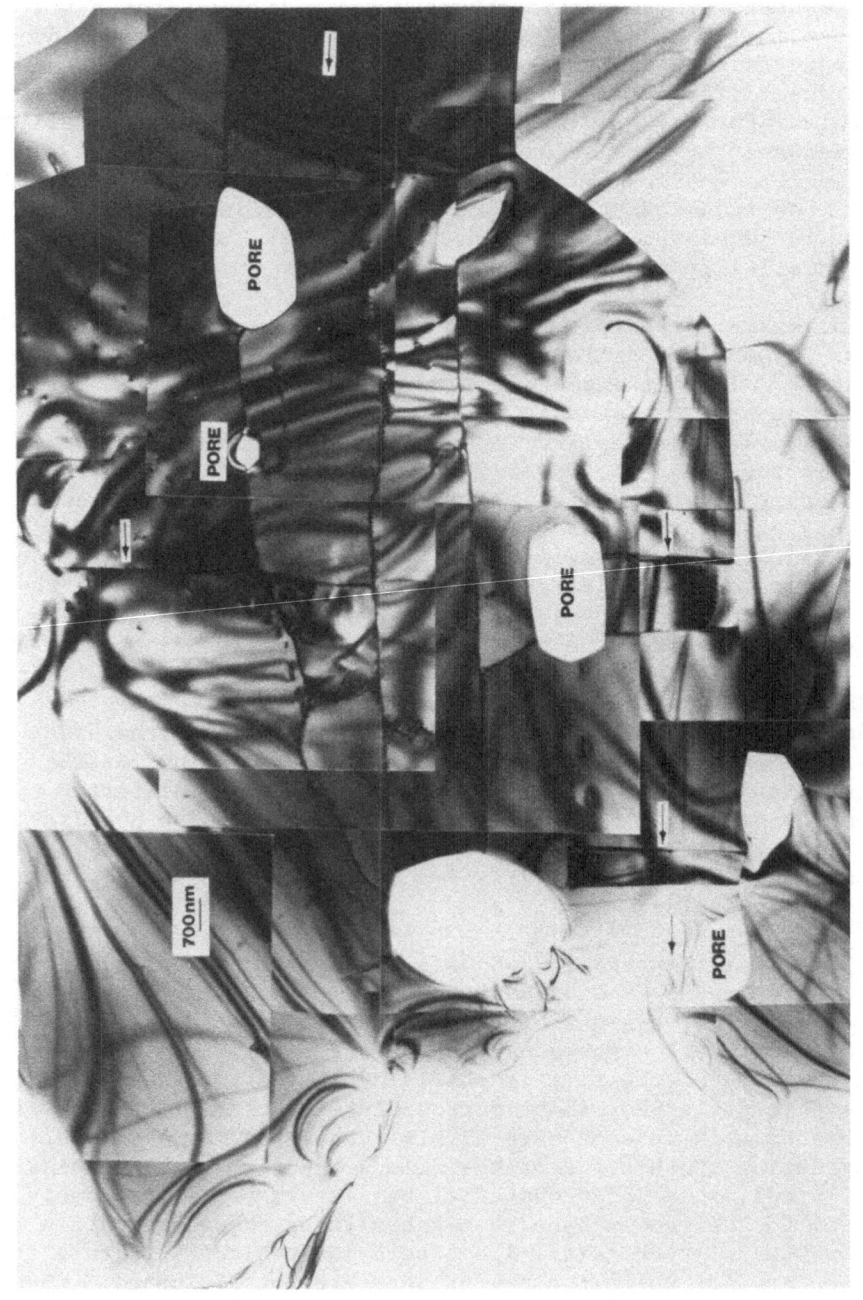

Fig. 2. A composite bright-field image of a grain which has grown much larger than the surrounding matrix. There are a number of basal twins, arrowed.

184

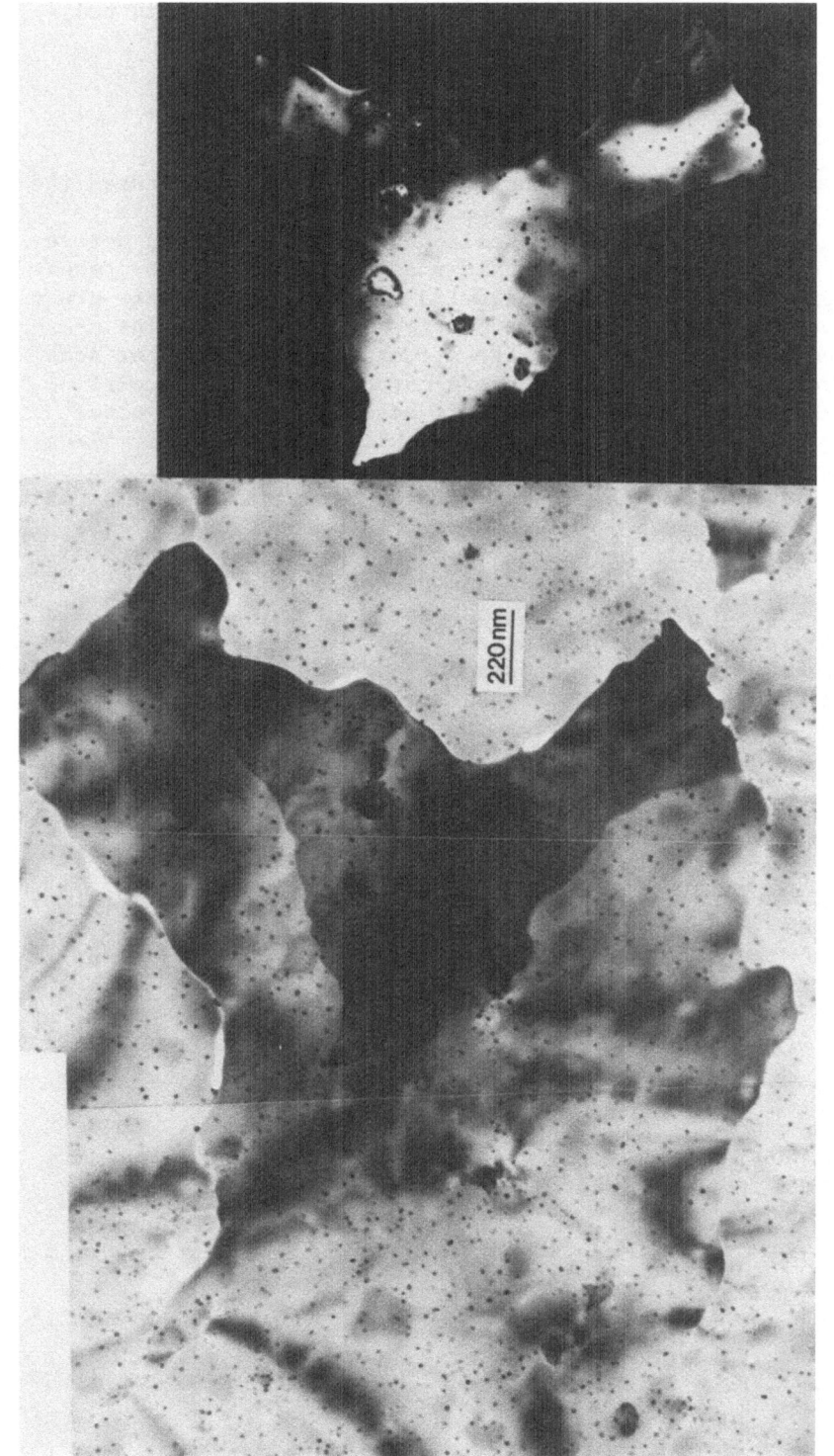

Fig. 3. An area from a thin film of alumina, a bright-field image and a dark-field image illustrating a twinned region. The twin is shown by the outline of the grain in the dark-field image. The black dots are due to Mo from the TEM grid.

220 nm

the study of the same type of defects as those observed in poly-crystalline compact.

SUMMARY

The early work on grain boundaries in α-alumina stressed the structural aspects of the grain boundaries and related this to similar analysis in metals and semiconductors, but it did not relate this structure to the morphology of the grains. The present study of grain boundaries in α-alumina has shown that there are a large number of special grain boundaries in polycrystalline α-alumina compacts and show that the structure of these interfaces can be directly related to the grain morphology. The special role of basal twins in such materials has been stressed, since they are often associated with grains which have undergone 'exag-gerated' grain growth. A number of other twin boundaries includ-ing the (10$\bar{1}$1) twin which was illustrated here and the (11$\bar{2}$3) and rhombohedral twins (see 15) have also been observed. The exis-tence of special grain boundaries and their interaction with pores and other grain boundaries has been illustrated in this paper by the exact $\Sigma = 13$ grain boundary. In this example grain boundary migration was shown to result in the formation of very large grains. Finally it has been shown that thin films of α-alumina which can be prepared by the oxidation of aluminum also contain special grain boundaries and provide a model system for the study of the structure and migration of such interfaces.

ACKNOWLEDGMENTS

The microscope used in this study is part of the Materials Science Center Facility which is maintained by Mr. Ray Coles and supported by the NSF. This research is supported by DOE under contract no. DE-FG02-84ER45092.

REFERENCES

1. R. Coble patent #3,026,210 (1962).
2. K. J. Morrissey, C. B. Carter, Adv. in Mat. Char., Plenum Press, pp. 297-307 (1983).
3. A. H. Heuer, et al., J.A.C.S., 63 (1980), pp. 53-58.
4. N. J. Tighe, A. Hyman, Anisotropy in single-crystal refractory compounds, ed. F. Vahldiek, SMersol (1967), pp. 121-136.
5. A. H. Heuer, Phil. Mag., 13 (1966), pp. 379-393.
6. W. D. Scott, Deformation of Ceramic Materials, ed. R. C. Bradt, R. E. Tressler, Plenum, NY (1965), pp. 151-166.

7. O. P. Bhandari et al., Acta Metall., 21 (1973), pp. 1515-1521.

8. M. L. Kronberg, Acta Met., 5 (1957), pp. 507-524.

9. B. J. Hockey, Deform. Ceram. Mater., Proc. Symp. (1974), pp. 167-169.

10. R. C. Pond, V. Vitek, Proc. Roy. Soc., A357 (1977), pp. 453-470.

11. B. Cunningham et al., Scripta Met., 16 (1982), pp. 349-352.

12. K. J. Morrissey, C. B. Carter, J.A.C.S., 67 [4] (1984), pp. 292-301.

13. W. R. Wagner, et al., Phil. Mag., 29 (1974), pp. 895-904.

14. K. J. Morrissey, C. B. Carter, Adv. in Ceramics 6 (1983), pp. 85-95.

15. C. B. Carter, K. J. Morrissey, to be published in Advances in Ceramics, 12, American Ceramic Society, Columbus, OH, 1984.

16. K. J. Morrissey, C. B. Carter, Mat. Res. Soc. Sym. Proc. 24 (1984), pp. 121-127.

17. S. C. Hansen, D. S. Phillips, Phil. Mag. A, 47 [2] (1983), pp. 209-234.

18. R. P. Fortunee, M.S. Thesis, Case Western Reserve Univ. (1981),

19. L. A. Bursill, R. L. Withers, Phil. Mag., A40 (1979), pp. 213-232.

20. Y. R. Shiue, D. S. Phillips, sub. for publication.

21. K. J. Morrissey, K. K. Czandera, C. B. Carter, R. P. Merrill, J.A.C.S., 67 (1984), pp. C88-C90.

SURFACE REACTIVITY OF SILICA AND ALUMINA CERAMIC POWDERS

Wayne C. Hasz* and Alan Bleier**

Massachussetts Institute of Tech., Cambridge, MA
*Sprague Electric Co., North Adams, MA 01247
**Oak Ridge National Laboratory, Oak Ridge, TN 37831

INTRODUCTION

Colloidal and surface properties of powders control aqueous ceramic suspension properties such as agglomeration, viscosity, and ordering. The important particle-particle interactions are controlled by the charge and potential distributions in the ion cloud surrounding particles in suspension, the electrical double layer (EDL). These interactions arise, in turn, from acid-base reactions that occur on the powder surface as a function of pH and electrolyte concentration. Consequently, understanding these mechanisms is critical for controlling ceramic suspensions during processing of high quality pieces (Bleier, 1983).

Dry oxide powders, by their insulating nature, are electrically neutral and have no net surface charge or potential. When such powders are placed in an aqueous solution, H^+ and OH^- ions and H_2O react with the surface to generate an amphoteric, hydroxylated layer. If the pH of the suspension is changed, H^+ and OH^- ions from solution re-equilibrate with the hydroxylated surface sites, SOH^O, changing their concentration (see Figure 1 and Table 1). These ionization reactions change the number of charged species on the surface, altering the surface charge and potential as indicated in Figure 2. Electrolyte species may also adsorb at the hydroxylated surface layer (termed complexation) and thereby change the charge and potential distributions in the EDL.

The triple layer model of Davis, James, and Leckie (DJL) (Davis, 1978a,b and James, 1982) specifically considers these surface reactions and provides a method to determine their corresponding equilibrium constants, K_a. This model considers two regions of adsorbed charge and a diffuse, unoriented layer. H^+ and OH^- ions react at the oxide surface, contribute to the surface charge, σ_o, and experience a surface potential, ψ_o (see Figure 2). Counterions situated in the first adsorbed layer, the inner Helmholtz plane (IHP), are considered bound to charged surface sites. These complexed ions contribute to σ_β and generate a potential, ψ_β. The outer Helmholtz plane (OHP), with potential, ψ_d, and diffuse layer charge, σ_d, is separated from the IHP by a region of thickness, Δ, and constant capacitance, C_2. Since the triple layer model considers ionization of surface sites and complexation of electrolyte ions in the EDL, chemical reaction constants can be derived for both types of reactions. These reaction constants apply over a wide range of electrolyte concentrations and may be used to predict ψ_d, and to estimate the ξ-potential at the OHP in ceramic suspensions.

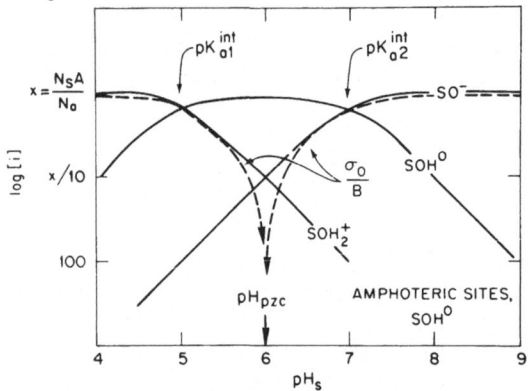

Figure 1 Concentration of surface species as a function of pH (after James, 1982).

Table 1   Surface reactions on oxide particle surface.

Ionization

$$SOH_2^+ = SOH^O + H_s^+$$

$$SOH^O = SO^- + H_s^+$$

Complexation

$$(SOH_2^+Cl^-)^O = SOH^O + H_s^+ + Cl^-$$

$$SOH^O + Na_s^+ = (SO^-Na^+)^O + H_s^+$$

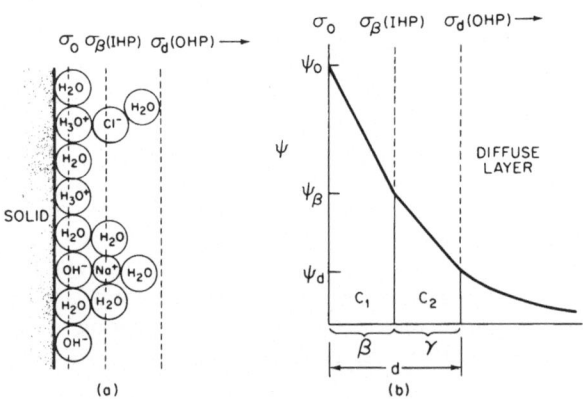

Figure 2 Distribution of a) charged ions and b) charge and potential in the electrical double layer (after James, 1982).

190

METHODS TO DETERMINE SUSPENSION PROPERTIES

Potentiometric, acid/base titration data for oxide powder surfaces can be used to calculate surface charge, potential, and reaction constants. Bolt (1957) and Parks and de Bruyn (1962) were among the first investigators to use this technique to study the interaction of acids and bases with the oxide-electrolyte solution interface.

In potentiometric titration, a control electrolyte solution, e.g. 10^{-3} to 1 N NaCl, is titrated with the corresponding acid or base, HCl or NaOH, and the volume of titrant required to attain various pH-values is recorded. A sample of the electrolyte solution is equilibrated with an oxide powder of known surface area. This dispersion is titrated and the two titration curves are compared; see Figure 3. The difference in volume of acid or base required to obtain a given pH for each sample, ΔV, relates to the net uptake or release of H^+ and OH^- ions at the powder's surface. If supporting electrolyte does not specifically absorb and if the H^+ and OH^- ions have equal affinity for the surface, then their net uptake is given by:

$$1 \qquad \Gamma' = [(C_a - C_b) - (C_{H+} - C_{OH-})] / A_s$$

where Γ' is the net uptake per cm^2 of surface; $C_{a(b)}$ is the concentration of acid(base) required to attain a given pH; $C_{H+(OH-)}$ is the concentration of $H^+(OH^-)$ species at a given pH; A_s is the specific surface area in $m^2 \ g^{-1}$. The relative surface charge, $R\sigma_o$, can be calculated:

$$2 \qquad R\sigma_o = F(\Gamma_{H+} - \Gamma_{OH-})$$

where F is the Faraday constant. If a series of $R\sigma_o$ versus pH isotherms are plotted for different electrolyte concentrations, the curves intersect at a unique pH called the point of zero charge, PZC, as in Figure 4a. The surface charge is independent of the concentration of nonadsorbing electrolyte at the PZC, implying the

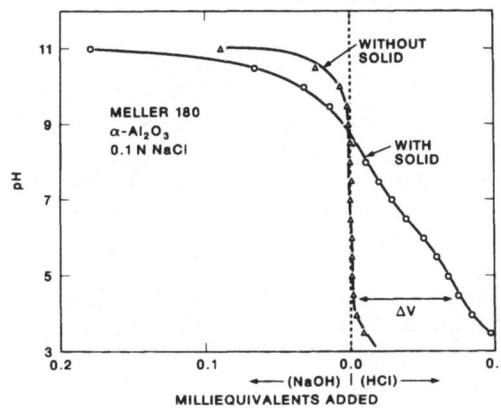

Figure 3 Potentiometric titration of electrolyte control and suspension to yield surface charge. (Hasz, 1983)

191

surface density of positive sites equals that of negative sites; ($\Gamma_{H+} = \Gamma_{OH-}$).

The intersection of the relative surface charge curves may not occur at $R\sigma_0=0$, see Figure 4a. To compensate for this effect, the axis is shifted by a value, $\Delta\sigma_0$, such that the intersection occurs at $\sigma_0=0$ as shown in Figure 4b. The second plot defines the actual value of the surface charge as a function of pH. The shift, $\Delta\sigma_0$, in the relative surface charge results from a release or uptake of charged species from the surface during solid-solution equilibration prior to titration. This release or uptake of species results from either dissolution of surface impurities or initial hydroxylation of the oxide surface if dry powder is placed in solution. The magnitude of $\Delta\sigma_0$ denotes the equivalent amount of surface impurity or the extent of hydroxylation.

The PZC corresponds to the average of the surface ionization constants. At this pH, the overall surface is uncharged since the concentration of neutral sites is high and those of positive and negative sites are small and equal ($[SOH_2^+] = [SO^-] \ll [SOH^O]$); see Figure 1. Titration data directly yield the area-average concentration difference between positive and negative groups on the surface; however, a quantitative distinction between the actual chemical sites is not possible from these data alone. Strongly acidic oxides generally have low PZC-values and strongly basic ones have high PZC-values (Parks, 1967). Examples are SiO_2 with a PZC of 2 and MgO with a PZC of 11. The PZC can be predicted from electrostatic considerations and is related to the cationic charge and radius (Parks, 1967 and Yoon, 1979). Shifts in the PZC reflect changes in the state of hydration, crystal cleavage habit, and

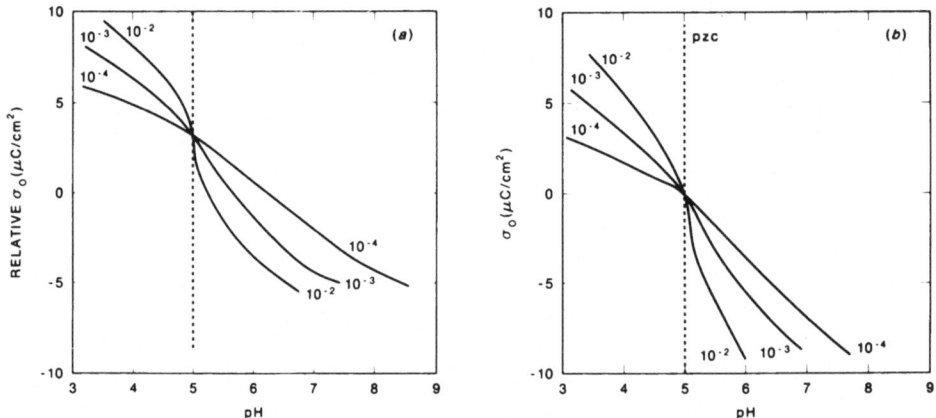

Figure 4 Method to determine the point of zero charge of the oxide surface. a) the pH at the intercept is the PZC. b) a correction factor, $\Delta\sigma$, is applied to obtain the absolute surface charge (after Hunter, 1981); Ionic strength in molarity.

crystallinity, making the actual value difficult to predict. Thus, the experimentally determined PZC-values may differ from that expected due to variations in sample preparation and impurity concentration.

Potentiometric titrations can also be used to measure the concentration of electrolyte species adsorbed at the oxide surface. Ions such as Na^+, Cl^-, Br^-, I^-, CN^-, F^-, Cu^{2-}, S^{2-}, Ag^+, Ca^{2+}, and K^+ can be studied with the use of specific ion electrodes. The procedure is similar to that for a pH-titration: the pX-titration for a suspension is compared with that of the control electrolyte, where X is the ion of interest.

A phenomenon known as the suspension effect may cause problems in measuring the pH of a suspension during titration (Cherbonowski, 1982). Errors result from an interaction between the junction potential at the glass electrode and the charged particles in suspension. The magnitude of this effect depends on particle size and concentration, being greatest for very concentrated, low ionic strength suspensions containing highly charged particles. The use of a second junction, reference electrode eliminates the suspension effect. The outer reservior is filled with the control electrolyte, eliminating the concentration gradient that normally exists between the suspension and the electrode's reservoir. Thus, the particles do not interact with the electrode during a titration.

Electrophoresis is another technique used to explore the electrical double layer at the oxide-solution interface. In this technique, an applied voltage interacts with the charged particles, creating a hydrodynamic shear plane in the EDL and causing the particle to move towards the oppositely charged electrode. The electrophoretic mobility of the particle, μ, is calculated and the electrostatic potential, ξ, at the shear plane is estimated. This ξ-potential is thought to be identical to the diffuse layer potential, ψ_d, even at high electrolyte concentrations (0.1 N). The ξ-potential is found to decrease as the pH approaches the isoelectric point (IEP- zero potential), since the concentration of charged surface sites and thus the induced OHP potential decrease. The ξ-potential also decreases as the electrolyte concentration increases since complexing ions screen the surface charge. The PZC equals the IEP when the electrolyte ions do not specifically adsorb. In this case, only the H^+ and OH^- ions contribute to the charge developed at the oxide surface.

Errors introduced in the measurement of surface charge, zeta potential, and other EDL properties may be significant and occur from several causes (Hasz, 1983): 1) Dissolved CO_2 produces carbonic acid which changes the pH and which may specifically adsorb at the powder surface; 2) Dissolution at extreme pH may consume titrant or induce surface reprecipitation; 3) Variability in measuring the surface site density (N_s); 4) Lack of independent experimental checks on the titration data, and 5) Impurities.

The model of Davis, James, and Leckie (James, 1982) may be used to calculate ionization (K_a) and complexation (K^*) constants. Graphs of surface charge, σ_o, vs pH are constructed and the effective ionization constants, pQ-values [$\equiv pH \pm log(\alpha/1-\alpha)$], where α is the fraction of surface sites ionized ($\alpha \equiv \sigma_o/eN_s$), are calculated and plotted versus α^* ($\equiv 10\alpha + c^{1/2}$). Extrapolation to conditions of zero charge and infinitely dilute solution yield the intrinsic ionization constants, $pK_{a1,a2}$. Plots of pQ^*_{ion} ($\equiv pH \pm log(\alpha/1-\alpha))C_{ion}$) versus α^*_{ion} ($\equiv 10\alpha - log(c)$), when extrapolated to the zero surface charge and log[NaCl] = 0 conditions yield the intrinsic complexation constants, $pK^*_{Na,Cl}$.

Once the ionization and complexation reaction constants have been estimated from the pH titration data, the calculation of the distribution of species in the EDL is possible. The simultaneous evaluation of expressions for the bulk solution and EDL chemical equilibria can be made using computer programs such as MINEQL (Westall, 1976) or MICROQL (Westall, 1979a,b). The incorporation of surface charge effects, which modify the species electrochemical potential, must be included and can be handled by a subroutine in the computer programs. These programs determine the concentration distribution of species in the EDL from which the charge and potential distribution may be estimated. These values are then compared with the experimentally determined surface charge and ξ-potential to test the model employed.

EXPERIMENTAL MATERIALS

The silica used in this study, Quso G30, was obtained from the PQ Corporation, Valley Forge, PA. It is a precipitated silica produced by depolymerizing a high purity sand to yield soluble silicate. The soluble silicate is then repolymerized to precipitate fine, particulate, amorphous silica (PQ, 1978). According to Yates (1976), most commercial precipitated silicas are usually prepared by the addition of acid to a sodium silicate solution under controlled conditions yielding a hydrous silica which is filtered, washed and dried. The alumina used in this study was obtained from Adolf Meller Co., Providence, RI. Alpha alumina is produced by the refining and thermal decomposition of a complex aluminum salt which is said to yield high purity alumina due to multiple recrystallizations and precise heating (Meller, 1983).

The commercial SiO_2 and Al_2O_3 powders were cleaned using soxhlet extraction. In this technique the powder is placed in a porous, filter paper cup which is placed in a siphon chamber. Freshly deionized (DI) water flows through the powder removing soluble ions and filling the chamber. When the siphon height is obtained, the siphon tube drains the dirty water into the solvent container where it is redistilled and the process repeated. This technique allows continuous washing with hot, DI water where the washing solvent is replaced every few minutes. Parfitt and Wharton (1972) found that continuous extraction for 48 h removes all traces of surface chloride from chloride-originated powders, e.g. SiO_2 from $SiCl_4$, TiO_2 from $TiCl_4$ and Al_2O_3 from $AlCl_3$. Titrations on nonsoxhleted Quso G30 powders yielded $R\sigma$ versus pH curves that did not reliably intersect at $R\sigma = 0$, whereas soxhleted powders did for both silica and alumina (Hasz, 1983). The $\Delta\sigma$ discrepency for SiO_2 apparently derives from the dissolution and then replacement of surface sodium ions with H^+ ions from solution in the uncleaned system. For the cleaned system, surface sodium was not present and hence no shift in the titration curves was evident, i.e. $\Delta\sigma = 0$. Soxhlet extraction also eliminated problems in the intersection of surface charge curves for Meller 180, α-alumina. The cleaned powders were then dispersed in DI water and stored in cleaned polypropylene containers.

Dilute solutions of 0.1 N HCl and 0.1 N NaOH were used for pH-control during all experiments. NaOH solutions were standardized using the phthalate technique of Vogel (1962). The acid solutions were then standardized against these primary standards. All solutions were standardized to 0.001% at 25 $^\circ$C. Deionized, 18 MΩ-cm water was used in all experiments. All chemicals used in this study were of analytical grade. NaCl was used to maintain constant ionic strength in all titration, electrophoresis, and dissolution experiments. Prepurified nitrogen, saturated with water vapor, was used as an inert atmosphere for all titration studies.

EXPERIMENTAL TECHNIQUES

The specific surface area of the powders was determined using the single-point BET method on a Quantasorb surface area instrument by Quantachrome. The solubility of alumina and silica in aqueous solutions at different pH-values was determined using a Bausch and Lomb Spectronic 21 UVD, matched cells, and the procedure of Vogel (1962).

Potentiometric Titration. Potentiometric titrations were carried out using a Radiometer RTS822 system. The electrodes included Radiometer's K701 double junction, G2040C glass, and T701 temperature sensor. Prior to titration, the burette was thoroughly cleaned by repeated flushings with fresh titrant. The second junction of the K701 electrode was rinsed with DI water and flushed with the electrolyte before the final filling. 1.0 N NaCl solutions were used for samples with 1.0 and 0.1 N electrolyte and 0.01 N for samples with 0.01 and 0.001 N electrolyte. The samples were N_2-bubbled for at least 40 min to ensure removal of any residual CO_2. At the end of the run, the final pH and volume added were recorded. The usual titration rate was 2 pH units per hour. The titrations for the control electrolyte and suspension were compared, yielding the net uptake of titrant by the powder's surface, as described earlier.

Electrophoresis. Electrophoresis was carried out using a Rank Brothers Particle Micro-Electrophoresis Apparatus Mark II. Measurements were made at 25 $^{\circ}$C in the cylindrical cell. At least ten measurements were made for each sample to ensure accuracy. Samples were titrated under nitrogen and allowed to equilibrate at the desired pH for two hours. The samples were removed and stored in polypropylene test tubes for 4 hours when electrophoresis measurements were made. Unfortunately the measurements had to be made in air, so pH and ξ-potential shifts due to CO_2 absorption from the atmosphere may have occurred.

RESULTS AND DISCUSSION

Silica. In the case of soxhleted Quso G30, the PZC is 4.25 and the corresponding surface charge profile shown in Figure 5. The σ_o curves have a unique intersection at $\sigma = 0$, however, for unsoxhleted powder, the intersection occurs at $R\sigma = 4.74 \ \mu C/cm^2$, requiring a $\Delta\sigma$-shift. Soluble sodium is apparently removed upon soxhletion, thereby removing the cause of the surface charge shift, $\Delta\sigma$.

The second ionization constant for Quso G30, pK_{a2}, is 6.6, determined by plotting pQ vs α^* and by extrapolating to the conditions of zero charge and infinite dilution; see Figure 6a. The complexation constant for sodium, pK^*_{Na}, is 5.4, determined by plotting pQ^* vs α^*_{ion} and by extrapolating to the conditions of zero charge and $\log[NaCl] = 0$; as seen in Figure 6b. The first ionization and complexation constants for silica cannot be directly determined from these data since the pH-conditions under which the

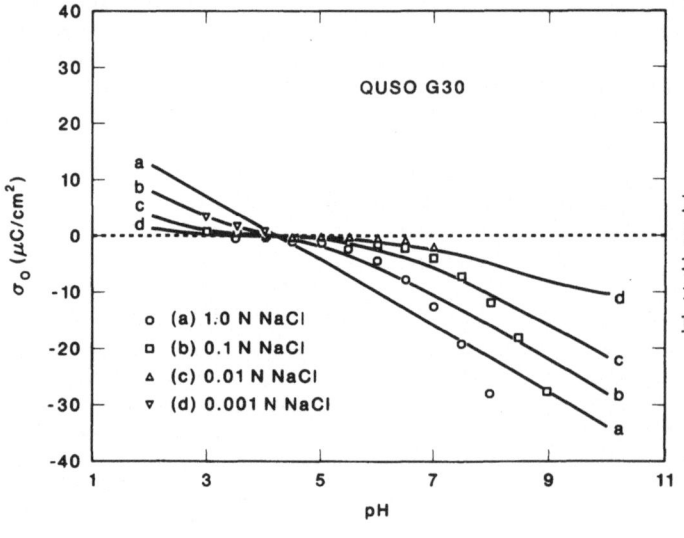

Figure 5 Experimental (symbols) and theoretical (lines) surface charge, σ versus pH for Quso G30.

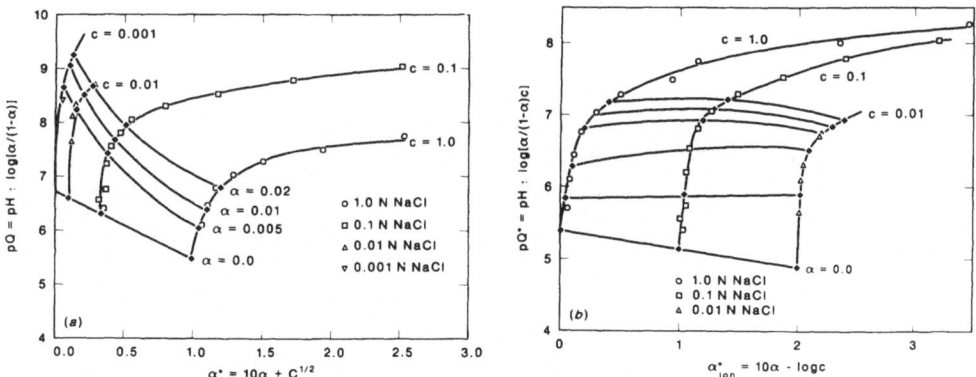

Figure 6 Double extrapolation plots for Quso G30 yielding the a) ionization constant, pK_{a2}, and b) complexing constant, $pK*_{Na}$.

surface species transitions occur are pH = 1.90 and 3.10, values that are too low to determine constants confidently.

The simultaneous surface reaction equations were evaluated once the reaction pK's had been determined. The theoretical surface charge curves agree well with the experimentally determined values for low electrolyte concentrations and pH values near the PZC of 4.25 (Figure 5). The experimentally determined curves appear to increase in the negative direction more sharply towards the PZC than do the theoretically calculated ones, possibly indicating surface dissolution during titration. However, dissolution experiments showed very low solubility, significantly below the saturation level, in the region of interest (Hasz, 1983).

Alumina. The experimental and theoretical surface charge profile for Meller 180, α-Al_2O_3, is shown in Figure 7; the PZC is 8.75. The theoretical curves compare favorably with the experimental ones, except for slightly lower values below the PZC. The experimentally determined ξ-potential is compared to the theoretical ψ_d curves in Figure 7b. The theoretical values lie in the range of the experimental data, however discrepencies are evident and are probably due to CO_2 adsorption phenomena occurring in the experimental system (Hasz, 1983). The initial $R\sigma$ vs pH data for the corresponding unsoxhleted material yield a nondistinct intersection, implying surface or solution contamination. After soxhletion, the $R\sigma$-curves all intersected at a unique pH, confirming impurity removal. The ionization constants for Meller 180, determined from plots of pQ vs $\alpha*$ in the same manner as for the silica are pK_{a1} = 5.0 and pK_{a2} = 10.7. Corresponding complexation constants are $pK*_{Cl}$ = 7.1 and $pK*_{Na}$ = 10.2.

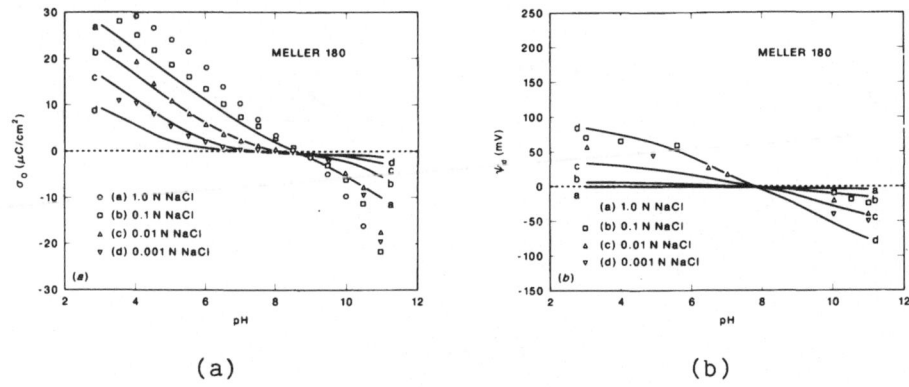

(a)   (b)

Figure 7 Experimental (symbols) and theoretical (lines) surface
         charge and IHP potential profile versus pH for Meller
         180, $\alpha-Al_2O_3$.

A comparison of the EDL properties for the silica and alumina
in this study to those of other works is shown in Table 2. The
variability in the properties of these materials is caused by the
synthesis and preparation techniques, leading to different crystal
habits at powder surfaces.

Table 2. Surface Properties of SiO_2 and Al_2O_3 in NaCl solutions.

Material	Reference	pK_{a1}	pK_{a2}	pK^*_{Cl}	pK^*_{Na}	PZC
Silica						
Quso G30	this study	–	6.6	–	5.4	4.25
SiO_2	Huang, 1977	–	6.7	–	–	2.0
SiO_2	Schindler, 1968	–	5.8	–	–	2.0
Alumina						
Meller(α)	Hasz, 1983	5.0	10.7	7.1	10.2	8.75
Al_2O_3 (α)	Yorpps, 1964	8.5	9.7	–	–	9.10
Cabot (γ)	James, 1982	5.2	11.8	7.9	9.2	8.5
Cabot (γ)	Huang, 1973	7.7	9.3	–	–	8.5
Cabot (γ)	Davis, 1978	5.7	11.5	–	–	8.6

CONCLUSION

Surface charge arising from interfacial reactions of
electrolyte ions with SiO_2 and Al_2O_3 suspensions are analysed using
potentiometric titration. Points of zero charge, PZC, are
calculated and surface charge profiles are plotted for these oxides.

The intrinsic surface ionization and complexation constants of electrolyte ions on the surface are determined from the surface charge data using the double extrapolation technique of Davis, James, and Leckie.   Theoretical equilibrium calculations are employed to solve for the interfacial reactions at the oxide/solution interface using the experimentally determined ionization and complexation constants to estimate the concentration of charged species in the electrical double layer, EDL. From this estimate, the charge and potential distribution in the EDL are calculated.   The theoretical surface charge curves for SiO_2 and Al_2O_3 are compared to those determined experimentally. ξ-potentials for Al_2O_3 are calculated from electrophoresis data and compared with the theoretical values of the diffuse layer potential, ψ_d. The theoretical charge distributions predict the experimental surface charge data well, implying that the DJL model applies to these systems. Theoretical diffuse layer potentials predict the experimentally measured ξ-potential well for pH-values greater than the isoelectric point.

ACKNOWLEDGEMENTS

The experimental work was funded by the Department of Energy, Contract No. DE-AC02-81ER1053.

REFERENCES

Bleier, A., 1983, Acid-Base Properties of Ceramic Powders, in: "Advances in Material Characterization," R.A. Condrate, D.R. Rossington, and R.L. Snyder eds., Plenum Publ. Co., New York.

Bolt, G.H., 1975, Determination of the Charge Density of Silica Sols, J. Phys. Chem., 61:1166.

Chernoberezhs'kii, Y.M., 1982, The Suspension Effect, in: "Surface and Colloid Science," v12, E. Matijevic, ed., Plenum Publ. Co., New York.

Davis, J.A., James, R.O., and Leckie, J.O., 1978a, Surface Ionization and Complexation at the Oxide/Water Interface 1. Computation of Electrical Double Layer Properties in Simple Electrolytes, J. Colloid Interface Sci., 63:480.

Davis, J.A. and Leckie, J.O., 1978b, Surface Ionization and Complexation at the Oxidation/Water Interface 2. Surface Properties of Amorphous Iron Oxyhydroxide and Adsorption of Metal Ions, J. Colloid Interface Sci., 67:90.

Hasz, W.C., 1983, "Surface Reactions and Electrical Double Layer Properties of Ceramic Oxides in Aqueous Solution," S.M. Thesis, Massachusetts Institute of Technology, Cambridge, MA.

Huang, C.P., 1977, The Determination of Surface Acidity by Alkalimetric Titration, Annual NSF Report.

Huang, C.P., 1982, The Surface Acidity of Hydrous Solids, in: "Adsorption of Inorganics at Solid-Liquid Interfaces," Anderson, M.A., and Rubin, A.J., eds., Ann Arbor Science Publ. Inc., Ann Arbor, MI.

Hunter, R.J., 1981, "Zeta Potential in Colloid Science- Principles and Applications," Academic Press, New York.

James, R.O., and Parks, G.A., 1982, Characterization of Aqueous Colloids by Their Electrical Double-Layer and Intrinsic Surface Chemical Properties, in: "Surface and Colloid Science," v12. Matijevic, E., ed., Plenum Publ. Co., New York.

Meller, 1983, "Alumina Powder for Ceramic Manufacturing," The Adolph Meller Co., RI.

Parfitt, G.D., and Wharton, D.G., 1972, J. Colloid Interface Sci., 38:431.

Parks, G.A., and de Bruyn, P.L., 1962, The Zero Point of Charge of Oxides, J. Phys. Chem., 66:967.

Parks, G.A., 1967, Aqueous Surface Chemistry of Oxides and Complex Oxide Minerals, in: "Advances in Chemistry Series," No.67, Amer. Chem. Soc., Washington. DC.

PQ, 1978, "Properties and Applications, Quso Micro-fine Precipitated Silicas," Technical Brochure #18-1, PQ Corporation, Valley Forge, PA.

Schindler, P.W., and Kamber, H.R., 1968, Die Aciditat von Silanolgruppen, Helv. Chim. Acta, 51:1781.

Vogel, A.I., 1962, "A Text-Book of Quantitative Inorganic Analysis, Including Elementary Instrumental Analysis," 3rd ed., John Wiley, NY.

Westall, J., Zachary, J., and Morel, F., 1976, "MINEQL: A Computer Program for the Calculation of Chemical Equilibrium Composition of Aqueous Systems," Technical Note No. 18, Ralph Parsons Laboratory, MIT, Cambridge, MA.

Westall, J., 1979a, "MICROQL I. A Chemical Equilibrium Program in Basic," Swiss Federal Institute of Technology EAWAG.

Westall, J., 1979b, "MICROQL II. Computation of Adsorption Equilibria in Basic," Swiss Federal Institute of Technology EAWAG.

Yates, D.E., and Healy, T.W., 1976, The Structure of the Silica /Electrolyte Interface, J. Colloid Interface Sci., 55:9.

Yoon, R.H., Salman, T., and Donnay, G., 1979, Predicting Points of Zero Charge of Oxides and Hydroxides, J. Colloid Intrface Sci., 70:483.

Yorpps, J.A., and Fuerstenau, D.W., 1964, The Zero Point of Charge of α-Al$_2$O$_3$, J. Colloid Interface Sci., 19:61.

DIFFUSION OF WATER IN SiO_2 AT LOW TEMPERATURE

W. A. Lanford and C. Burman

Joseph Henry Physics Building
Department of Physics, SUNY Albany
Albany, New York 12222

R. H. Doremus
Materials Engineering, R.P.I.
Troy, New York 12181

ABSTRACT

^{15}N hydrogen profiling is used to study diffusion of water in fused silica at 90°C. Two different diffusion mechanisms are observed, one with $D \sim 6 \times 10^{-18}$ cm^2/sec and one with $D = 10^{-15}$ cm^2/sec. These results are compared to measurements of others and the possible origin of these two diffusions are briefly discussed.

INTRODUCTION

SiO_2 exhibits a remarkable combination of optical, mechanical, electrical and chemical properties. As a result, it is widely used in such diverse technologies as optical fibers, microelectronics and fine particle composite materials. However, all of the properties listed above are well known to be sensitive to the amount of water (or number of OH groups) in the silica. Because of the nearly universal presence of water in most environments, prediction of the long term properties of devices made with silica depends on knowledge of the rate of diffusion of water in SiO_2.

There exists a long history of measurements of the transport of water in SiO_2 from the benchmark work of Moulson and Roberts (1) who worked at high temperatures (600-1200°C) to the very recent work of Nogami and Tomozawa (2) who worked at ~ 100°C. Since in most applications SiO_2 will be at temperatures between room temperature

and 100°C, measurements of lower temperature diffusion coefficients
are most important in predicting long term properties. Remarkably,
extrapolation of the high temperature data (1) and the low temperature measurements (2) agree to within an order of magnitude and
both indicate that D at 90°C is of order 10^{-17} - 10^{-18} cm^2/sec, and
D at 30°C would be ∿ 100 times smaller.

These are small diffusion coefficients implying diffusion of
water is not a serious limitation in most applications. To put this
in perspective, if D = 10^{-19}, using x = $\sqrt{2DT}$ as a measure of diffusion distance, it takes 1600 years for water to diffuse 1 µm or
two months to diffuse 100 Å.

Because of the importance of this diffusion coefficient and
because of the indirect nature of most previous experimental methods,
it seemed worthwhile to use the ^{15}N nuclear reaction hydrogen profiling technique to make a simple direct measure of this diffusion.
As will be seen below, these measurements indeed confirm that there
is a diffusion process occuring with D ≃ 10^{-18} cm^2/sec at 90°C but
that there is also a much more rapid diffusion occuring with
D = 10^{-15} cm^2/sec. <u>This is not a small diffusion coefficient</u>.
Using the same extrapolation as before to estimate D at 30°C implies
water takes only of order 16 years to diffuse 1 µm or of order 14
hours to diffuse 100 Å. This may present serious limitations to
several technologies, perhaps most importantly to optical fibers.
It may also be important in the detailed understanding of fracture
of silica fibers.

^{15}N HYDROGEN PROFILING

^{15}N hydrogen profiling has now been used for a number of years
to quantitatively measure hydrogen concentration <u>vs</u> depth

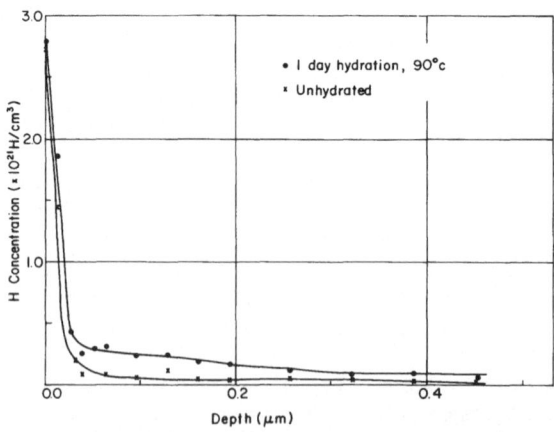

Figure 1: Hydrogen concentration profiles of fused silica:
one sample exposed to distilled water for 24 hours,
one sample untreated.

variety of insulating, semiconducting and superconducting materials (3). The first detailed application to glass involved the study of the ion exchange mechanism responsible for the hydration of alkali silicate glasses (4). Descriptions of this technique exist in the literature (including the Proceeding of this Conference held two years ago at Alfred) and will not be repeated here (5).

For our purposes it seems sufficient to summarize the properties of this method: It is capable of quantitatively measuring hydrogen concentration <u>vs</u> depth in any material (without use of a reference standard) with a depth resolution of order 100 Å and a sensitivity of order 100 ppm (atomic). It works equally well in insulators and conductors. Because this method makes use of a nuclear reaction to probe for hydrogen, it measures only H/cm^3 regardless of chemical bonding of the hydrogen. As a result, it is often useful to combine nuclear reaction analysis with a technique sensitive to bonding (such as infrared absorption spectrometry) to give a more complete description of the material. For the present study, infrared studies are underway and will be reported shortly (6).

SAMPLE PREPARATION AND ANALYSIS

Flat plate samples of fused silica glass (optosil) were etched in HF (3% for 5 minutes) and then exposed to temperature controlled distilled water ($90°C$) for varying times. The water was agitated using a teflon coated magnetic stirrer. Subsequently the hydrogen concentration <u>vs</u> depth for these samples was measured.

Typical hydrogen profile results are shown in Figure 1. Because there are two diffusion mechanisms acting with very different <u>diffusion coefficients,</u> it is difficult to display the effects of both on a single graph. However, in Figure 1, one sees the two stages of diffusion: a shallow profile extending of order 100 Å with a high hydrogen concentration and a deep profile extending beyond 0.4 μm with a low hydrogen concentration.

Because the diffusion which causes the shallow profile is so slow and because of the substantial amount of hydrogen on the unhydrated sample, it is difficult to extract a diffusion coefficient for this slow component without exposing samples for times much longer than one day. However, taking the difference between the treated and untreated sample in Figure 1 gives a diffusion distance of order 100 Å which implies an effective diffusion coefficient D ($90°C$) ~ $6x10^{-18}$ cm^2/sec, in reasonable agreement with the data of Moulson and Roberts (1), and Nogami and Tomozawa (2) cited above.

For the rapidly diffusing component, time dependence can be more easily measured. Figure 2 shows the hydrogen profiles of a series

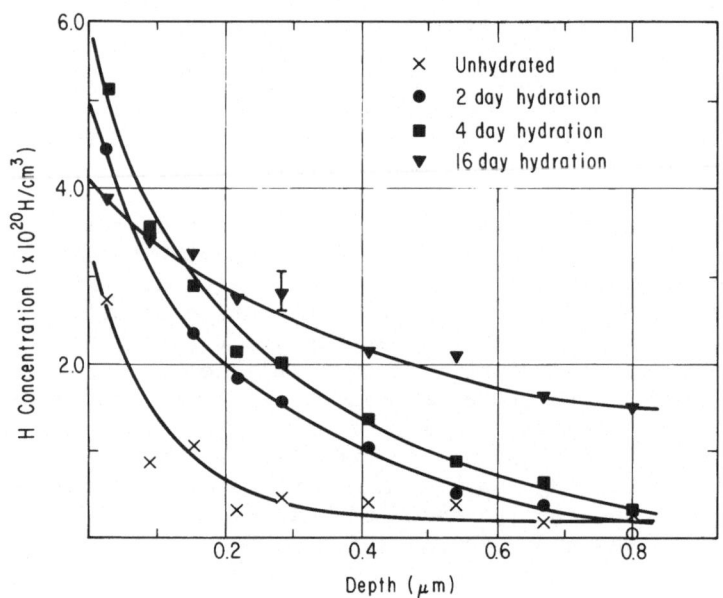

Figure 2: Hydrogen concentration profiles of a series of samples of SiO_2 exposed to water for varying lengths of time.

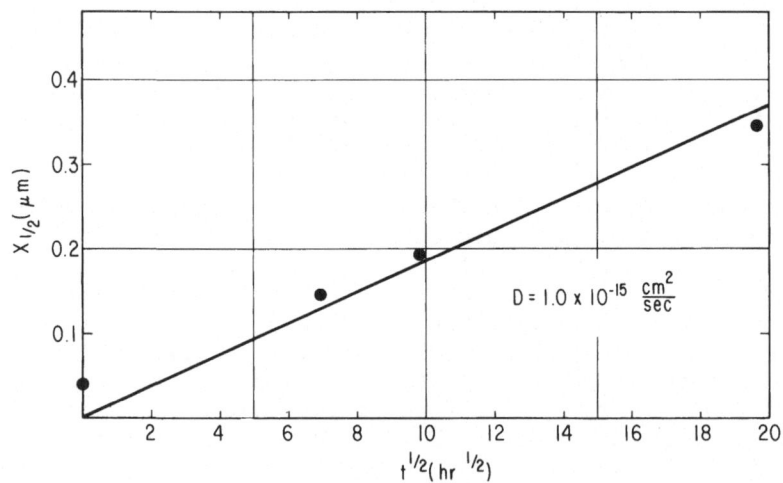

Figure 3: Depth of penetration of hydrogen <u>vs</u> square root of exposure time for data in Figure 2.

of samples exposed at 90°C for 2, 4 and 16 days. Figure 3 shows the depth of penetration (measured from the surface to the depth where the concentration is one half the surface value) vs the square-root exposure time. As can be seen, a good square root of time dependence is seen with D $(90^\circ$C$) = 1.0 \times 10^{-15}$ cm^2/sec.

DISCUSSION

The present report is intended to make only two central points: First, in the spirit of this conference, we wish to emphasize that the simple application of nuclear reaction analysis to unambiguously measure hydrogen concentration profiles can greatly clarify problems, even ones which have been extensively studied by other methods. Second, in the case of diffusion of water in fused silica, there are at least two distinct transport mechanisms, with one being three orders of magnitude faster than the other.

We presently have underway experiments designed to determine the mechanisms responsible for both types of hydrogen transport. These include studies of the pressure, temperature, and time dependence, the effects of water vapor vs liquid water, infrared studies to determine bonding and a more careful detailed determination of the shape of the diffusion profile. While for more definitive conclusions we must wait for the results of these measurements, it seems likely that the two mechanisms observed correspond to: A) the rapid transport of hydrogen in the form of molecular water diffusing through the Si-O-Si network followed by reaction with the network and B) the reaction of H_2O with Si-O-Si bridges at the glass surface followed by the slow diffusion of the resulting Si-OH groups.

From the perspective of various technologies, the rapid component of diffusion should be considered in both materials manufacturing methods and in prediction of long term characteristics.

ACKNOWLEDGMENTS

This research was supported by grants from the Army Research Office (work at SUNY Albany) and National Science Foundation (work at R.P.I.)

REFERENCES

1. J. Moulson and J. P. Roberts, Trans. Far. Soc. 57 (1961) 1208.
2. M. Nogami and M. Tomozawa, Physics and Chemistry of Glasses 25 No. 3 (1984) 82.
3. W. A. Lanford, H. P. Trautvetter, J. Ziegler and J. Keller, Applied Physics Letters 28 (1976) 566.

4. W. A. Lanford, K. Davis, D. LaMarche, T. Laursen, R. Groleau and R. H. Doremus, Journal of Non-crystalline Solids 33 (1979) 249.

5. W. A. Lanford, C. Burman, R. H. Doremus, Y. Mehrotra and T. Wassick, <u>Advances in Materials Characterizations,</u> Ed. D. R. Rossington, R. A. Condrate and R. L. Snyder, Plenum Press, New York (1983)

6. W. A. Lanford, C. Burman, R. H. Doremus and M. Tomozawa private communication (1984).

SPECTROSCOPIC ANALYSIS OF MOLECULAR SURFACES BY ION, PHOTOELECTRON

AND VIBRATIONAL SPECTROSCOPY

Joseph A. Gardella, Jr.[*], Robert L. Schmitt[*],
Joseph H. Wandass[*] and Roland L. Chin[**]

[*]Chem. Dept., State University of New York at Buffalo
Buffalo, New York 14214
[**]Allied Chemical Corporation Buffalo Research Laboratory
Buffalo, New York 14210

INTRODUCTION

The study of structure, bonding and composition in near surface and interfacial regions of molecular and polymeric solids involves the development and application of many techniques (1). Among those which are finding broad use for these analyses are X-Ray Photoelectron Spectroscopy (XPS or ESCA) (2) and Fourier Transform Infrared (FT-IR) Spectroscopy (3). More recently Ion Beam techniques like Ion Scattering Spectroscopy (ISS) and Secondary Ion Mass Spectrometry (4,5) have been making inroads into polymer structure analysis, complementing the high surface sensitivity of ESCA and rich bonding information from FT-IR, which samples over a much greater depth. Development of these methods and others like High Resolution Electron Energy Loss Spectroscopy (HREELS) involve the study of well characterized model systems, where surface bonding and composition is well known. ESCA and FT-IR, alone and in conjunction are more established in their information content and capabilities for polymer analysis.

The goals of this report are twofold. We will describe our research in the development of SIMS and HREELS for molecular and macromolecular structure analysis (4-7) of model systems. The second set of studies described have as their goal the development of a picture of surface bonding and composition for controlled morphology of highly crystalline homopolymers, where folding and looping of polymer chains yields a surface or interface with very different bonding and reactivity than in the bulk (8). Extending this approach to block copolymers involves the analysis of the contributions of morphology and bulk block composition in order to provide a quantitative and qualitative description of structure and bonding of molecular and polymeric surfaces.

DEVELOPMENT OF NOVEL METHODS FOR POLYMER ANALYSIS

Secondary Ion Mass Spectrometry (SIMS)

Many researchers have pointed out that SIMS should have unique applicability for the analysis of surface molecular structure of polymeric solids (1,4,9-11). All of these researchers have pointed out that SIMS has a high information content which includes bonding information from molecular, quasi-molecular, and fragment ions, compositional data, isotopic analysis, sensitivity to hydrogen, and low detection limits. In discussing the further development of SIMS for polymer research, it is necessary to review the conclusions of previous work, to understand the limitations of SIMS.

Gardella and Hercules (4,5) first pointed out that SIMS can be sensitive to structure and bonding of structurally similar surfaces which ESCA cannot differentiate. Specifically homologous and isomeric sidechains in a series of poly(methacrylates) were distinguished by SIMS and it was shown that structure related polyatomic ions would be generated from the pendant sidechain ester group. The most significant result was that isomeric butyl sidechain, indistinguishable by their ESCA core level spectra, were distinguished based on the intensity of the carbonium ion generated from the butyl sidechain group. This reflected the stability of the carbonium ion series: $3° > 2° > 1°$ (4). Based on this work Gardella, Novak and Hercules (5) were able to quantitate the surface chemical hydrolysis of poly(t-butyl methacrylate) with SIMS. Work at the Naval Research Laboratory (10) and by Briggs (11-13) has also pioneered SIMS in polymer analysis, detecting platicizers in polymers and refining the characterization of the methacrylates (10), and characterizing sample preparation and analysis conditions (11), analyzing working polymers (12) and developing some ion imaging characteristics (13) of SIMS.

Two conclusions can be drawn from this work. First of all, static primary ion beam conditions, which are important in minimizing damage and charging and retaining high surface sensitivity while increasing the probability of molecular ion generation (14), are necessary in SIMS of polymers (4,11). Gardella (4) and Briggs (11) have done careful studies noting that different polymers suffer from different damage rates during ion beam sputtering. Briggs has also noted (11,12) that even with this preliminary work much work needs to be done to describe the surface sensitivity of static molecular SIMS. Gardella, Novak and Hercules' results (5) address the detection of a very mild surface reaction, but no measurement of the depth of reaction was made, so sampling depth cannot be evaluated. No real account of static SIMS sampling depth for a macromolecular solid exists.

Thus, studies underway in our laboratory seek to define the surface sensitivity for molecular SIMS under static conditions (6). These studies involve the use of model systems which are mono-molecular assemblies produced via Langmuir-Blodgett techniques (15). These systems provide a wide variety of characteristics and controllable variables which make them uniquely suited for studying the SIMS experiment. Among these are: i) a highly regular and pure model system can be prepared, ii) molecular orientations for large molecules are precisely determined via control over surface tension during preparation and transfer of the monolayer, iii) modifications such as substrate changes, added cations and chromophores can be effected, iv) multilayer systems can be produced with specific orientations and most importantly, v) the "physical state" of the monolayer can be controlled. Ancillary techniques such as ellipsometry, FT-IR and thermodynamic measurements can confirm the nature of the monolayer orientation and thickness. This system provides a means to prepare an "ideal" model system for static molecular SIMS.

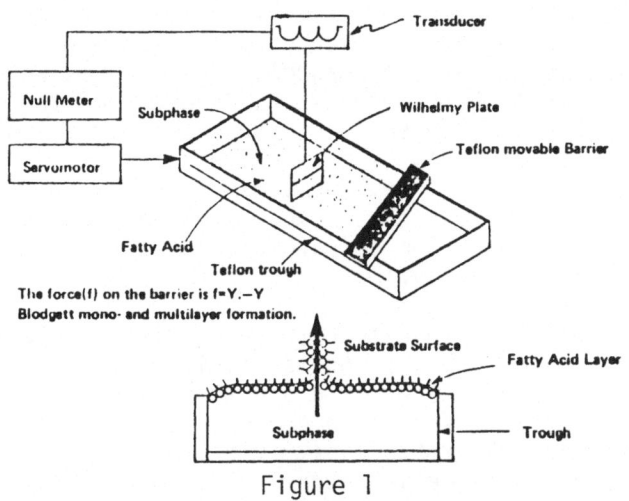

Figure 1

Figure 1 shows a diagram of the LB apparatus. The molecular systems which form insoluble monolayers at gas-liquid interfaces and can be transferred to suitable substrates run the chemical gamut (15). For our initial work we have chosen a set of fatty acids whose LB characteristics are well understood, and produced monolayers for transfer to prepared silver surfaces, since silver is often used as a support for compounds analyzed by molecular SIMS (14), the interaction of the monolayer with the silver substrate was of interest. Additionally, we were interested in evaluating the sensitivity of static SIMS to a quantified monolayer of molecules well known for difficulty in SIMS analysis (16).

Structure Models for Aromatic Siloxane Homo and Copolymers: a) TMpS/DMS Copolymer-low DMS concentration, high crystallinity, b) TMpS/DMS copolymer-high DMS concentration, c) crystal structure of pure TMpS single crystal.

The study of changes in the morphology of the fractional homo-polymer TMpS was accomplished by examining solution grown crystal mats (I), melt cast mats (II), melt quenched mats (III) and solution cast films (IV) with ESCA and FT-IR, correlated with differential scanning calorimetry (DSC), crystallographic (Small Angle X-Ray Scattering) and gas phase HF reaction studies. The results provide a complete picture of the surface structure of TMpS demonstrating for the first time the use of ESCA to analyze the morphology of an ordered homopolymer. ESCA analyses of the various degrees of crystallinity (I>II>III>IV) were correlated with FT-IR results yielding the conclusion that the surface (amorphous) region of each material consists of surface segregated $Si-C_6H_4-Si-O$ linkages, exposing siloxane bonds in the folds of the long chains back into the rod-like crystalline "core" formed by the polymer chains. ESCA analyses showed higher surface oxygen concentration while FT-IR changes in the Si-O-Si stretches in the higher amorphous content materials. Chemical etching studies with gas phase HF (diluted with He 10/30 v/V), which selectively reacts with and removes the amorphous surface regions, increased the overall crystallinity. This process of chemical and morphological changes were followed spectroscopically by both methods. ESCA detected the formation of Si-F, and the decrease in surface oxygen as crystallinity increased. FT-IR showed the growth in Si-OH and the shifting of Si-O-Si stretches with reaction time. Combining these results, we deduced the addition of HF across the folded Siloxane bond to form Si-F and Si-OH, etching away the surface amorphous region to increase the relative crystallinity.

We then set about to extend these studies to the TMpS/DMS copolymers. Considering the models in Figure 5 is very important in understanding and interpreting spectroscopic results. The series of block copolymers is varied in concentration of TMpS (see structure in Fig. 5) and in the crystallinity via preparation. For this study, the samples listed in Table 1 below were analyzed, and can be viewed by considering the models. Copolymers of high TMpS content and high crystallinity will show segregation of the highly amorphous short blocks of DMS (12 monomer units long) at the surface. The amorphous nature of the DMS blocks "force" the TMpS blocks into a thick surface region in the case illustrated in Figure 5b, the high DMS concentration. Basing ESCA analysis on the previous work (8), Figure 6 shows the results of the ESCA analysis of this series of copolymers. As in the analysis of the TMpS homopolymer, the interpretation of the C/O ratio is the key piece of data. While Table 1 shows the gradual shift of the Si2p binding energy

Table 1. ESCA Binding Energy Analysis of TMpS/DMS Copolymer Series

Carbon 1s (eV)	Shakeup (Δ)	Oxygen 1s (eV)	Shakeup (Δ)
285.0 (1.9)	6.3	532.7 (1.9)	6.2

SAMPLE	WT % RATIO	SILICON 2p (+0.2ev)	Shakeup (Δ)
TMpS xtal	100/0	102.0 (2.1)	6.1
TMpS melt	100/0	101.9 (2.0)	6.2
TMpS/DMS xtal	90/10	102.1 (2.2)	6.3
TMpS/DMS melt	90/10	102.2 (2.1)	5.9
TMpS/DMS soln	90/10	102.1 (2.1)	6.0
TMpS/DMS xtal	75/25	102.2 (2.3)	6.0
TMpS/DMS melt	80/20	102.2 (2.2)	6.0
TMpS/DMS soln	80/20	102.1 (2.2)	6.1
TMpS/DMS xtal	50/50	102.3 (2.4)	6.3
TMpS/DMS melt	50/50	102.5 (2.2)	6.0
TMpS/DMS melt	30/70	102.6 (2.2)	5.8
DMS soln cast	0/100	102.6 (2.0)	---

from pure TMpS to pure DMS, the plot of C/O ratio shows a deviation from a homogeneous mixture based on the straight line interpolation between the pure compound values only at the low concentration DMS copolymer of high crystallinity. Based on the model in Figure 5a it is clear that higher crystallinity would only affect the surface concentration where all the short DMS blocks could be segregated in the surface. Given the sampling depth of ESCA, higher concentrations of DMS in the copolymers with greater than 10% DMS do not affect the surface concentration of DMS. Crystallinity of the bulk TMpS in the copolymer drastically affects the degree of segregation of the short DMS block for the 90% TMpS case as sampled by ESCA.

ISS and FT-IR (18) provide complementary results, ISS showing that shadowing effects from atoms in different bonding situations can be affected by the sample morphology and the FT-IR/ATR sampling showing the segregation of the DMS over a thick layer in the lower crystallinity samples. In this unique case, microphase unhomogeneities can be affected by two controllable factors in these copolymers: concentration and crystallinity. Given the shortness of the DMS blocks, the ESCA results confirm the model of surface composition in Figure 5, and these spectroscopic method, yield a firm picture of crystallinity-structure relationships.

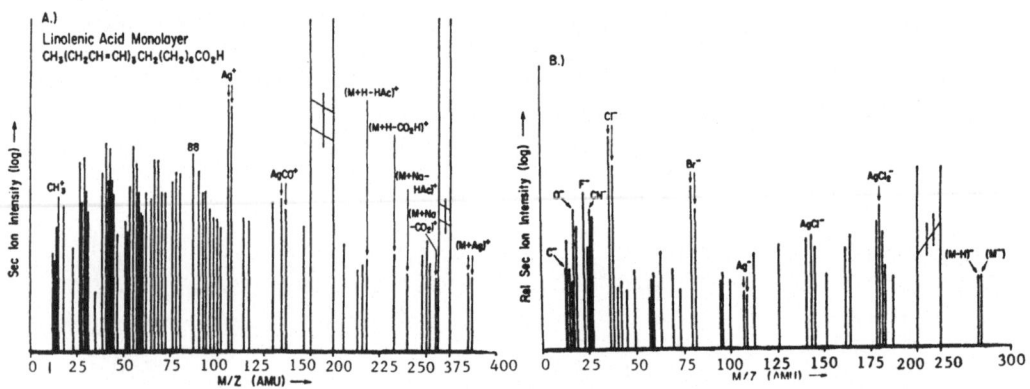

Figure 2

Results in Figure 2 (linolenic acid) illustrate the type of information available from static SIMS of LB monolayers of fatty acids. Molecular ions were detected in the negative ion spectrum, and can be explained by charge delocalization over the acid head carboxyl group, thus stabilizing the (M-H)⁻ ion. Some cationization by silver and sodium occurs to give quasi-molecular ions observed in the positive ion spectrum. Additionally, fragmentation patterns at lower mass yield characteristic ions from bond breaking phenomena. The presence of a strong silver signal indicates the sampling depth exceeds the monolayer length even under these mild static conditions. Thus, the further study of multilayers should yield a way to describe the true sampling depth for static SIMS of molecular systems. The extension of this approach to polymerizable LB systems promises to further describe the capability of understanding the SIMS process for macromolecular systems.

High Resolution Electron Energy Loss Spectrometry (HREELS)

Extension of the same arguments for utilizing LB monomolecular layers as model systems can apply in our development work for new methods of vibrational spectroscopy. HREELS (17), outlined in Figure 3, is a technique which has been used extensively to probe small molecules adsorbed on single crystals and model catalytic systems.

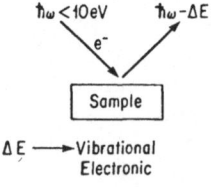

Figure 3

HREELS offers several advantages over optical techniques like
IR and Raman in that it has single instrument vibrational and
electronic spectral capabilities, high sensitivity (<0.1% of a
monolayer), utility of application, i.e. to various surface
topographies, and perhaps the most useful feature - orientation
sensitivity. If one considers a long range dipole scattering
mechanism (17) to be operational in the HREELS, only normal
vibrational modes that are orthogonal to the sample surface will
be accessed by HREELS. This selection rule is analogous to that
in IR reflection spectroscopy (17). We have demonstrated that
successful HREELS experiments can obtain spectra from long chain
fatty acids "adsorbed" on polycrystalline silver via LB methods
with enough resolution and sensitivity to adequately make vibrational assignments (7). These results illustrated HREELS
sensitivity to LB orientations and importantly, to determine the
degree of unsaturation in various assemblies, discriminating among
them.

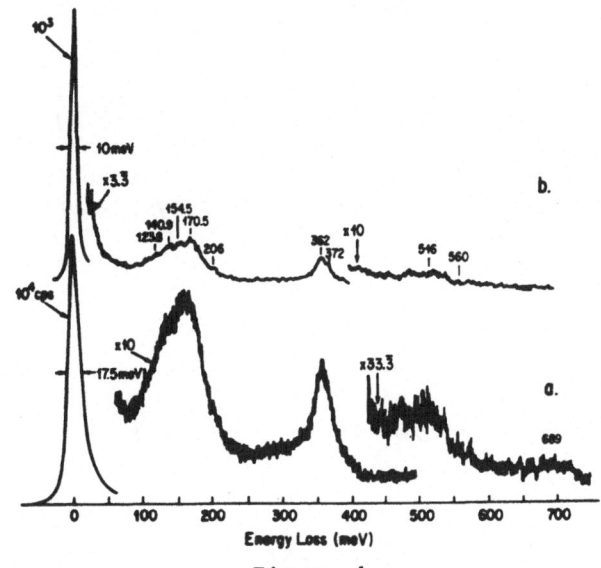

Figure 4

HREELS Spectra of Linolenic Acid Monolayer on Silver: a) low
resolution survey scan, b) high resolution ΔE = 10 meV.

Results in Figure 4, the HREELS spectrum of Linolenic acid,
shows the relatively high scattering intensity of the loss features,
vibrations due to CH stretches, some characteristic of unsaturation,
and a very weak feature due to the normally dominant (in the IR)
carbonyl stretch. This is explained by the distance between the
acid head (oriented down) and the molecular interface in this
system, about 24 Å away from the scattering process. These
results indicate that HREELS can be developed as a surface sensitive

vibrational method, and show promise in analyzing samples which model macromolecular surfaces, with surface sensitivity limited to 10-20 Å.

APPLICATION OF ESCA, FT-IR AND ISS TO HOMO AND COPOLYMERS

In this portion, we will discuss results from studies based around ESCA analysis of controlled morphology polymers. By complementing results with FT-IR and/or ISS, for more bonding information or higher surface sensitivity than ESCA, respectively, these studies illustrate how crystallinity affects the surface composition of homo and block copolymers, how bulk and surface composition are related, how speciated blocks are arranged in the surface morphology of block copolymers and what a depth profile of the changes in morphology and composition looks like. The multi-technique approach is important, ESCA has proven capability for deducing bonding and composition (2) and high surface sensitivity (∼20-50 Å). FT-IR has more detailed bonding information, but samples much deeper, even with surface sensitive sampling methods like Attenuated Total Reflectance (ATR) (1,3). ISS offers elemental analysis with higher surface sensitivity (∼3-5 Å) than ESCA, but limited (and not well understood) chemical information (1). The combination of these three methods is one approach our laboratories have taken to accomplish the study of polymer surfaces.

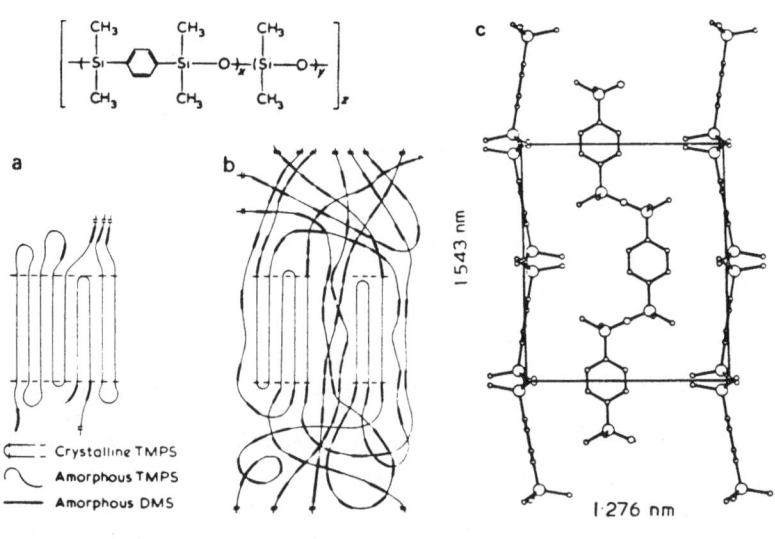

Figure 5

Studies of Aromatic Siloxane Polymers

We have reported results from the study of the homopolymer poly(tetramethyl-p-silphenylene siloxane) (TMpS) (8) and block copolymers of TMpS with poly(dimethyl siloxane) (DMS) by ESCA, FT-IR and ISS (18). These spectroscopic results are correlated with bulk structure information about the unique TMpS polymer systems, which achieve a high degree of crystallinity (see Figure 5). These studies allow the investigation of various models of surface bonding and looping of polymer chains in crystalline macromolecular networks.

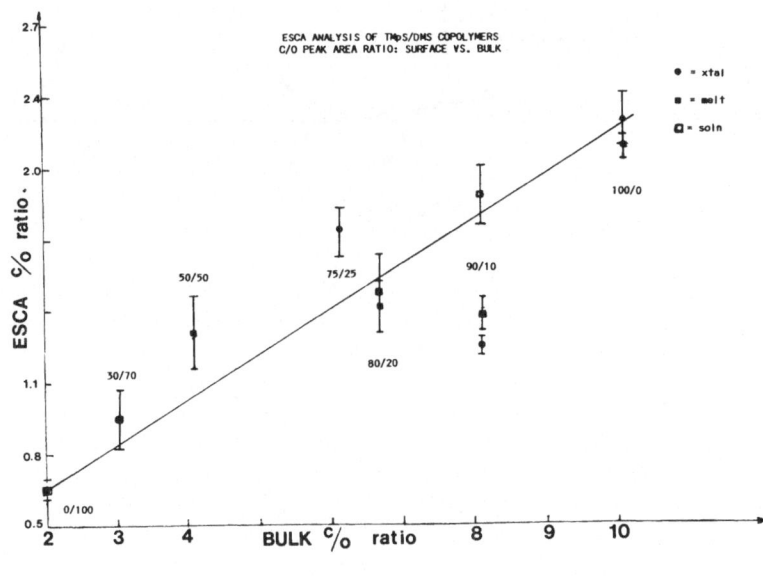

Figure 6

ESCA and ISS Studies of Polycarbonate/DMS Copolymers

The final set of studies (19) involves the application of ESCA and ISS to the study of the series of copolymers whose structure is given in Figure 7. Bisphenol A polycarbonate (BPAC)/DMS copolymers are important thermoplastics, where long chain DMS blocks (average = 18 monomer units) can segregate freely to the surface. Unlike the TMpS/DMS system, BPAC does not achieve a high degree of crystallinity. In these studies the ESCA and ISS analyses are calibrated by the pure homopolymers as endpoints (Figure 7), a straight line is interpolated as a homogeneous mixture, and surface composition

217

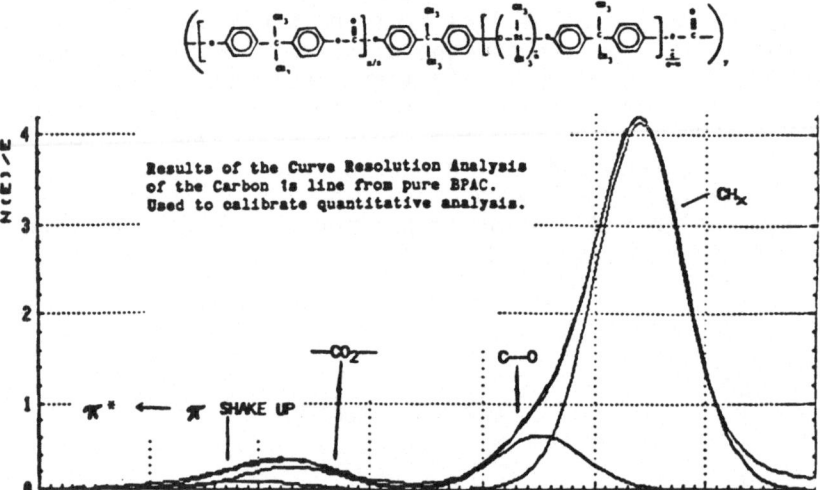

Figure 7

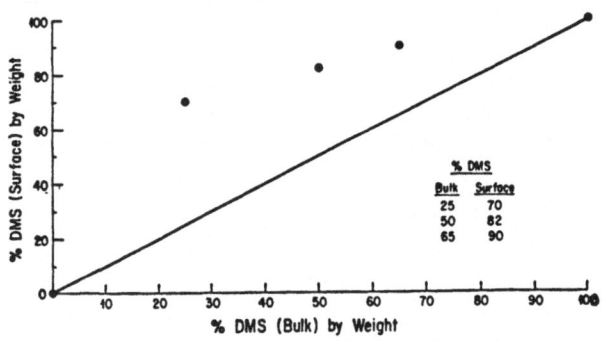

Figure 8   ESCA Analysis of DMS/BPAC Copolymers

is calculated. The degree of segregation is illustrated as a
positive deviation in the plots of Figures 8 and 9. These show the
quantified segregation of DMS to the near surface layer sampled by
the method. The ESCA results given in Figure 8 show how drastic
the segregation is, the 65% bulk DMS copolymer having a surface
concentration of 90% DMS. The ISS results in Figure 9 show the
effects of the higher surface sensitivity. These results show
that the 50/50 and 65/35 copolymers are both almost entirely DMS
at the top-most layer, the 65/35 case equivalent to pure DMS in
Si/O ratio.

218

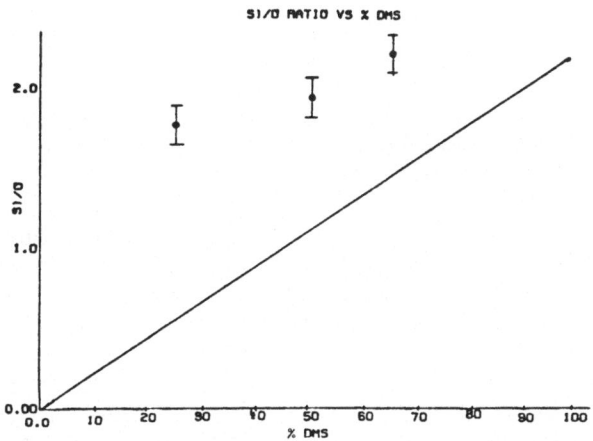

Figure 9   ISS/Analysis of DMS/BPAC Copolymers

These studies illustrate the power of the multitechnique approach to the determination of polymer bonding, structure and morphology at surfaces. The development of new techniques and the application of better known methods to solve surface structure of polymeric materials can help the long term development of property specificity in materials science. These studies also illustrate the analytical power of these methods and open new areas of application to complex non-crystalline materials.

Acknowledgement

The direction and assistance of Professor J. H. Magill of the University of Pittsburgh is greatly appreciated. Dr. Lawrence Salvati Jr. of Perkin Elmer Physical Electronics has provided instrumental support and help in the development and interpretation of the ISS results. Financial support for this work from SUNY/ Buffalo, US Public Health Service, IBM Corporation and the Donors of the Petroleum Research Foundation, administered by the American Chemical Society is greatly appreciated. HREELS Analyses were performed at the NSF Center for Research in Surface Science and Submicron Analysis (CRISS) at Montana State University. We are grateful for the continued NSF support for this facility.

References

1.   J. A. Gardella, Jr., Multitechnique Spectroscopic Analysis of Polymer Surfaces, Trends in Analytical Chemistry, 3(5): 129 (1984).
2.   D. T. Clark, Chapter 6 in "Handbook of X-Ray and Ultraviolet Photoelectron Spectroscopy", D. Briggs, Ed., Heyden: London, 1978.

3. J. L. Koenig, "Chemical Microstructure of Polymer Chains", Wiley: New York, 1980.
4. J. A. Gardella, Jr. and D. M. Hercules, Comparison of Static Secondary Ion Mass Spectrometry (SIMS), Ion Scattering Spectroscopy and X-Ray Photoelectron Spectroscopy for Surface Analysis of Acrylic Polymers, Anal. Chem., 53(12): 1879 (1981).
5. J. A. Gardella, Jr., F. P. Novak and D. M. Hercules, Analysis of the Surface Hydrolytic Degradation of Poly(t-butyl methacrylate) by Static SIMS, Anal. Chem., 56(8): 1371 (1984).
6. J. H. Wandass and J. A. Gardella, Jr., Secondary Ion Mass Spectrometry of Fatty Acids Prepared by Langmuir-Blodgett Techniques, J. Am. Chem. Soc., submitted.
7. J. H. Wandass and J. A. Gardella, Jr., High Resolution Electron Energy Loss Spectroscopy of Fatty Acids Prepared by Langmuir-Blodgett Techniques on Polycrystalline Silver, Surface Science, submitted.
8. J. A. Gardella, Jr., J. S. Chen., J. H. Magill and D. M. Hercules, Spectroscopic Studies of Polymer Surfaces and Interfaces: Poly(tetramethyl-p-silphenylene siloxane), J. Am. Chem. Soc., 105(14): 4536 (1983).
9. J. W. Rabalais, Chapter 16 in "Photon, Electron and Ion Probes of Polymer Structure and Properties", D. W. Dwight, T. J. Fabish and H. R. Thomas, Eds., American Chemical Society: Washington, D.C. (1981), p. 237.
10. J. E. Campana, J. J. DeCorpo and R. J. Colton, Characterization of Polymeric Thin Films by Low Damage SIMS, Appl. Surf. Sci., 8: 337 (1981).
11. D. Briggs and A. B. Wootton, Analysis of Polymer Surfaces by SIMS, 1. An Investigation of Practical Problems, SIA, Surf. Interf. Anal., 4(3): 109 (1982).
12. D. Briggs, Analysis of Polymer Surfaces by SIMS, 2. Fingerprint Spectra from Simple Polymer Films, SIA, Surf. Interf. Anal., 4(4): 151 (1982).
13. D. Briggs, Analysis of Polymer Surfaces by SIMS, 3. Preliminary Results from Molecular Imaging and Microanalysis Experiments, SIA, Surf. Interf. Anal., 5(3): 113 (1983).
14. A. Benninghoven, Some Aspects of Secondary Ion Mass Spectrometry of Organic Compounds, Int. J. Mass Spectrom. Ion Physics, 53: 85 (1983).
15. G. L. Gaines, "Insoluble Monolayers at Liquid Gas Interfaces", Wiley: New York (1966).
16. R. Colton, Naval Research Laboratory, Chemistry Division, Personal Communication.
17. H. Ibach and D. L. Mills, "Electron Energy Loss Spectroscopy and Surface Vibrations", Wiley: New York, 1982.
18. R. L. Schmitt, J. A. Gardella, Jr., R. L. Chin and J. H. Magill, in preparation.
19. R. L. Schmitt, J. A. Gardella, Jr., R. L. Chin, J. H. Magill, and L. Salvat, in preparation.

THERMAL WAVE IMAGING OF DEFECTS IN OPAQUE SOLIDS

D.N. Rose, D.C. Bryk and D.J. Thomas
US Army TACOM, Warren, Michigan

R.L. Thomas, L.D. Favro, P.K. Kuo, L.J. Ingelhart
and M.J. Lin
Department of Physics
Wayne State University, Detroit, Michigan

K.O. Legg
School of Physics
Georgia Tech University, Atlanta, Georgia

ABSTRACT

Thermal wave imaging of defects in opaque solids is carried
out by focusing the periodically modulated intensity of a heat
source (conventionally a laser, electron beam, or ion beam) at the
surface of the solid. Solution of the heat equation shows that the
ac temperature in the solid is wave-like (thermal waves), and
critically damped. These waves can be used to probe the thermal
properties of the subsurface of the solid, and their reflections
from discontinuities in thermal impedance can be used to image
subsurface defects such as inclusions, voids, cracks, and
delaminations. Various techniques have been developed to detect
the resulting ac temperature variations at the sample surface,
including photoacoustic (gas-cell) detection, thermoacoustic
(piezoelectric transducer) detection, optical probe beams, and ac
infrared detection. Thermal wave imaging is particularly useful to
probe subsurfaces from depths of about 1 micron to 300 microns,
with a maximum depth of about 2 mm. The effective depth is about
one thermal diffusion length, and can be varied systematically by
varying the heat source modulation frequency.

Examples will be given of thermal wave images of ceramics and metals, as well as an image of a turbine blade. These images were taken with laser beam excitation and gas cell or optical probe beam detection. An additional "ion-acoustic" image of implanted areas in a single crystal slab of zirconia will be presented as an example of the use of ion beam excitation of thermal waves.

INTRODUCTION

Thermal wave imaging, or photoacoustic microscopy as it is also called, offers a different view of materials through its dependence on thermal properties. Its history began with a report by Alexander Graham Bell in 1880. There was an initial flurry of work by scientists of the day including Roentgen and Lord Rayleigh, but after that, the effect lay dormant for 50 years until it was revived for work with gases by Viengerov in the Soviet Union. It was not reapplied to solids until 10 years ago [1] and the first photoacoustic image was produced only 5 years ago using a borrowed laser and an O-ring for a drive belt between a motor and a stage micrometer [2].

PHOTOACOUSTIC PRINCIPLE

Figure 1 schematically describes the effect as applied to solids. Periodic light, which was chopped sunshine in one of Bell's later experiments, shines on a material in a closed vessel. As the surface of the material heats up, it heats the air above it. As this air is heated, the pressure in the vessel rises. When the light is off, the material's surface cools, so the air above it cools, so the pressure drops. This periodic pressure variation or sound was detected using a listening tube in Bell's day. It was found that the sound was stronger when the sunshine was stronger and was also stronger when the material had a darker, more absorbent color [1].

The first and most widespread use of the effect has been in spectroscopy where the interest lies in the variation of the signal with excitation light frequency. This has permitted spectra to be taken of absorbing and highly scattering materials. While the interest here is microscopy, where point to point differences over a surface are imaged, there are good illustrations from spectroscopic applications. The photoacoustic signal has a phase with respect to the light modulation as well as a magnitude. The phase can be related to depth. For example, the red in an apple peel shows up in a photoacoustic spectrum of the peel taken with a phase delay because the natural wax coating on an apple delays the appearance of the thermal response at the surface [3]. Utilizing this, it was shown that a lobster shell which is black optically had different colored dyes at different depths [4]. The phase is also dependent to an extent on surface profile though it is relatively insensitive to surface reflectivity.

222

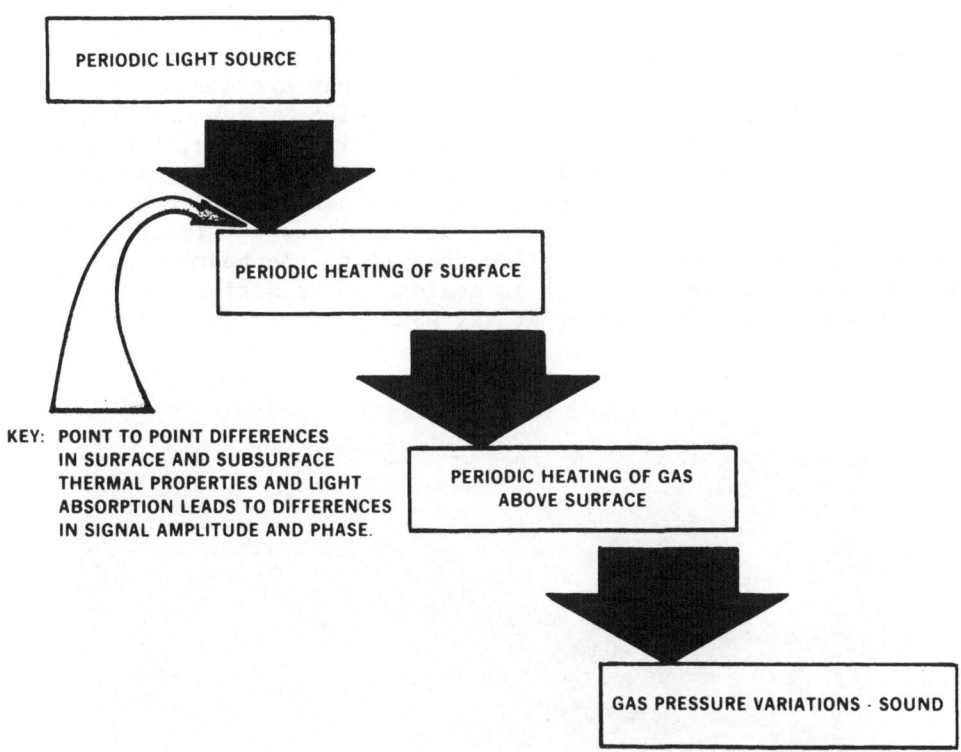

Figure 1.   Principle of the photoacoustic effect using gas cell
            detection.

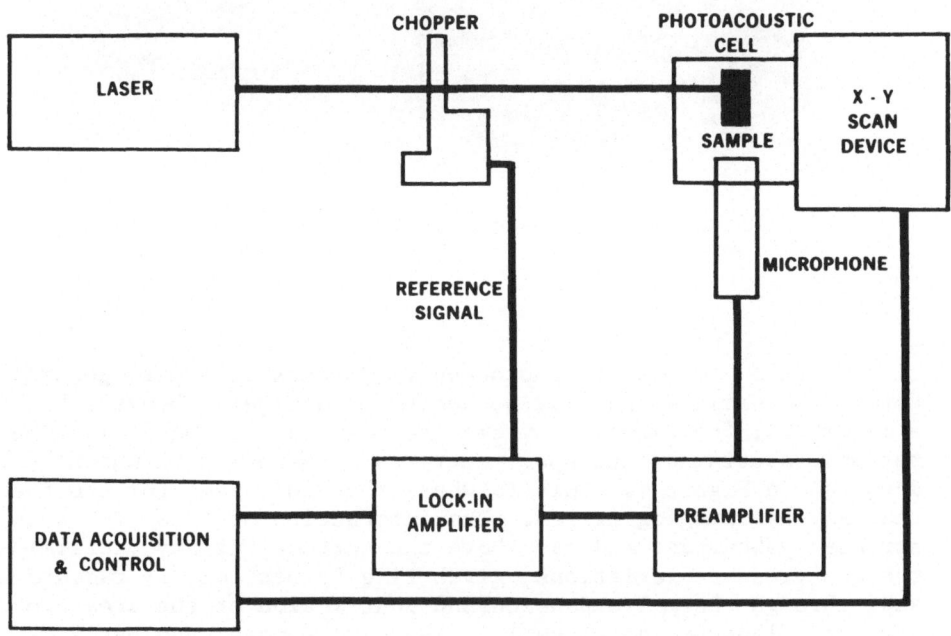

Figure 2.   Schematic of a photoacoustic microscope using gas cell
            detection.

Figure 2 shows a schematic layout of a modern imaging system utilizing gas cell detection. The sound is detected with a sensitive microphone and the electronics that are now available to extract small signals from considerable noise. Lack of these hindered nineteenth century researchers. In addition, the advent of the laser gave easily directed beams with which to interrogate different spots on a surface. Figure 3 shows top and bottom views of a modern gas cell. Note the window to allow the radiation to reach the surface and the shallow channel to the hearing aid microphone. In use, the cell is sealed to the surface to be examined with a thin bead of sticky wax.

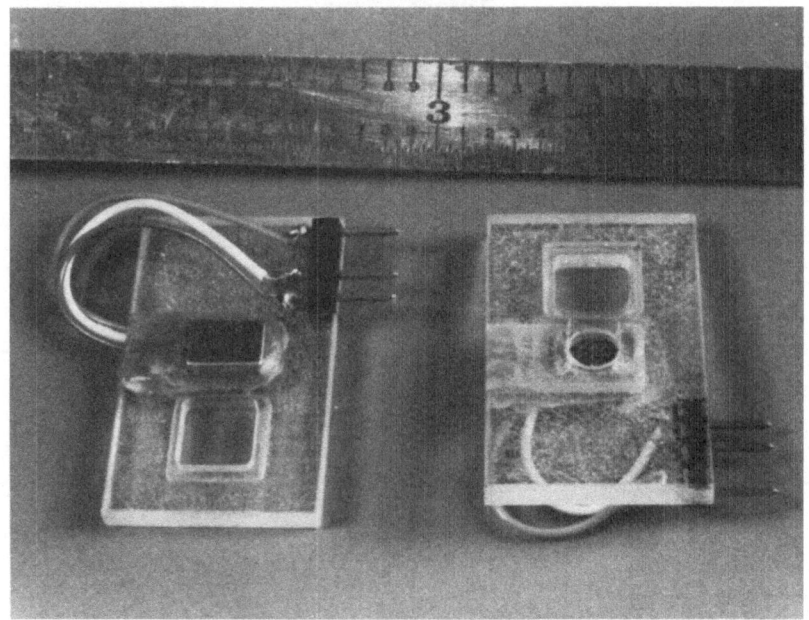

Figure 3.   Photoacoustic microscopy gas cell.

Figure 4 shows the photoacoustic effect in a more general form. The excited spot may be heated by energetic beams other than a laser [5], for example, by electrons [6] or by ions [7]. The periodic heating of the spot generates a thermal wave which is not depicted in Figure 4. This is where the other name for the field, thermal wave imaging arises. The interaction of this wave with the surface, subsurface and gas above the surface, if any, determines the gas pressure variations, gives rise to periodic IR radiation and, through thermal expansion and contraction of the area around the spot, launches an acoustic wave. If we model the basic situation in an opaque material as a one dimensional case of the heat equation, on a semi-infinite solid, a solution for the temperature T into the material can be written as

$$T = T_O \exp [i (qx - \omega t)]$$

where $q = (1 + i) (\omega \rho c/2K)^{\frac{1}{2}}$, $f = \omega/2\pi$ is the modulation frequency, ρ is the mass density, c is the specific heat capacity and K is the thermal conductivity. This solution is periodic in space and time but is highly damped, dropping by a factor of 535 in one thermal wavelength which, in turn, equals $2\pi(2K/\omega\rho c)^{\frac{1}{2}}$. It is customary to take a thermal diffusion length, defined as $(2K/\omega\rho c)^{\frac{1}{2}}$, as a measure of the depth for thermal wave imaging for a particular material at a particular modulation frequency f. At this point, the wave has dropped to 1/e of its initial amplitude. This corresponds to skin depth for electromagnetic waves. This illustrates that this technique is a near surface tool being useable in aluminum down to roughly 1 mm maximum using the magnitude of the signal and 2 mm using the phase. More details of the theory using gas filled cells are given elsewhere [8], [9], [10].

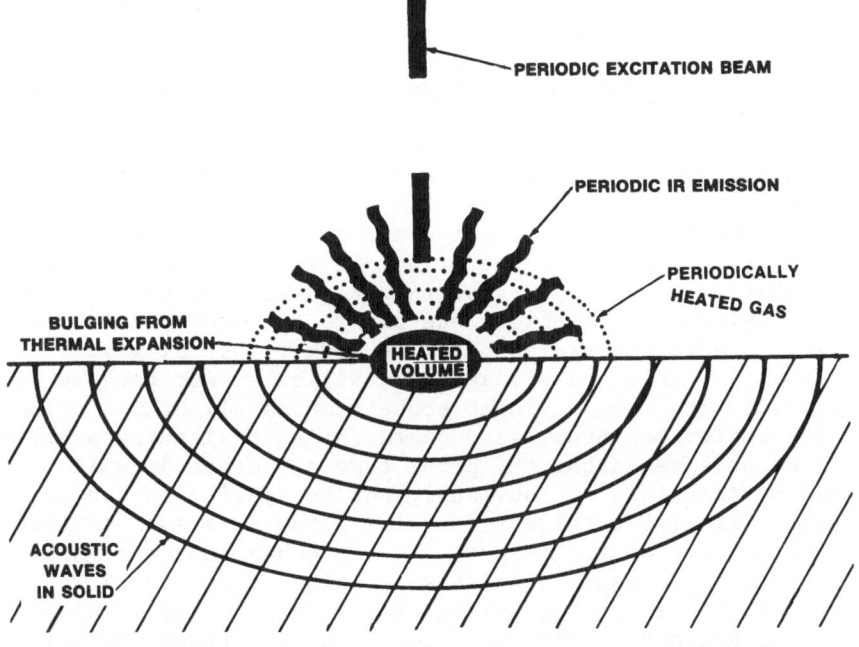

Figure 4.   Effects at a periodically heated spot.

DETECTION TECHNIQUES

Figure 5 shows 5 of the most common methods for detecting the thermal wave signal. The first one, gas cell detection, is what we now call the approach of Bell which has been described.

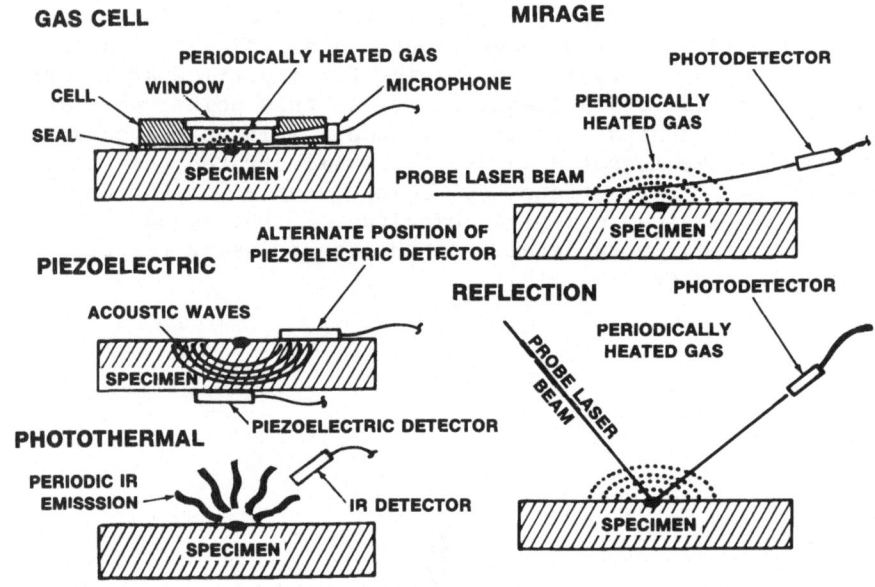

Figure 5. Common photoacoustic detection techniques.

The "mirage" technique, also called optical beam deflection [11], was pioneered by the French [12]. In this approach, a second laser beam called the probe beam skims within 50 microns of the surface. The periodically heated air above the excited spot deflects the probe beam. The principle is the same as that responsible for what appears to be water ahead of us on a hot highway. With this technique one can go to lower modulation frequencies and thus see deeper with longer thermal waves. Unfortunately, this makes image acquisition slower because of the longer diffusion time. If the probe beam is slightly displaced from the excitation beam, very characteristic signatures are seen of cracks oriented along the probe beam direction as will be shown. The mirage technique can obviously not be applied to concave surfaces. It also requires maintaining the relative position of the excitation beam focal spot, the probe beam and the sample surface during a scan. With this technique local variations in surface height can produce signal variations in the magnitude and phase that are not easily separated from subsurface thermal wave interactions [13]. It is, however, a non-contact method.

Piezoelectric detection relies on the acoustic waves that are generated as the excited spot on the sample that is effected by the thermal waves periodically expands and contracts. The conversion efficiency from thermal to acoustic energy is typically quite low but this technique can be faster since one does not have to wait for the thermal information to come back to the surface. Thermal wavelengths are typically 0.3 microns in insulators and 3 microns in metals at 1 MHz dropping to 0.03 microns and 0.3 microns respectively at 100 MHz. This approach can be used where the exciting beam is a chopped electron or ion beam in which case the sample must be in a vacuum. Using this technique and a modified electron microscope, images have been obtained in the same length of time that it takes to get an electron microscope picture. Subsurface flaws in integrated circuits, the doping pattern in semiconductors, the ion implantation pattern in GaAs and the crystalline structure in welds and also in wire bonds to integrated circuit chips have been imaged in this way [14]. This technique does require bonding a transducer to the sample. Theoretically, the interpretation is more complicated since the material's elastic properties, generally expressed as a tensor, become important and acoustic wave reflections must be considered. Theoretically, by themselves, thermal waves are nice because they die out so quickly that reflections, if they come into consideration are simpler and wave mechanics approximations give accurate descriptions [15].

In photothermal imaging, the periodic infrared emissions from the heated spot or the emissions from the back of thin specimens from thermal waves propagating through the specimen are detected. This is a point sampling approach but it relies on thermal diffusion so it is not as inherently fast as piezoelectric detection. It is however, a non-contact method and can also be used in a vacuum. Using this approach, 0.5 micron changes in a 50 micron layer of paint on a 0.5 mm thick sheet of metal have been detected [13] [16]. Also the thickness of diffusion hardened surface layers in steel has been measured to an accuracy of 0.15 mm [17]. It is adaptable to complex geometries and its resolution is at least 0.001 Celsius degrees.

In the reflection approach, a probe laser is reflected directly off the excited spot. It is felt now that changes down to at least 10 nm can be detected (for reference, green light has a wavelength of 550 nm). While this is very good, state of the art ultrasonic transducers can do better. Both thermal lensing, as in the mirage technique, and thermal expansion, as in the piezoelectric approach, affect the signal [18]. This technique does not require a particularly well polished surface. It does require maintaining the relative positions of the excitation beam focal spot and the spot at which the probe beam is reflected which is usually slightly displaced from the excitation beam's focal spot. It is a non-contact method and high modulation frequencies are possible giving fine resolution.

A sixth approach not shown in Figure 5 uses interferometry to measure the height variation of the heated spot [19]. If this technique, also known as photodisplacement, were used, it should be possible to simultaneously profile the surface.

Yet another approach utilizes an ultrasonic beam to interrogate the periodic gas pressure variations by detecting the periodic phase shift in the ultrasonic wave in the gas as it is beamed down to the heated spot and reflected to another transducer [20]. A variation of this is to monitor the phase of an ultrasonic surface wave in the solid as it traverses the heated area. Changes corresponding to 0.001 Celius degree have been detected. This approach was used to show poor bonding between a silicon wafer and an aluminum backing.

This variety in detection approaches has been given to emphasize the generality of the effect and to show the range of options for applying thermal wave imaging in particular situations. Examples of gas cell, mirage and piezoelectric detection are given in the following examples.

NEAR SURFACE RESOLUTION

Figure 6 shows how another sample was constructed which illustrates the photoacoustic principle further. Two slots, approximately 70 microns deep, were milled in aluminum and came together forming a sharp point. The slots were then filled with plastic and the surface was polished smooth. Next a coat of silver

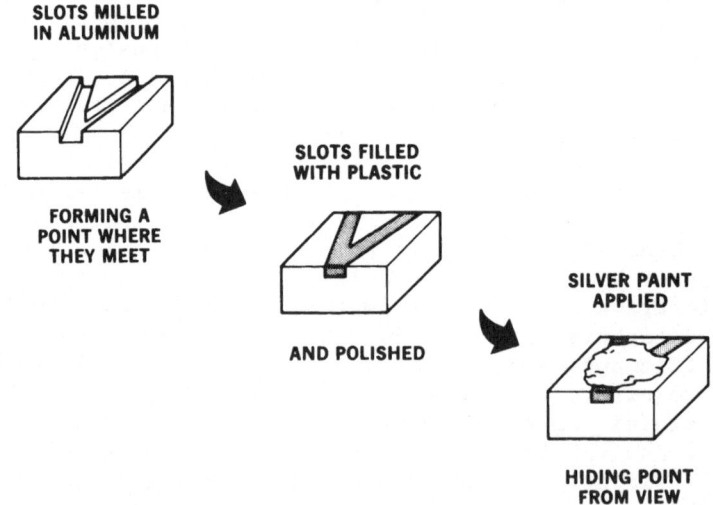

Figure 6.   Construction of a photoacoustic point resolution test specimen.

228

paint was applied, hiding the structure from optical view. Figure
7 shows the photoacoustic picture at two modulation frequencies.
This presentation is in inverse video as shown by the reference bar
at the bottom so that higher thermal responses are darker. The
thermal wavelength in the upper image is 2.2 mm, while the thermal
wavelength of 1 cm at 11 Hz for the lower image is longer than the
entire image and yet the point is sharp. This feature is in the
near field for thermal waves so geometrical shadowing holds [21].
Under these conditions though, the phase shift is essentially zero
so no depth informaton can be obtained from the phase.

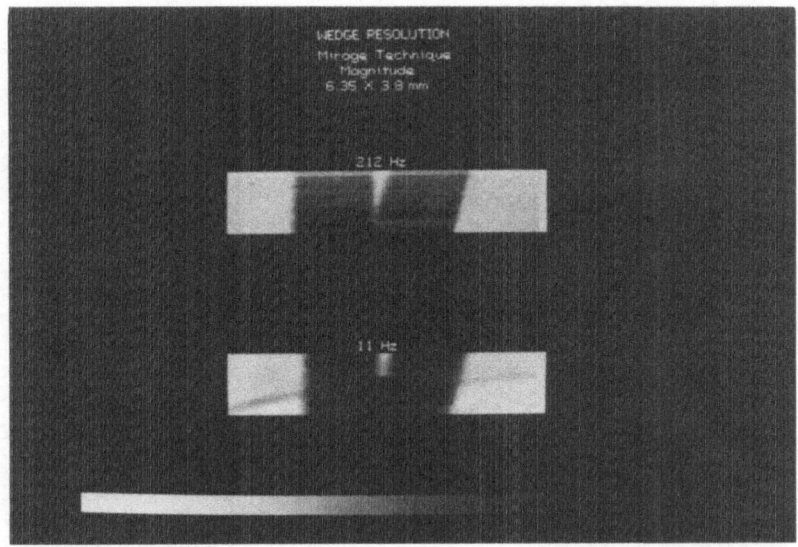

Figure 7. Mirage image of point resolution test specimen.

BURN-IN CARRIER

     A burn-in carrier for integrated circuits made of glass
bonded mica was imaged for Figure 8. The metal connecting leads
are the broad vertical strips angling in at the top to where the
integrated circuit would be. Optically, the matrix was white.
Under a microscope, the matrix had small, irregular clear glassy
areas in a white matrix. Included with the magnitude and phase
images is a concurrently collected image formed by monitoring the
light reflected from each point sampled. This was inverted to
serve as a visual reference of the relative amount of light energy
absorbed. As expected, more of the light was reflected from the
metal leads and the thermal wave signal from the leads is

correspondingly lower. However, the matrix shows areas of strong
thermal response for which there is no corresponding feature in the
absorbed light. This illustrates that thermal wave imaging is a
different way of seeing. A crack can be seen slanting vertically
on the right side of the thermal wave images. This crack could be
seen optically under the right conditions. With gas cell
detection, perfectly vertical, closed cracks can not be seen [10];
however, "vertical" cracks in practice, such as this one, are
frequently neither perfectly vertical nor perfectly closed so they
can be detected.

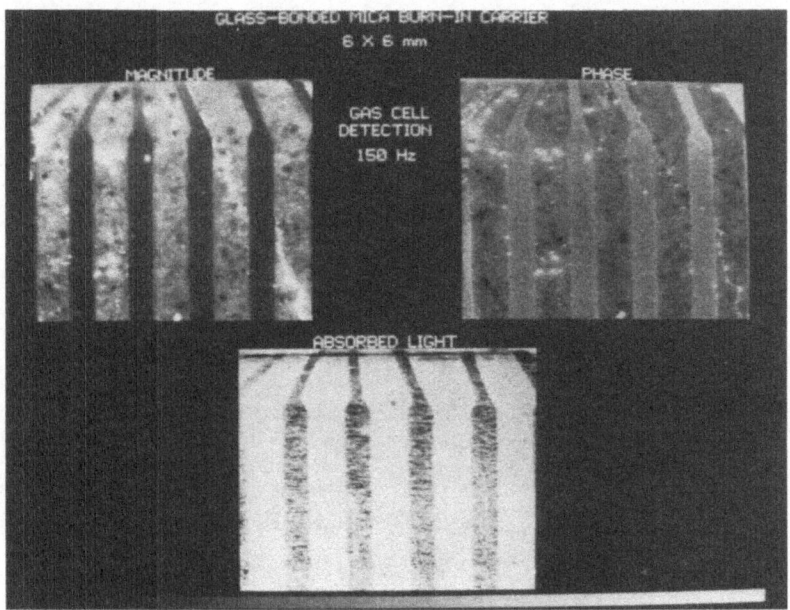

Figure 8. Gas cell and corresponding scanned optical absorption
images of a glass-bonded mica burn-in carrier. A crack
may be seen slanting vertically on the right side of the
thermal wave images.

OPEN FATIGUE CRACK IN ALUMINUM

Figure 9 demonstrates the distinctive signature that can be
obtained from a crack using mirage detection. The upper image
shows the crack using gas cell detection while the lower is the
same crack using mirage detection with the probe beam slightly
displaced from the heating beam and also parallel to the crack
[13]. As a first approximation, one can think of the crack as

230

blocking the heat flow when the crack is between the heated spot and the probe beam so that the thermal response the probe beam sees is low in this case, but, when the probe beam is between the crack and the heated spot, the crack reflects the heat so that the probe beam is more strongly deflected.

When the probe beam is displaced from the excitation beam, there is a transverse deflection of the probe beam as well as a deflection normal to the surface. This transverse deflection can be used for extracting the thermal diffusivity of the spot under examination [22].

In the mirage image, there is a crown-like structure over the crack and a vertical line in the upper half of the image. We first noticed this in a pseudocolor image. We have found a pseudocolor capability useful for revealing subtle detail in the images which we subsequently verify in the gray scale images.

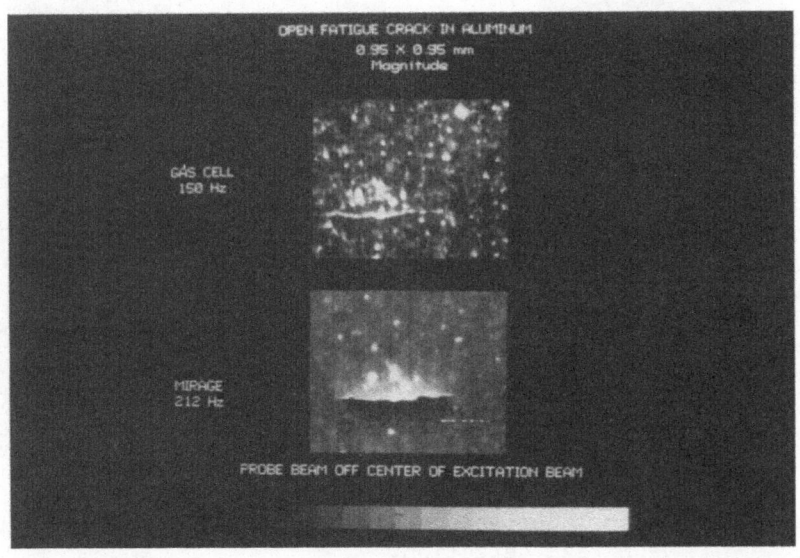

Figure 9. Fatigue crack in aluminum: Gas Cell vs mirage detection.

KNOOP INDENTS ON SiC

Figure 10, in inverse video, shows a comparison of thermal wave images of Knoop indentations on silicon carbide obtained using gas cell detection and mirage detection. Though the images are of different indentations, they illustrate the two approaches. The visible indentation, 140 x 20 microns, is somewhat less than half the height and 6 % of the width in the field of view in these images. There is a background signal, minimal for the 90 degree images and highest for the 0 degree images that has been subtracted from the data.

According to the theory, the effect of cracks that are shallow compared to the thermal wavelength should show up at − 45 degrees phase angle [9], [23]. Accordingly, the smallest indications can be seen in the + 45 degree images and the biggest indications in the − 45 degree images. The residual indication in the + 45 degree gas cell image corresponds in size to the visible surface deformation of the indentation. In contrast, mirage detection responds to vertical cracks and a much larger response is seen for mirage detection at + 45 degrees. The + 45 degree mirage response however, is still less than the − 45 degree mirage response as can be seen most clearly in Figure 11 which is a

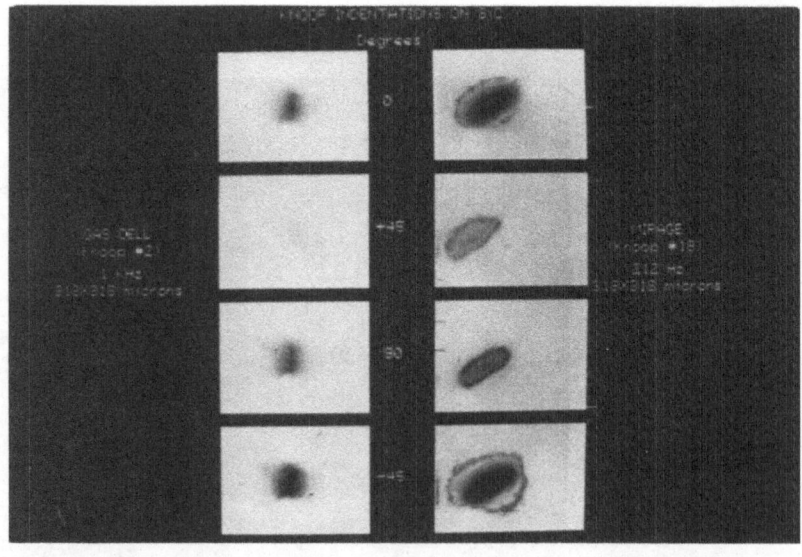

Figure 10. Knoop indentations in sintered silicon carbide: Gas Cell vs Mirage detection.

232

comparison of these two images. Figure 11 also illustrates that a combination of data presentations helps to bring out all of the features of the data.

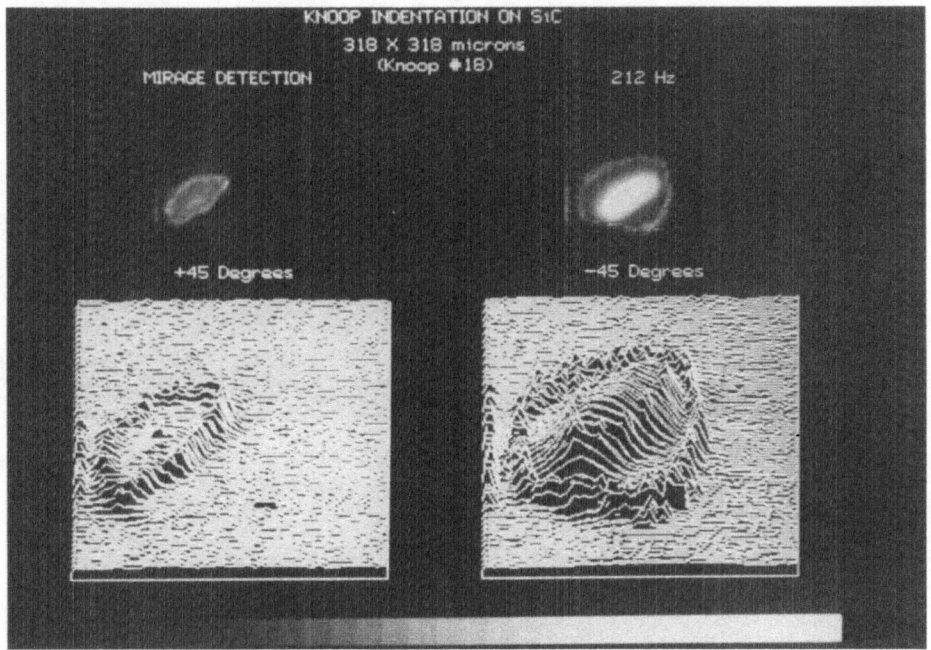

Figure 11. Knoop indentation in sintered silicon carbide: Mirage images at + 45 degrees and - 45 degrees

TIGHT VERTICAL CRACK IN SiC

For this demonstration, a Knoop indentation was again used. The indentation was made in hot pressed silicon carbide with a 2.7 Kg load but then, the visible indentation was machined off, leaving only the vertical crack formed during indentation, estimated as 140 microns long and 70 microns deep. Figure 12 shows a comparison of a reflected light image obtained as in Figure 8 and a mirage image. The crack was not detectable by optical microscopy, by scanning laser acoustic microscopy (SLAM) or by an advanced eddy current probe. In this case, weaker bands can be seen on either side of the strong pair delineating the crack.

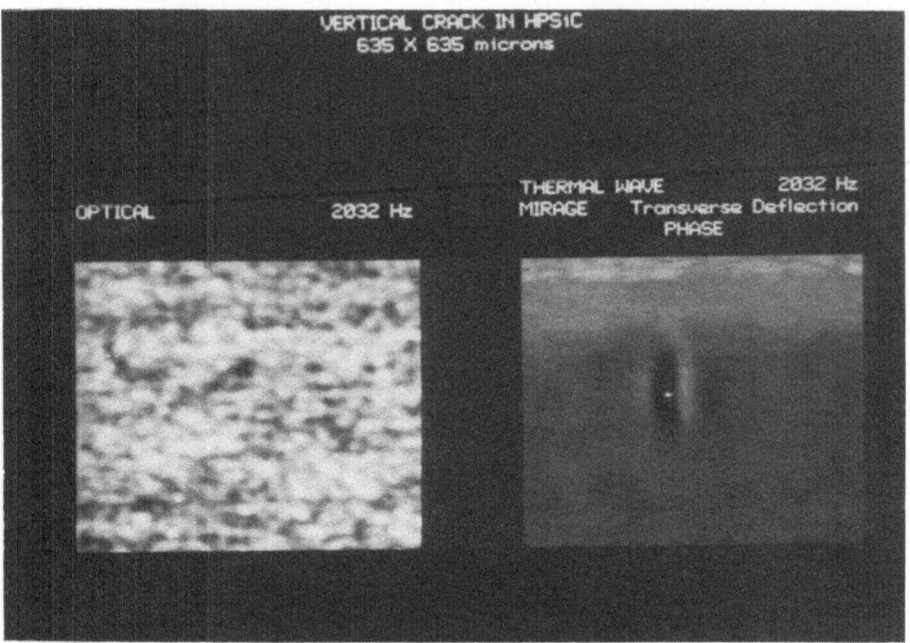

Figure 12. Tight vertical crack in hot pressed silicon carbide:
Scanned optical and mirage images.

ION ACOUSTIC IMAGE OF ZIRCONIA

For Figure 13, a modulated ion beam was used for excitation
with piezoelectric detection to image two nitrogen implanted
regions in a slab of single crystal zirconia. Only very faint
yellowish areas were visible to the eye at the implanted areas
using glancing light. We see the best potential for "ion
acoustics" in on-line control of ion implantation on complex or
inhomogeneous surfaces [7]. Similarly, the photoacoustic effect
can be useful for control of materials processing via lasers.

SUBSURFACE COOLING PASSAGES

Figure 14 is an image at the trailing edge of a TF30 first
stage turbine blade which has through cooling air holes of which
two are visible as broad horizontal bands in the image [24]. As
expected, the thermal signal is stronger where the heat flow is
blocked by the air holes as shown by the whiteness of the bands
compared to the reference bar at the bottom of the picture. This,
also illustrates that an absolutely flat surface is not required
for thermal wave imaging, even with mirage detection.

234

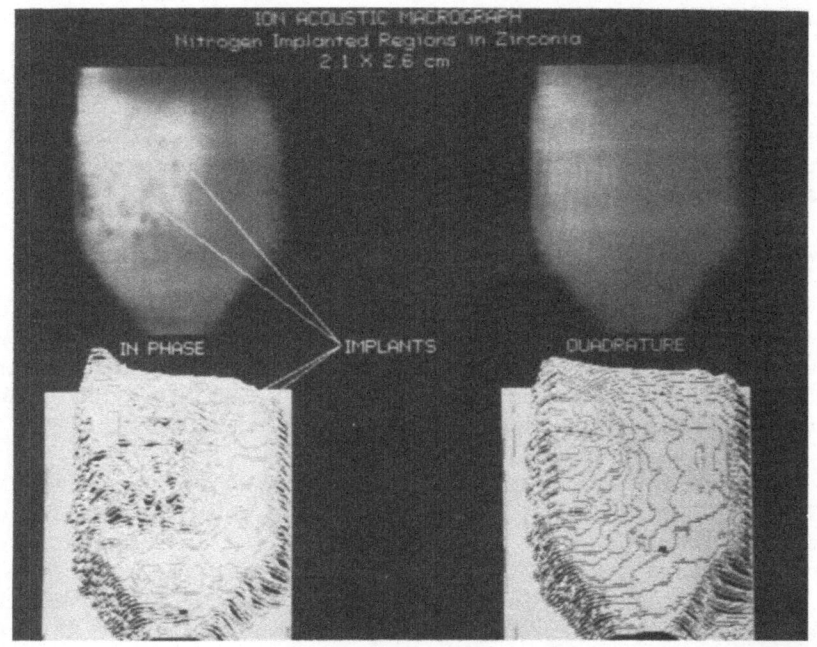

Figure 13. Ion acoustic image of nitrogen implanted areas in a
zirconia crystal.

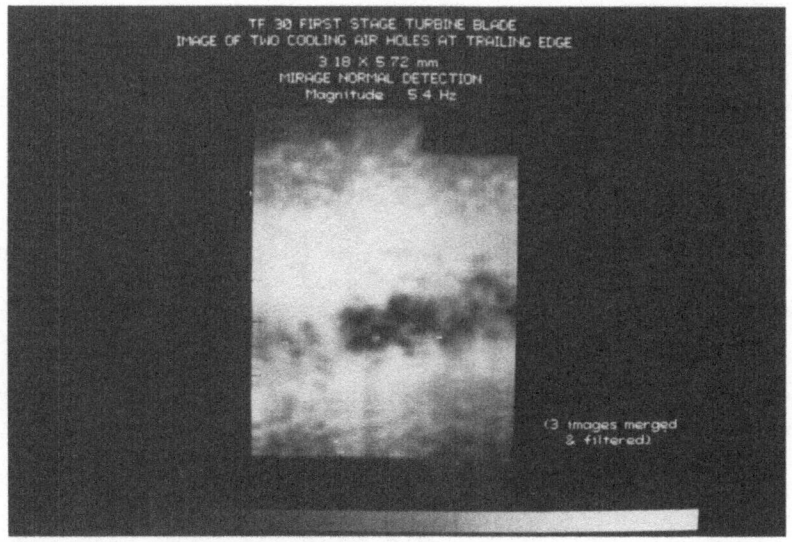

Figure 14. Image of subsurface turbine blade air cooling holes.

CONCLUSION

Thermal wave imaging offers a unique view of materials through its dependence on thermal properties. It is capable of probing opaque materials. It is better adapted to surfaces which can withstand more heat such as ceramics and metals but rubber has been successfully imaged [25].

ACKNOWLEDGEMENTS

This work was partially funded by Dr. Fred Schmideshof of the Army Research Office, the Track Elastomer program under Mr. Jacob Patt at the US Army TACOM, the Reliability for Readiness Program under Mr. Paul Doyle and Mr. Jim Kidd at the US Army AMMRC, and an MTT program under Mr. Robert Brockelman of the US Army AMMRC.

The burn-in carrier sample was furnished by Mr. Jack Liker of MYKROY/MYCALEX Division of Monogram Industries. The fatigue crack sample was part of the DoE Ames Laboratory NDE development program. The hot pressed silicon carbide sample was prepared by Dr. George Quinn of the US Army AMMRC. Mr. Kenneth Fizer of the Naval Air Rework Facility, Norfolk supplied the turbine blade and Xavier Mann prepared the zirconia crystal sample.

REFERENCES

1. A. Rosencwaig, "History of Photoacoustics", Chapter 2 in Photoacoustics and Photoacoustic Spectroscopy, John Wiley and Sons, New York, New York (1980).

2. Y. H. Wong, R. L. Thomas and G. F. Hawkins, "Surface and Subsurface Structure of Solids by Laser Photoacoustic Spectroscopy", Appl. Phys. Lett. 32, 538 (1978).

3. J. A. Noonan and D. M. Munroe, "What is Photoacoustic Spectroscopy", Optical Spectra, February (1979).

4. T. A. Moore, R. Tom, E. P. O'Harra, D. M. Anjo, and D. Benin, "In Vivo and In Vitro Studies of Biological Materials by Photoacoustic Techniques", presented at AAAS meeting, 26-31 May 1983, Detroit, Michigan.

5. A. Rosencwaig, Photoacoustics and Photoacoustic Spectroscopy, op. cit, p.297.

6. E. S. Cargill, III, "Electron-acoustic Microscopy", Phys. Today, 34, 27 (1981).

7. K. O. Legg and D. N. Rose, "Scanning Ion Acoustic Microscopy for Analyzing Ceramic Surfaces", presented at the 8th Annual Conference on Composites and Advanced Ceramic Materials, 15-18 Jan 1984, Cocoa Beach, Florida.

8. R. L. Thomas, J. J. Pouch, Y. H. Wong, L. D. Favro, P. K. Kuo and A. Rosencwaig, " Subsurface Flaw Detection in Metals by Photoacoustic Microscopy", J. Appl. Phys. 51, 1152 (1980).

9. P. K. Kuo, L. D. Favro, L. J. Inglehart, R. L. Thomas and M. Srinivasan, "Photoacoustic Phase Signatures of Closed Cracks", J. Appl. Phys. 53, 1258 (1982).

10. P. K. Kuo and L. D. Favro, "A Simplified Approach to Computations of Photoacoustic Signals in Gas Filled Cells", Appl. Phys. Lett. 40, 1012 (1982).

11. J. C. Murphy and L. C. Aamodt, "Signal Enhancement in Photothermal Imaging Produced by Three Dimensional Heat Flow", Appl. Phys. Lett. 39, 519 (1981); L. C. Aamodt and J. C. Murphy, "Photothermal Measurements Using a Localized Excitation Source:, J. Appl. Phys. 52, 4903 (1981).

12. A. C. Boccara, D. Fournier and J. Badoz, "Thermo-optical Spectroscopy: Detection by the Mirage Effect", Appl. Phys Lett., 36, 130 (1980)); A. C. Boccara, D. Fournier, N. Jackson and N. Amer, "Sensitive Photothermal Deflection Technique for Measuring Absorption in Optically Thin Media", Opt. Lett. 5, 377 (1980).

13. R. L. Thomas, L. D. Favro, K. R. Grice, L. J. Inglehart, P. K. Kuo, J. Lhota and G. Busse, "Thermal Wave Imaging for Nondestructive Evaluation", p 586, Proceedings of the 1982 IEEE Ultrasonics Symposium, B. R. McAvoy, ed., IEEE Press, New York, New York (1982).

14. A. Rosencwaig, "High Resolution Thermal Wave Imaging" presented at AAAS meeting 26-31 May 1983, Detroit, Michigan.

15. L. D. Favro, K. R. Grice, L. J. Inglehart, P. K. Kuo and R. L. Thomas, "Gas-Cell Scanning Photoacoustic Microscopy (SPAM)", presented at AAAS meeting 26-31 May 1983, Detroit, Michigan.

16. G. Busse, "Photothermal Remote Nondestructive Material Inspection", presented at AAAS meeting 26-31 May 1983, Detroit, Michigan.

17. M. Luukkala, J. Jaarinen and A. Lehto, "Photothermal Measurement of the Thickness of Diffusion Hardened Surface Layers in Steel", Proceedings of the 1983 IEEE Ultrasonics Symposium, B. R. McAvoy, ed., IEEE Press, NY, NY (1983).

18. J. C. Murphy and L. C. Aamodt, "Photothermal Deflection Imaging and Microstructural Characterization of Solids", presented at AAAS meeting 26-31 May 1983, Detroit, Michigan.

19. Y. Martin, H. K. Wickramasinghe and E. A. Ash, "Thermo and Photo Displacement Microscopy", p 563, Proceedings of the 1982 IEEE Ultrasonics Symposium, B. R. McAvoy, ed., IEEE Press, New York, New York; L. C. M. Miranda, "Photodisplacement Spectroscopy of Solids: Theory", Appl. Optics, 22, 2882 (1983).

20. R. G. Stearns, B. T. Khuri-Yakub and G. S. Kino, "Measurements of Thermal-Elastic Interactions with Acoustic Waves", p 595, Proceedings of the 1982 IEEE Ultrasonics Symposium, B. R. McAvoy, ed., IEEE Press, New York, New York. R. Stearns, B. T. Khuri-Yakub and G. S. Kino, "Phase Modulated Photoacoustics", p 649, Proceedings of the 1983 IEEE Ultrasonics Symposium, B. R. McAvoy, ed., IEEE Press, New York, New York (1983).

21. L. J. Inglehart, M. J. Lin, L. D. Favro, P. K. Kuo and R. L. Thomas, "Spatial Resolution of Thermal Wave Microscopies", Proceedings of the 1983 Ultrasonics Symposium, B. R. McAvoy, ed., IEEE Press, New York, New York (1983). L. J. Inglehart, D. J. Thomas, M. J. Lin, L. D. Favro, P. K. Kuo and R. L. Thomas, "Resolution Studies for Thermal Wave Imaging", Review of Progress in Quantitative NDE, Vol. 4, D. O. Thompson and D. Chimenti, ed., Plenum Press, New York, New York, to be published. L. J. Inglehart, K. R. Grice, L. D. Favro, P. K. Kuo and R. L. Thomas, "Spatial Resolution of Thermal Wave Microscopes", Appl. Phys. Lett. 43, 446 (1983).

22. R. L. Thomas, L. J. Inglehart, M. J. Lin, L. D. Favro and P. K. Kuo, "Thermal Diffusivity in Pure and Coated Materials", Review of Progress in Quantitative NDE, Vol. 4, D. O. Thompson and D. Chimenti, ed., Plenum Press, New York, New York, to be published.

23. P. K. Kuo, L. J. Inglehart, L. D. Favro and R. L. Thomas, "Experimental and Theoretical Characterization of Near Surface Cracks in Solids by Photoacoustic Microscopy", p 837, Proceedings of the 1981 IEEE Ultrasonics Symposium, B. R. McAvoy, ed., IEEE Press, New York, New York (1981).

24. D. N. Rose, "Photoacoustic Microscopy - An Emerging Tool", p 201, Proceedings of the 32nd Defense Conference on Nondestructive Testing, 1-3 November 1983, Wright Patterson Air Force Base, Ohio.

25. R. L. Thomas, L. D. Favro, P. K. Kuo and D. N. Rose,
    "Scanning Photoàcoustic Microscopy of Aluminum with Aluminum
    Oxide, Roughness Standards and Rubber", US Army
    Tank-Automotive Command Research and Development Center",
    Warren, Michigan, Technical Report no. 12668 (1982). R. L.
    Thomas, L. D. Favro, P. K. Kuo, D. N. Rose, D. Bryk, M.
    Chaika and J. Patt, "Scanning Photoacoustic Microscopy of
    Aluminum with Aluminum Oxide, Roughness Standards and
    Rubber", US Army Tank-Automotive Command, Research and
    Development Center, Warren, Michigan, Technical Report No.
    12957 (1984).

STUDIES OF ADHESION FAILURE MECHANISMS AT METAL-HgCdTe

AND ZnS-Ge INTERFACES BY SEM/SAM/EDX

Gerald A. Garwood, Jr. and Michael Ray

Santa Barbara Research Center
A Subsidiary of Hughes Aircraft Company
75 Coromar Drive
Goleta, CA 93117

INTRODUCTION

Advanced techniques for electron microscopic[1] and electron spectroscopic[2] characterization of surfaces and interfaces have ushered in an era of rapid progress for thin film adhesion studies.[3] These modern techniques have great potential usefulness in technologies involving multilayer films such as microelectronic infrared (IR) detector arrays and optical coatings for IR windows.[4] In particular, scanning electron microscopy[5] (SEM) provides detailed views of microcircuitry and optical coatings from which structural defects such as metal delamination and microcracks can be studied. Furthermore, scanning Auger microscopy (SAM) is particularly appropriate for analyzing thin metal films and optical coatings from a chemical point of view.[4] Auger spectroscopy is sensitive to the top 10 to 50 Å of the sample surface and capable of detecting all elements except hydrogen and helium.[6] By combining SAM with ion beam sputtering, a depth profile can also be obtained which reveals the elemental composition not only of the layers themselves but also of interfacial impurities.[4] Additionally, energy dispersive X-ray spectroscopy (EDX) provides data regarding the elemental constitution in approximately the top 1 to 2 μm of material analyzed.[5,7] All elements from sodium and above are detectable by EDX. An increasing number and variety of other techniques also highly useful for investigating thin films have become available, although they were not exploited for the present study.[8-11]

Studies directed toward improving our understanding of factors influencing adhesion are important from both fundamental and practical perspectives. A number of metals and other materials can be uniformly deposited on a variety of substrate surfaces, and the strength of adhesion is inherently sensitive to the nature of the interaction between the adjacent materials. Adhesion can be greatly affected by interdiffusion and subsequent compound formation and alloying at the interface.[4] Surface cleanliness prior to deposition is also known to be a major factor in adhesion. The strength of adhesion can also be altered due to the effects of exposure to thermal, mechanical, or chemical stresses. Accordingly, investigations of thin film systems such as metal-semiconductor devices and windows with optical coatings by electron spectroscopies and related methods will yield facts concerning structures, chemical composition, and stability which will be informative to process engineers and material scientists.

In the present article, results are reported on the application of SEM, SAM, and EDX to the investigation of adhesion problems encountered with two important thin film-substrate systems. In the first case, examples and causes of peeling of electroplated and thermally evaporated diode contact metallization layers from an important IR detector material, HgCdTe, are described. The outgassing of volatile inorganic contaminants from an alumina-coated tungsten crucible used in the thermal deposition process is discussed. A technique which was implemented to minimize electron-beam desorption of Hg from the HgCdTe lattice during SAM analysis of the detector array is explained together with a family of depth-composition profiles. In the second case, an interesting blistering phenomenon is illustrated along with the associated failure mechanism for a ZnS-Ge multilayer optical coating on a Ge substrate. In both cases the structures of the delamination sites were characterized by SEM, surface elemental composition by SAM, and bulk chemical composition by EDX and depth-profile SAM.

EXPERIMENTAL ASPECTS

SEM and EDX investigations were conducted using an in-house Cambridge Stereoscan Model 250 Mark 2 with a cryocooled Li-drifted silicon detector (Model No. 5194) and a Link Systems Model 860 processor. Samples were mounted on an aluminum substrate with quickly drying conductive cement and analyzed at room temperature. A primary beam energy of 20 keV was typically employed.

SAM information was obtained either with a Physical Electronics (PHI) Model 595 Microprobe at Hughes Research Laboatory, Malibu, California or with a Jeol Scamp system at Charles Evans and Associates, San Mateo, California. Specimens were held in place on metal carriers with a conductive adhesive and/or metal clips and analyzed at ambient temperature. An incident electron beam having a diameter of $>500 \text{Å}$ and an energy of either 5 or

10 keV impinged the sample at 30° relative to the surface. The beam was operated either in the spot mode for point analyses, or was rastored over areas up to about 1 mm^2 for area surveys. Argon ion sputtering was utilized for the depth profiles.

Deposition of metal layers by thermal evaporation was accomplished via a NRC thermal evaporator equipped with a quartz crystal thickness monitor. High purity palladium metal was melted in an alumina-coated tungsten basket and heated to about 1500°C. The base pressure of the evaporator was typically 10^{-6} Torr and depositions usually occurred at ~10^{-5} Torr. Wafers were attached with metal clips to the bottom of a planet which rotated about its own axis which in turn revolved about a central axis. These combined circulatory motions randomized the angle of incidence of incoming metal atoms and resulted in a more uniform metal coating. Electroplating was conducted in a special electroplating tank equipped with its own recirculating pump, 0.2 μm filter, and temperature control. Palladium was plated at room temperature (RT) using a commercial RT plating solution. Metal thickness was determined by a Dektak surface profilometer.

The $Hg_{(1-x)}Cd_xTe$ substrates used in the first part of this study were grown by liquid phase epitaxy on CdTe single crystals [(111) orientation] at Santa Barbara Research Center (SBRC) and had x values in the 0.2 to 0.4 range. All optically-coated Ge windows discussed in part two of the present investigation were obtained from a supplier external to SBRC.

ADHESION STUDIES: METAL-$Hg_{(1-x)}Cd_xTe$ INTERFACE

Background on IR Detectors and Metallization

Infrared detector systems play a vital role in deep space exploration, weather monitoring, and fire sensing and suppression, as well as in heat-seeking systems and devices for night vision. The "eyeball" in many of these applications requires an IR sensitive "retina" which is often made of mercury cadmium telluride, $Hg_{(1-x)}Cd_xTe$, where x is the mole fraction of CdTe. $Hg_{(1-x)}Cd_xTe$ (HgCdTe) is a compound semiconductor comprised of the alloy system, HgTe and CdTe, and its bandgap, which determines its long wavelength cutoff, can be varied continuously from 0 to 1.6 eV by changing the Cd/Hg ratio.[12]

Figure 1 shows a simplified schematic of the "retina," a two-dimensional matrix of diode contact metallization pads on a $Hg_{(1-x)}Cd_xTe$ epitaxial layer. The metal pattern is fabricated using photolithographic techniques, and is herein briefly described. Typically, the HgCdTe surface is chemically cleaned and coated with a thin layer of insulator. Photoresist (PR) is spun

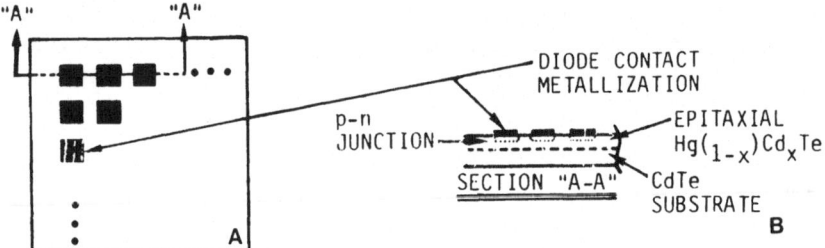

Figure 1. Schematic of $Hg_{(1-x)}Cd_xTe$ IR detector array showing:
(A) plan view of diode contact metallization,
(B) cross-sectional view. This schematic is greatly
simplified and not to scale. Features not shown
include ground contacts and insulator layer.

on, baked, and exposed to UV light through a mask containing the
desired pattern. The UV-stimulated regions of PR (an array of
squares in this case) become acidic and are dissolved in a basic
developer. Windows are then chemically etched in the insulating
layer thereby exposing bare HgCdTe. Metal is deposited onto the
sample by a thermal evaporation technique. To remove (or reject)
extraneous metal from the areas surrounding the "windows," the
wafer is exposed to solvent. This process step is known as the
rejection step. The solvent attacks the PR causing it to swell as
it dissolves. The swelling action literally shears the metal
which is on the PR from the metal which is bonded to the HgCdTe,
leaving a sharply delineated array of contact pads.

Peeling Problem and Approach Taken

Ideally, during rejection, the metal pads rigidly adhere to
the HgCdTe and form ohmic contacts. Occasionally, one or more
metal pads peel from the substrate resulting in a low production
yield. The occurrence of a number of peeling incidents prompted
an investigation of the underlying cause(s). Due to the fact that
the peeling could conceivably be attributable to a number of hypo-
thetical causes (e.g., to residual photoresist, incompletely
removed insulator, inclusion of evaporator impurities, excessively
thick metal, etc.) the investigation took on a multi-faceted
approach over a period of several months. The various aspects of
the approach fell into six categories:

1. Evaluation of alternate deposition techniques.

2. Compilation of accurate and complete adhesion yield statistics and correlation to process/material parameters.

3. Studies aimed at assessing and improving the degree of pre-metallization clean-up.

4. Evaluation of alternate diode contact metals.

5. Efforts directed toward identifying, monitoring, and eliminating contaminants introduced during the deposition.

6. Investigations of metallurigical interactions at the metal-HgCdTe interface.

The limited scope of this paper does not permit a detailed description of all these studies and their results. Consequently, only a brief overview of the first four categories will be given here. Items 5 and 6 will be covered in the following subsections entitled "Crucible Condensates" and "Pd-HgCdTe Interactions," respectively.

First of all, deposition by electroplating was pursued as an alternative to thermal evaporation. Palladium was chosen as the candidate metal for these studies because it had been widely used as the base layer metal for diode contact metallizations on HgCdTe. Care was taken to assure that the plating solution was regularly monitored and kept clean. To compare the effectiveness of electroplating to that of thermal evaporation, wafers of epitaxial HgCdTe were split, pre-cleaned and processed identically up to the deposition step. Following rejection, the arrays of contact metallization pads were closely examined under an optical microscope for adhesion yield. Some of the first electroplated layers were unusually thick (~2000Å) and tended to peel simply due to the tremendous stresses inherent in such thick films. Closer examination of the peeled pads by SEM (see Figure 2A) revealed that in those instances for which the Pd-to-HgCdTe bond was tenacious, chunks and layers of HgCdTe were literally pulled up "by the roots," rigidly clinging to the underside of the Pd pad, leaving craters and striations at the substrate surface. This phenomenon was found to be virtually independent of the deposition method, as is substantiated by the rough striated surface seen in Figure 2B for a thick layer of thermally evaporated Pd. On the other hand, in cases for which the adhesion between the metal layer and the substrate was relatively weak, portions of Pd pads peeled and broke away from the HgCdTe surface, leaving the characteristically smooth surface shown in Figure 2C. The "smooth" mode of peeling was observed with about the same low

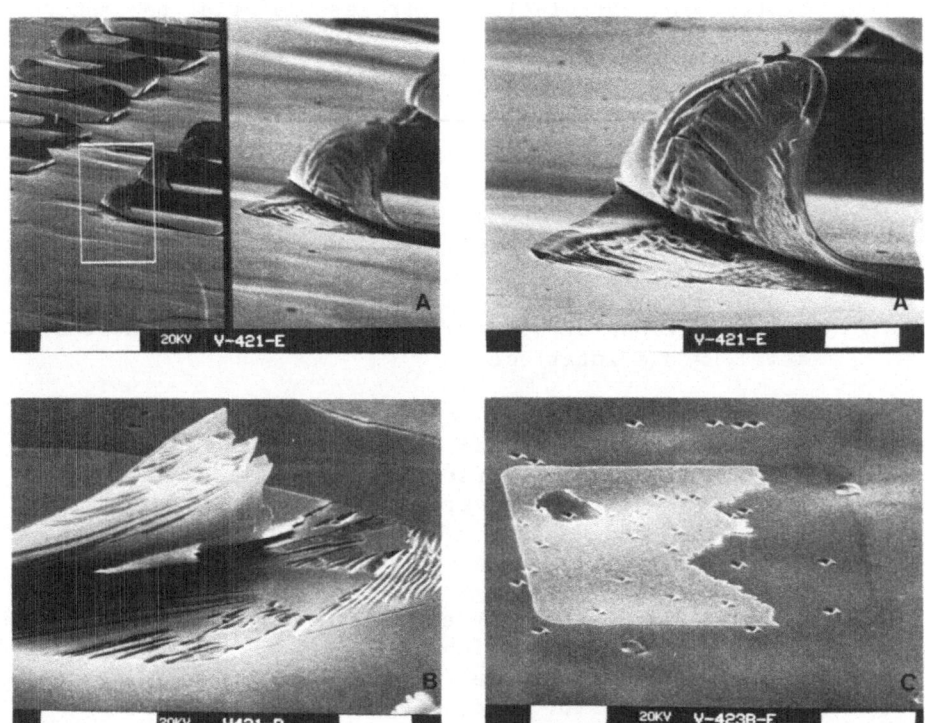

Figure 2. SEM photos showing two modes of peeling of Pd metal pads from $Hg_{(1-x)}Cd_xTe$ substrates: (A) "striated": 2200Å electroplated Pd film ripping up $Hg_{(1-x)}Cd_xTe$; (B) "striated": 1600Å thermally evaporated Pd film ripping up $Hg_{(1-x)}Cd_xTe$; (C) "smooth": 1500Å electroplated Pd film peeling smoothly (also occurs for thermally evaporated Pd).

frequency for both deposition techniques. Furthermore, regardless of technique, mixed modes of both "striated" and "smooth" delamination could often be observed within the same pad; individual pads in a 32 × 32 test array were 75 μm × 75 μm square and located on 4 mil (~100 μm) centers. These observations suggest that electroplating is neither better nor worse than thermal evaporation for depositing Pd diode contact metallization in terms of adhesion quality. In addition, the occurrence of mixed mechanisms of peeling within the same small regions suggests that some sort of lateral material-related or process-related inhomogeneity may exist on a metallized HgCdTe substrate.

Secondly, adhesion statistics were recorded for all metal/substrate combinations, and attempts were made to find correlations between adhesion yield for a given metal and other parameters such as $Hg_{(1-x)}Cd_xTe$ composition, array size, contact pad spacing, and predeposition clean-up. The only statistically significant correlation that could be found was that between adhesion yield and thoroughness of removal of insulator material. Over a period of about eight months, the overall adhesion yield for Pd increased from ~65% to over 90%.

Thirdly, the study to assess and improve the degree of pre-metallization clean-up was undertaken essentially in parallel to the statistics category and can be invoked to explain this improvement in adhesion yield. During the investigation, a micro-spot attachment became available for use with the ellipsometer in the process lab. The ellipsometer[13] is an instrument which measures the change in the state of polarization of elliptically polarized light when it reflects off the surface of a sample. From such optical measurements one can determine certain characteristics of the sample. If the index of refraction of the substrate is known and if the surface is reasonably smooth, ellipsometry can be used to determine the thickness of a thin film from several thousand angstroms to <10Å. Accordingly, the ellipsometer can be used to identify the presence of residual insulator films on HgCdTe that are too thin to see with the naked eye or even with a microscope. The micro-spot feature allowed processors to "look" inside small diode "windows" prior to metallization and stop a wafer from continuing in the process if residual insulator material was detected. As a result, as time went on and as the technique became more widely accepted, fewer arrays having incompletely removed insulator films were metallized. These arrays were typically re-etched until they passed ellipsometric inspection, so metallization yields naturally went up.

Fourthly, several different metals were deposited via thermal evaporation onto like epitaxial HgCdTe substrates having identical process histories. Test arrays (32 × 32) of contact pads were tape-tested and inspected for adhesion yield and failure mechanism. All three metals (including Pd) exhibited >95% adhesion yield under these controlled conditions. The few incidences of peeling displayed random mixed mode behavior from pad-to-pad and intra-pad.

Crucible Condensate

When diode contact metallization adhesion yield became an issue, one of the first areas of concern was that of the deposition process, itself. The reason for the suspicion was two-fold. Firstly, some HgCdTe wafers whose pre-metallization clean-up was not considered questionable suffered from peeling problems. Secondly, a dark band of discoloration along certain external surfaces of the white, alumina-coated tungsten crucibles was noted to appear after they had been fired once. The black/grey (and sometimes brownish) stain would usually darken to a maximum after two or three repeated usages. The parts of the crucible that became the most discolored were consistently the "upper" edge of the lower arm and the exterior of the funnel having line-of-sight proximity to the lower arm. These stains suggested that some material was outgassing from the ceramic coating.

To address this possibility, a brand new empty integral tungsten/alumina crucible was installed in the thermal evaporator with a clean sheet of copper foil (~5 mm × 5 mm square, ~100 μm thick) suspended midway between the lower arm and funnel exterior as shown below:

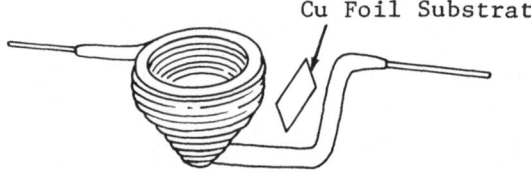

Cu Foil Substrate

Integral Tungsten/Alumina Crucible

The piece of foil was suspended by a "gallows"-shaped Cu hanger not shown. The purpose of the Cu foil was to act as a substrate to collect any condensable materials that might be released by the empty crucible during a dummy 10 minute run at ~1500°C. By removing the Cu sample after the first 10 minute run and replacing it with a fresh Cu foil, a second 10 minute run could be conducted to simulate the crucible's second usage and the associated outgassing products independently collected. Similarly, a third foil was used to collect condensate during a third 10 minute increment. These three Cu samples had been arbitrarily labeled numbers 3, 4, and 7, respectively.

Both SAM and EDX analyses were performed on these samples to determine the elemental composition of the condensates. Auger electron spectra for the Cu foil reference and for the three 10 minute increments are shown in Figures 3A-D, respectively. These spectra allow several conclusions to be made. The reference sample contains ambiently adsorbed oxygen and carbon with a trace of Cl as might be expected. Several metallic elements and/or

their oxides were deposited on the three Cu foil samples, including W, Ca, Mg, Al, and Si. The relative proportions of these
materials varied markedly with time: in particular, Ca increased
dramatically, Mg peaked out at ~20 minutes, Al grew steadily. The
thicknesses of the condensates were not directly measured, but

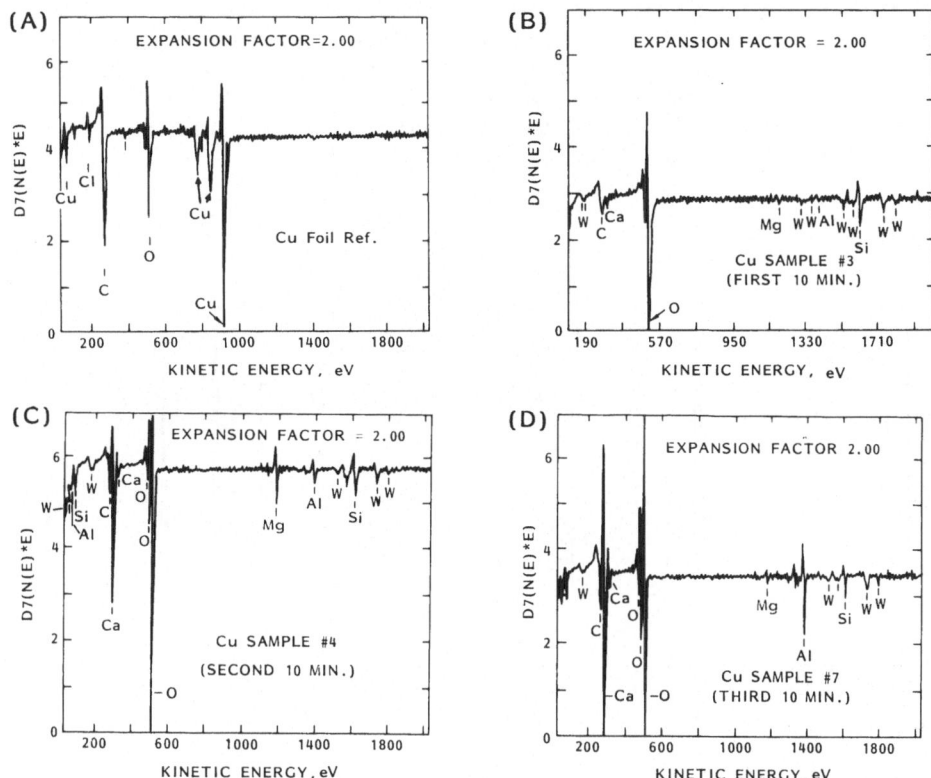

Figure 3. SAM Auger electron spectra of contaminants outgassed
from integral tungsten/alumina crubible in thermal
evaporator during various time increments: (A) Cu
foil reference, (B) first 10 minutes, (C) second 10
minutes, (D) third 10 minutes. Experimental
conditions: Crucible was heated at 1500°C and sample
was collected on fresh Cu foil for each time increment.

must be appreciable, because the dark stains on the Cu foils were
clearly visible, even for the third increment. Also, note that
the huge Cu Auger signal just above 900 eV in Figure 3A is completely "buried" in Figures 3B-D. Obviously, the condensates are
thicker than ~50Å; otherwise the 900 eV Cu signal would be seen.

EDX spectra for these same four specimens are given in Figures 4A-D, respectively. These spectra nicely corroborate the
Auger data, with peaks showing up for W, Ca, Mg, Al, and Si. In
addition, a peak can be seen for Fe, with a maximum intensity in
the spectrum for the first 10 minute increment of outgassing (Figure 4B). As stated earlier, EDX spectra can be representative of
elements as deep as 1 to 2 μm into the material. The thicker a Cu
surface is coated with a different material, the less intense will
be the Cu signal. One may observe that the Cu signal increases
with the number of time increments. Thus, the thickness of the
condensate decreases with time increments. In view of the tremendous attenuation of the Cu peaks in the spectrum for the first
increment (Figure 4B), one may estimate that the thickness of that
condensate is on the order of ~1 μm.

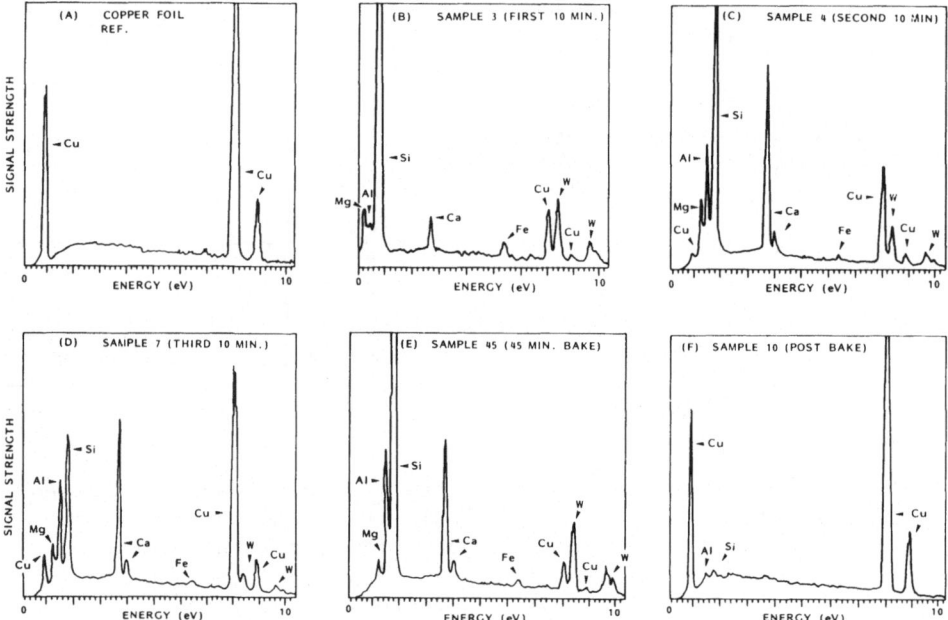

Figure 4.   EDX spectra of crucible condensates collected during
various time increments: (A) Cu foil reference;
(B) crucible No. 1, first 10 minutes;   (C) crucible
No. 1, second 10 minutes; (D) crucible No. 1, third
10 minutes; (E) crucible No. 2, 45 minute bake;
(F) crucible No. 2, 10 minute post-bake.   Experimental
conditions: same as in Figure 3.

To determine the effect of a long "bake-out" on the thoroughness of outgassing from these crucibles, a second new empty crucible was heated at 1500°C in the thermal evaporator and a sample of condensate was collected on a fresh Cu foil (Sample 45) during the entire bake time of 45 minutes. Then, the same crucible was allowed to cool, a fresh Cu foil was positioned (Sample 10), and the crucible was heated to 1500°C for a 10 minute increment. EDX spectra for Samples 45 and 10 are illustrated in Figures 4E and 4F, respectively. Note that the spectrum for the 45 minute bake is approximately a superposition/summation of the spectra for the three previously described 10 minute outgassings. In sharp contrast, however, is the spectrum for the 10 minute post-bake. Except for very small peaks at the Al and Si energies, this spectrum looks just like that of the Cu foil reference sample. These results show that a 45 minute bake-out at 1500°C is sufficient for removing nearly all of the outgassing species from the crucible.

Inspection of a typical chemical analysis of the alumina coating (see Table 1) reveals that essentially all of the materials (except W) observed in the condensate are, in fact, present as oxides in the coating. The manufacturer of these crucibles has confirmed that most of these oxides are added as binders to help facilitate the bonding of the alumina-based ceramic coating to itself and to the tungsten basket. These ceramic oxides tend to volatilize during the 1500°C treatment and apparently outgas at different rates depending on their boiling points. The most volatile, Na_2O, might be expected to come off first or most rapidly. An EDX spectrum (not shown) taken for a third crucible which was heated for <10 minutes displayed a rather strong Na peak in addition to peaks for the other materials. The fact that no TiO_2 was observed in any of the EDX or SAM spectra is due to its very small (0.04 wt %) abundance in the coating.

Table 1. Typical Chemical Analysis* of Coating - Percent by Weight

Al_2O_3	SiO_2	Fe_2O_3	TiO_2	Na_2O	CaO	MgO
98.55	0.58	0.10	0.04	0.31	0.19	0.23

* Data obtained from product description sheet from manufacturer.

Studies conducted subsequent to the present investigation showed that traces of several of these binders were incorporated into the deposited metal film.[14] No conclusive relationship between these impurities and adhesion yield was found. Nevertheless, once it was determined that the crucibles were a source of contamination, the crucibles were pre-baked for 45 minutes at 1500°C before their first usage. Unfortunately, only about one run could be performed per crucible after the pre-bake because of the severe cracking of the coating. Eventually, usage of the alumina-coated crucibles was discontinued in favor of a different style crucible.

Pd-HgCdTe Interactions

In order to determine whether or not any interfacial impurities were causing metal-to-HgCdTe delamination and to examine the elemental nature of the Pd-HgCdTe interface, a set of experiments was devised whereby SAM data was obtained while argon ion sputtering through the interface. Before these systematic experiments were begun, a decision was made to assess the degree of Hg outgassing from HgCdTe induced by the incident electron beam from the Auger electron gun. For each beam current value, the 500 Å diameter electron beam (5 keV) was rastored over a separate 5 μm × 10 μm rectangular area of the bare HgCdTe at RT, and Auger electron spectra were recorded over the 50 to 165 eV range. The peak-to-peak intensity of the Hg line at 76 eV was monitored as a function of incident beam current, I_p. As can be seen in Figure 5, the intensity is nearly at noise-level at $I_p = 1.07$ μA and asymptotically approaches a RT equilibrium maximum value at $I_p =$

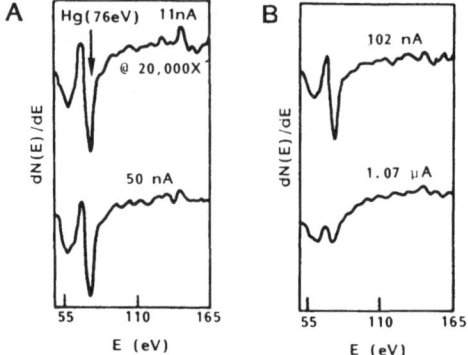

Figure 5.   Auger electron spectra of HgCdTe at various incident beam currents: (A) 11 nA and 50 nA; (B) 102 nA and 1.07 μA. Hg outgassing from the lattice increases with beam current. Experimental conditions: PHI Model No. 595 SAM; 5 μm × 5 μm rastor at 30° angle to surface; beam energy, 5 keV.

11 nA. Because HgCdTe is such a heat-sensitive material, the Hg can relatively easily be driven out of the lattice due to a combination of thermal and beam damage effects.

Accordingly, a beam current density of 10 nA/50 μm^2 (20 mA/cm^2), or lower, was chosen for the subsequent SAM experiments. Incidentally, other workers[15] have recently determined that the amount of electron beam stimulated Hg outgassing can be further decreased by chilling the HgCdTe during the Auger analysis. Their data suggests that between two and three times more Hg may be present on the surface when the analysis is done at ~180K than at RT.

Many SAM depth profiles were obtained for numerous HgCdTe samples containing Pd diode contact pads that had been deposited by either electroplating or thermal evaporation. A selection of four profiles is shown in Figure 6 to illustrate two main points:

1. The degree of Hg depletion at the HgCdTe surface was significantly greater underneath the Pd pads than on bare HgCdTe between the pads (compare Figures 6B and 6A).

2. Extremely different interfacial compositional profiles were observed but were not statistically correlatable to any material or process variables (compare Figures 6C and 6D).

The variety of profiles suggests that there are complex metallurgical interactions occurring at the Pd-HgCdTe interface. Particularly noteworthy in this regard, is the profile in Figure 6D which shows an accumulation of Hg at the same position in the interface as the Cd peak and Pd knee. The possibility exists that an alloy compound analogous to $Hg_{(1-x)}Cd_xTe$ such as $Pd_{(1-y)}Cd_yTe$ might have formed at the interface whereby some of the Hg^{2+} was reduced by Pd to elemental Hg. This possibility might be explored at a later date by a technique sensitive to chemical valence states, such as electron spectroscopy for chemical analysis (ESCA). Before concluding this part of the article, one last interesting observation must be cited. Although there were exceptions, the vast majority of Auger spectra which were acquired from Pd pad surfaces gave large Hg signals along with the usual peaks for ambiently adsorbed S, Cl, C, and O.

In conclusion, it would appear that the "smooth" and "striated" adhesion failure mechanisms are related to different metallurgical interactions between the Pd and HgCdTe which may be occurring as early in processing as the deposition itself. Different process steps which follow deposition could alter the chemistry. The exact mechanisms are nontrivial although Hg outdiffusion seems to play a decided role.

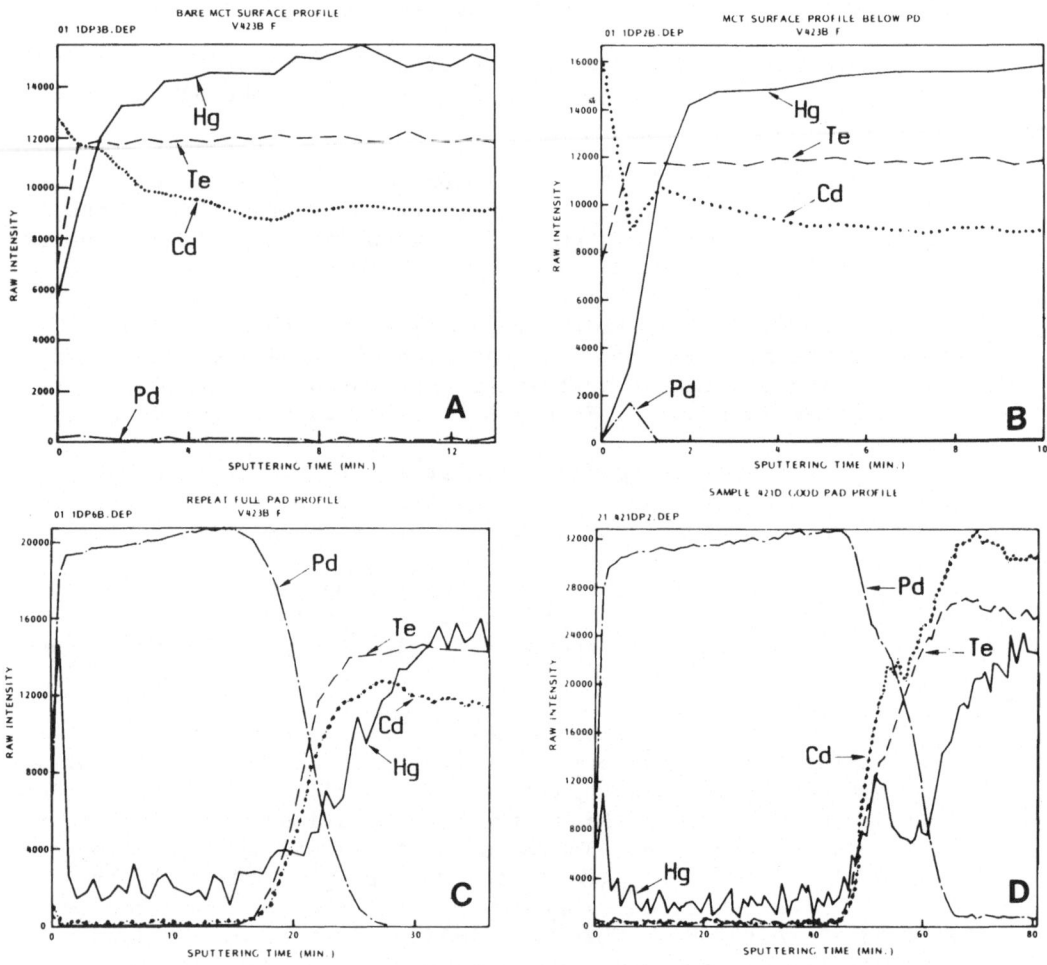

Figure 6. SAM depth profiles for various Pd–HgCdTe samples:
(A) bare HgCdTe surface, between Pd pads, sample
V423B–F; (B) beneath Pd pad, sample V423B–F;
(C) through Pd pad, sample V423B–F; (D) through Pd
pad, sample V421–D. Experimental conditions: Jeol
Scamp system; $I_p \approx 60$ nA at 10 keV and 4500X; sample
V423B–F, electroplated Pd (1500Å); sample V421–D,
thermally evaporated Pd; peak scaling factors:
sample V423B–F (Hg, 10.00; Pd, 1.00; Cd, 3.00;
Te, 2.00), sample V421–D (Hg, 10.00; Pd, 1.00;
Cd, 4.00; Te, 2.00).

ADHESION STUDIES: ZnS-Ge INTERFACE

Background on IR Windows

IR windows are an integral part of the overall IR detector package. An IR detector array is typically mounted on a cold-finger inside a metal vacuum dewar fitted at the light-incident end with a hermetically sealed IR transparent Ge window, as shown in the exploded view in Figure 7. The Ge window often has optical coatings on both sides which function together as a narrow band pass IR filter. The window is attached by circumferentially soldering it to the recess in the dewar porthole. To prepare the window for soldering, the optical coatings are temporarily masked and the outer periphery of the Ge is metallized, including the rim and the narrow annulus-shaped margins adjacent to the coatings. The mask is removed and the window is cleaned and inspected prior to installation. Acceptable pre-metallized windows have a uniform and tightly adhering metal film along the entire rim and annuli surfaces. Because the masks are slightly undersized the metal also typically follows the contour of the outer edge of the optical coatings, slightly stepping over their circumferential edge, as shown in Figure 7.

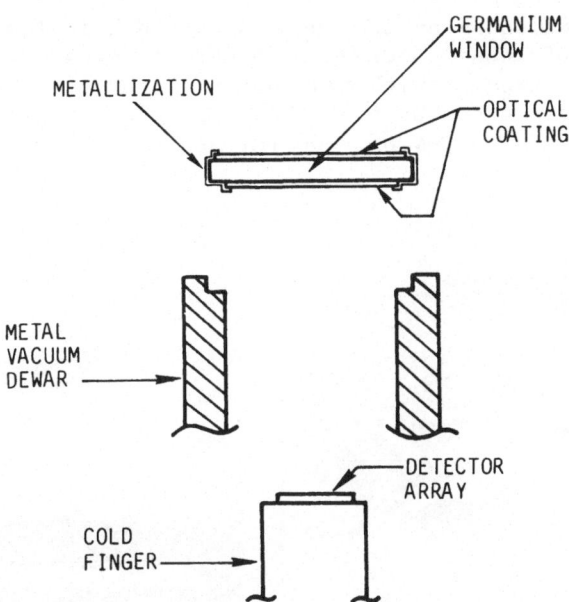

Figure 7. Simplified exploded-view schematic showing relative locations of IR window, vacuum dewar, IR detector array, and coldfinger.

Blistering Problem and Approach Taken

For no apparent reason, the normally high yield window-mounting process was marred by the appearance of small semicircular cracks and bubble-like blistered regions along the periphery of the optical coatings immediately following premetallization of the window. These cracks and blisters were first noticed upon removal of the protective masking material. Sometimes the mask would peel the "lid" from a blister, leaving a rough open crater near the edge of the optical coating. Blisters varied in size from tens of microns to 1 to 2 mm "diameter" and numbers ranged from one to several dozen per coating per wafer.

To gain some insight as to the cause of this problem, a rather extensive but systematic analytical approach was taken using complementary instrumental tools. SEM was used to characterize the morphology of cracks and blisters and to search for any microstructural evidence that might betray the cause. Blisters were opened by mechanical abrasion and by tape-testing. EDX was utilized to identify constituents of the optical coating. SAM was performed to analyze blister composition and to search for surface residues and interfacial contaminants.

Characterization of Cracks, Blisters, and Optical Coatings

Figure 8A shows an SEM photo of a crack in the optical coating typical of those observed on other windows. Note that the crack is semicircular with its center of radius at the point where the metal layer steps up over the upper edge of the IR coating. This crack has a radius of ~100 µm, and the window diameter is ~1 inch.

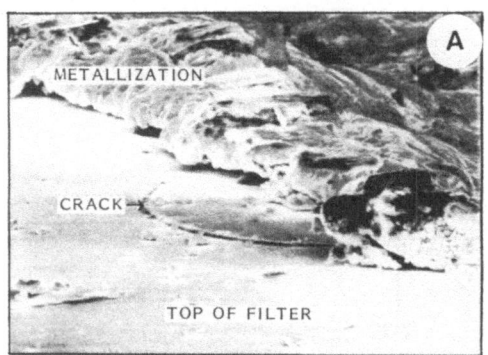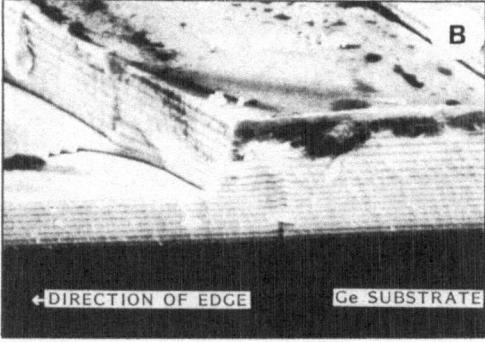

Figure 8.   SEM photos of defects in ZnS/Ge optical coatings on Ge windows: (A) crack next to big blister; (B) cross section of big blister base.

Adjacent to this crack was a large blister which had had its top
peeled off during mask removal. To obtain a closer look at the
base of the blister and determine where in the bulk of the optical
coating the plane of the blister base was located, the window was
carefully split in half such that the fracture zone bisected the
blister. The edge view of this blister is given in Figure 8B and
possesses features common to most other blisters. The base of the
blister is not a single plane, but rather a set of steps and terraces which tend to step upward away from the edge of the window.
Only one step-terrace can be seen in Figure 8B. Furthermore, the
location of the base happens to be a little over half way into the
optical coating in this case, whereas generally, blister bases can
be found at random distances from the top surface. One may
readily observe that the optical coating is a multilayered structure consisting of many alternate sublayers which show up as dark
and light stripes in Figure 8B. Almost in every case, the upper-
most surface of the various terraces in a blister base is the
light-colored material.

The top-most surface of the optical coating was analyzed by
SAM before and after ion milling, and corresponding Auger electron
spectra are shown in Figures 9A and 9B, respectively. The ion
milling succeeded in removing the carbon and oxygen contaminants
normally found on samples which have been exposed to air. The
relative intensities of the Zn and S signals are as expected for
ZnS.

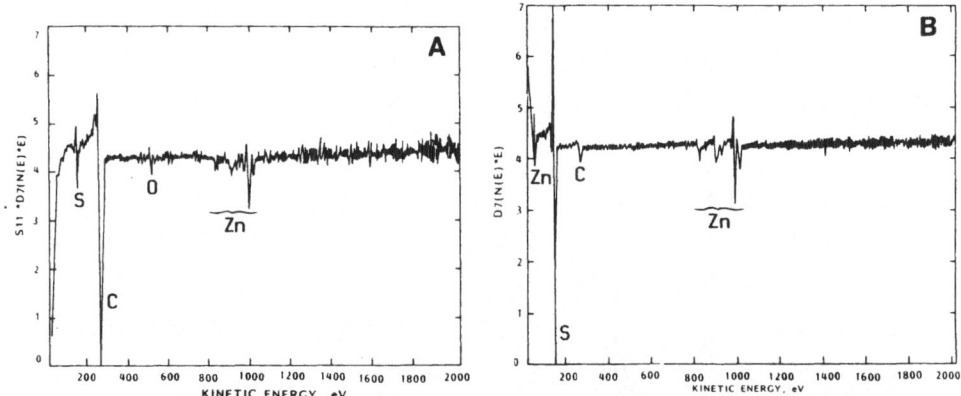

Figure 9. Auger electron spectra of top surface of optical
coating on Ge window: (A) before ion-mill cleaning;
(B) after ion-mill cleaning. Experimental
conditions: PHI 595 Microprobe; ~100Å material
removed with Ar$^+$ ion beam.

The alternating light and dark layers were analyzed by EDX and depth-profile SAM and found to be composed of ZnS (light) and Ge (dark). A SAM depth-profile for three of the sublayers is shown in Figure 10 and reveals "well-behaved" ZnS-Ge interfaces.

The upper dome of a blister was analyzed by SEM and EDX. Figure 11 shows a cross sectional view of a blister "lid" with the various ZnS and Ge sublayers marked. This particular optical coating was comprised of a stack of relatively thick ZnS/Ge sublayers on top of a stack of thin sublayers. Note that the underside of the dome is also terraced but consists of Ge in the bottom-most sublayer. Essentially all blisters had ZnS "floors" and Ge "ceilings."

Evidence of Chemical Penetration

Upon careful scrutiny by SEM of the entire surface area of the blister bases, a startling observation was made. As is vividly illustrated in Figure 12A, distinct semicircular stains showed up on the blister base at the outer region nearest the metallization layer. Some sort of chemical had seemingly penetrated an interlayer space of the optical stack at the outer edge and left a corrosion mark or chemical residue. Elemental analyses of these stains by SAM (Figure 12B) showed conclusively that

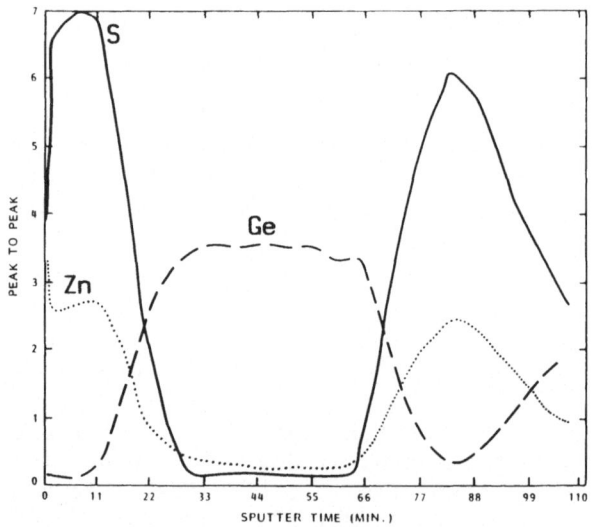

Figure 10. SAM depth profile through top three sublayers of ZnS/Ge optical coating on Ge window. Experimental Conditions: Ar^+ ion sputtering, PHI system.

species containing K, Cl, C, and O had somehow penetrated the optical stack. An inquiry of the metallization process revealed that these elements could have originated from constituents in the plating bath used in the electroplating operation. The issue of why the bath penetrated the optical stack was addressed next and is discussed in the next subsection.

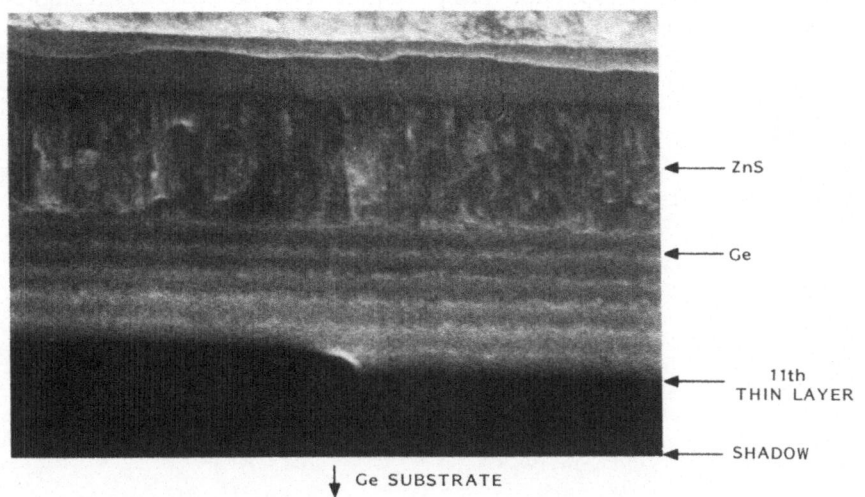

Figure 11. SEM photo showing cross sectional view of dome of blister.

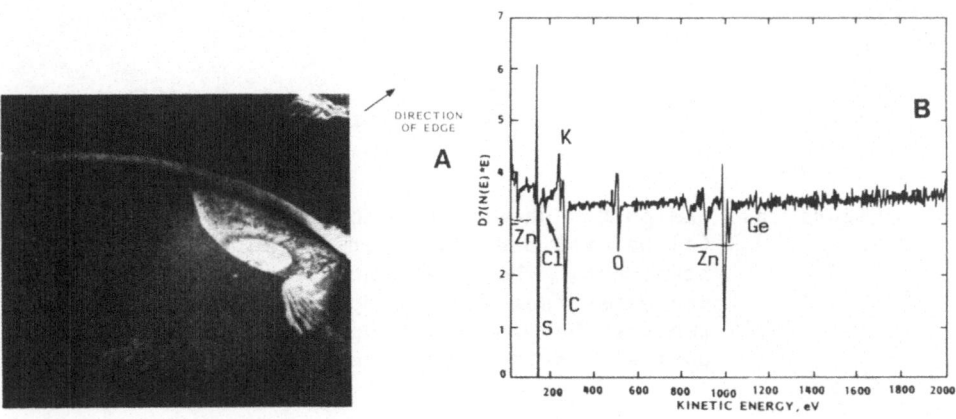

Figure 12. Evidence of chemical penetration: (A) SEM photo of semicircular stain on base of blister, (B) Auger electron spectrum of bright area in middle of stain.

Suggested Mechanism

SEM photos were taken of the outer periphery of a Ge IR window as received from the vendor and just prior to metallization (see Figures 13A and 13B, respectively). In Figure 13A, one may see that before the window is masked and prepared for metallization, the edge of the optical multilayer coating is smooth, rounded-off, and appears to have a top protective layer which tends to "seal" off the multilayer edges. However, a pre-metallization surface treatment (Figure 13B) not only leaves the Ge annulus roughened, but also removes the protective layer of the optical coating and mechanically perturbs the multilayer edges. The extent of physical degradation of these multilayers is much more apparent in the two high-magnification SEM photos in Figure 14. Not only are the edges of the multilayer coating exposed, but numerous ragged pits, pores, and microcracks are produced, any one of which could easily serve as an entrance path for plating bath solution.

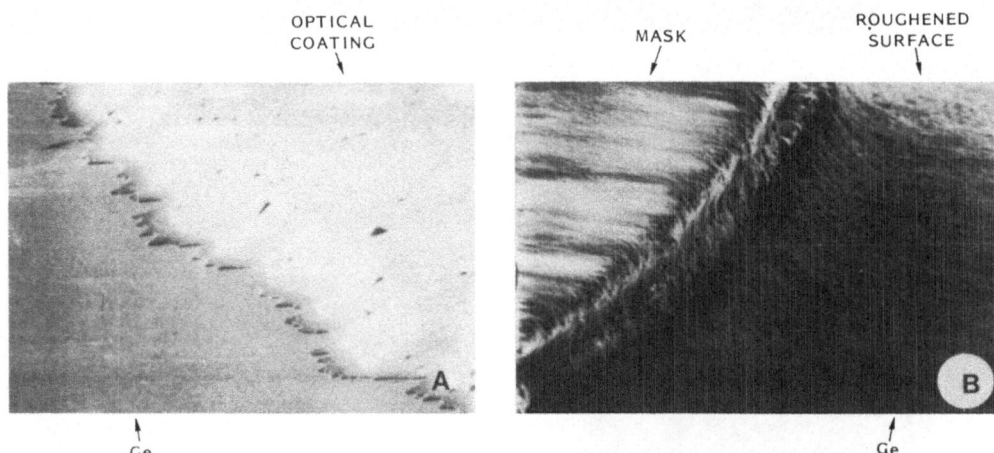

Figure 13.   SEM photos showing morphology of optical coating (A) before, and (B) after pre-metallization treatment. The treatment not only roughens both the Ge surface and the optical coating edge, but removes the protective upper-most layer of the optical coating multilayers, exposing their edges.

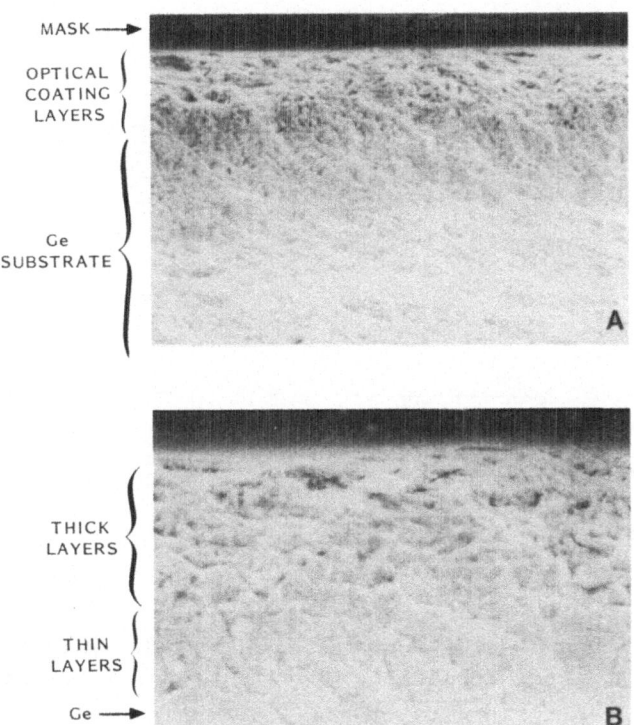

MASK →

OPTICAL
COATING
LAYERS

Ge
SUBSTRATE

A

THICK
LAYERS

THIN
LAYERS

Ge →

B

Figure 14. SEM photos showing close-up views of roughened
edges of optical coating multilayers: (A) high
magnification (B) higher magnification.

In accordance with the aforementioned observations, the following mechanism is proposed to account for cracks and blister
formation and is illustrated in Figure 15. Just prior to initial
electroplating, plating solution enters a porous defect in the
edge of the optical stack, such as a microcrack. During the initial stages of electroplating, the microcrack propagates radially
inward allowing more plating solution to enter. As the metal
layer thickens, the original microcrack entrance is effectively
sealed by the metal, entrapping process chemicals and solvent.
During the final stages of metallization, liquids volatilize and
gases are evolved, causing a build-up of pressure. Finally,
cracks and/or blisters form to relieve the pressure.

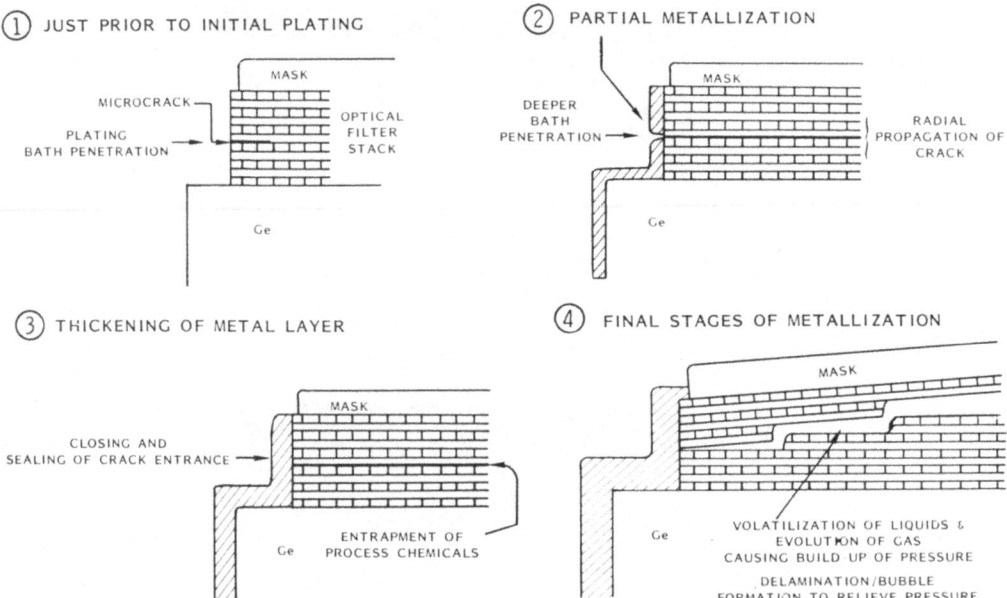

Figure 15. Suggested mechanism for ZnS-Ge delamination and blister formation. Sublayers are represented by symbols: ⊏⊏⊏⊏⊏⊏⊏ = ZnS, ⊏⊏⊏⊏⊏⊏ = Ge.

SUMMARY

The ability to control the mechanical integrity and chemical resistance of tightly adhering thin films is critical to the fabrication and performance of both high density microelectronic IR detector arrays and optical coatings for IR windows. In this paper, results are reported on the application of SEM, SAM, and EDX to the investigation of adhesion problems encountered with two important thin film-substrate systems. In the first case, examples and causes of peeling of electroplated and thermally evaporated diode contact metallization layers from the important IR detector material, HgCdTe, are described. Metallization adhesion yield was greatly improved by more tightly controlling the pre-metallization clean-up. The detection of volatile inorganic ceramic binders outgassing from alumina-coated tungsten crucibles used in the thermal evaporator is explained. Steps taken to minimize electron-beam desorption of Hg from the HgCdTe lattice during SAM analysis of the detector array are also discussed. Several SAM depth-profiles are given and shown to lead to the conclusion that some types of adhesion failure mechanisms are likely related to different metallurgical interactions between the Pd and HgCdTe, with Hg outdiffusion playing a decided role. In the second case, an interesting blistering phenomenon is illustrated along with a description of the associated failure mechanism for a ZnS-Ge multilayer optical coating on a Ge substrate. Physical defects were induced in the edges of the optical coating multilayers by a pre-metallization treatment. The blister formation mechanism involves four stages: (1) plating bath penetration into micro-cracks; (2) radial propagation of crack; (3) entrapment of process chemicals; and (4) volatilization, pressure increase, and delamination/blister formation to relieve pressure.

ACKNOWLEDGMENT

We extend our thanks to SBRC for support both of this research and of the publication of this paper. Much appreciation is offered to all process personnel who contributed effort toward various sample preparations and the statistical study. We are grateful to Richard J. Blattner, Charles Evans and Associates, San Mateo, California, and to Merrynell Colborn, Hughes Research Laboratory, Malibu, California for their SAM analyses. Betty Zuck and Ginny Harper assisted with the SEM/EDX analyses, Lupe Martinez typed the manuscript, and Kirk Gardiner provided the illustrations; to these four we say, "Thanks." Acknowledgment is also made to SBRC for supporting Dr. Garwood in the presentation of this paper. Lastly, Dr. Garwood offers a kind acknowledgment to the Conference Chairmen and Attendees who graciously voted in favor of nominating this paper as the "Official 'Advances in Materials Characterization II' Presentation of the 1984 Olympics."

REFERENCES

1. "Introduction to Analytical Electron Microscopy," J.J. Hren, J.I. Goldstein, and D.C. Joy, Plenum Press, New York (1979).
2. G.A. Somorjai, "Principles of Surface Chemistry," Prentice-Hall, Englewood Cliffs, NJ (1972); M. Prutton, "Surface Physics," Clarendon, Oxford (1975).
3. "Methods of Surface Analysis," A.W. Czanderna, ed., Elsevier Scientific Publishing Company, Amsterdam (1975); K.L. Chopra, "Thin Film Phenomena," McGraw-Hill Company, New York (1969).
4. P.E. Luscher and K.C. Ray Chiu, Technology Vectors 29:110 (1984).
5. J.I. Goldstein, D.E. Newbury, P. Echlin, D.C. Joy, C. Fiori, and E. Lifshin, "Scanning Electron Microscopy and X-Ray Microanalysis," Plenum Press, New York (1981); "Electron Microscopy of Materials," W. Krakow, D.A. Smith, L.W. Hobbs, ed., North-Holland, New York (1984).

6. D.T. Hawkins, "Auger Electron Spectroscopy; A Bibliography, 1925-1975," Plenum, New York (1977); C.C. Chang, Surface Sci. 25:53 (1971); G.A. Somorjai and F.J. Szalkowski, Advan. High Temp. Chem. 4:137 (1971); N.J. Taylor, p. 117, in: "Techniques of Metals Research," Vol. 7, R.F. Bunshah, ed., Wiley Interscience, New York (1971).

7. K.F.J. Heinrich, "Electron Beam X-Ray Microanalysis," Van Nostrand Reinhold, New York (1981); "Energy-Dispersive X-Ray Microanalysis, An Introduction," D. Vaughan, ed., Kevex Corporation, Foster City, California (1983).

8. "Electron Spectroscopy for Surface Analysis," H. Ibach, ed., Springer-Verlag, Berlin (1977).

9. T.A. Carlson, "Photoelectron and Auger Spectroscopy," Plenum, New York (1975).

10. C.A. Evans, Jr., Anal. Chem. 47(a):818A (1975).

11. T.E. Madey and J.T. Yates, Jr., J. Vacuum Sci. Technol. 8:525 (1971).

12. G.L. Hansen, J.L. Schmit, and T.N. Casselman, J. Appl. Phys. 53(10):7099 (1982); J.L. Schmit and E.L. Stelzer, J. Appl. Phys. 40(12):4865 (1969).

13. D.R. Rhiger and R.E. Kvaas, Second Interim Report, Contract No. F33615-80-C-5084, "Exploratory Development on PV HgCdTe Surface Leakage," Santa Barbara Research Center, 29 November 1982; D.E. Aspnes, p. 799, in: "Optical Properties: New Developments," B.O. Seraphin, ed., North-Holland, Amsterdam (1976).

14. A. Adams (unpublished work at SBRC).

15. T.P. Massopust and L.L. Kazmerski (private communication).

264

ELECTRON MICROSCOPIC STUDIES OF PLASMA-SPRAYED COATINGS

Christopher C. Berndt* and Reginald McPherson

NASA-Lewis Research Center, MS 105-1
Cleveland, Ohio 44135, USA
Monash University, Materials Engineering
Clayton, Victoria 3168, Australia

ABSTRACT

This work characterizes the coating surface and profile (cross-section perpendicular to the substrate surface) of plasma-sprayed deposits. Metal (mild steel and a composite of Ni-Al) and ceramic (alumina and zirconia based materials) deposits were produced and examined by scanning electron microscopy. The ceramic coatings were also examined by transmission electron microscopy. Structural observations from the controlled fracture of coatings are also included.

The ceramic coatings exhibited similar morphological features to the metal coatings in the form of flattened particles, dendritic artefacts on the particle surface, a columnar crystal structure and porosity of various forms. The columnar structure that was observed in some flattened particles may be explained by assuming efficient heat extraction through the underlying structure (substrate or coating). In a similar manner the dendritic artefacts would be expected to preferentially nucleate on the surfaces of large particles.

INTRODUCTION

Plasma-sprayed coatings are produced by injecting powder of approximately 25-125 microns in diameter into the tail flame of a

_____

*Fellow of the Joint Institute for Aerospace Propulsion and Power with Cleveland State University

DC thermal plasma (1, 2). The particles are accelerated and heated by the plasma effluent so that they may flow on impact against the substrate to form a saucer-shaped deposit. The deposition of many particles builds up an integral coating which has unique structural and materials properties. The physical state of the coating particles at their moment of impact and any interactions with the substrate may be determined by microscopy. Much work has previously been carried out on the metallography of coatings (3); for example to study how coating structure is controlled by plasma spraying variables such as the particle size and distribution (4) the type of arc gas (5), the plasma device used during the spraying procedure (6), the substrate surface topography (7), and the torch to substrate distance (8).

This work presents some electron microscopic studies of plasma-sprayed coatings. These structural observations may be related to the physical processes which occur during particle deposition. It should also be remembered that the materials properties of the coating will be influenced by the coating microstructure.

EXPERIMENTAL

Table 1 details the chemistries and size distributions of the powders used in this study. The powder morphologies were not the same and reflected the methods used in their preparation. The mild steel powder was produced by an atomization process, the nickel-aluminum was a composite of the individual components, the alumina was produced by grinding, and the yttria-zirconia material was prepared by crushing and then reaction at a high temperature.

The preparation of the substrate surface influences the coating adhesion (9). The substrates were prepared in the standard manner (10) by grit blasting with alumina (-59 to +0.58 mm) at a pressure of 0.15 to 4.00 MPa. The center line average surface roughnesses of mild steel, brass and aluminum grit-blasted substrates were approximately 8 μm. The roughness increased to values greater than 10 μm upon spraying metal powders whereas sprayed ceramic coatings exhibited values of about 3 μm.

The samples which were procured from industry were plasma-sprayed at power levels of either 18 kW (for metal coatings) or 31 kW (for ceramic coatings). Samples which were prepared for the controlled fracture experiments were plasma sprayed in the laboratory with a low power (7kW) N_2-Ar DC plasma. These coatings were incorporated into a loading fixture so that the coating could be subjected to a tensile load.

266

Table 1. Powder Specifications

Powder	Composition (in weight percent)		Particle Size (microns)
Mild Steel	C	0.20	-106 to +45
	Mn	0.50	
	Si	0.04	
	P	0.04	
	Fe	balance	
Alumina*	Al_2O_3	94	-53 to +15
	TiO_2	2.5	
	SiO_2	2.0	
	Other	balance	
Nickel -Aluminum**	Al	4.5	-90 to +45
	Ni	balance	
Yttria -Zirconia	ZrO_2	92	-90 to +40
	Y_2O_3	8	

* US Patent 3,607,343
**US Patent 3,305,326; 3,338,688; and 3,436,248

The surface topography of the coating and regions normal to the surface were examined by SEM. Taper sections of the coating profiles were produced for metallographic examination by mounting in bakelite and polishing in the usual manner. These samples were placed within brass tubing to prevent preferential abrasion of the coating and they enabled the "splat" appearance of the coating to be distinguished as well as oxide boundaries and porosity. The smooth surface of the metallographic specimens also permitted X-ray analysis for a qualitative description of elemental distribution within the coating and at the coating-substrate interface. Samples of plasma-sprayed zirconia were prepared for transmission electron microscopic examination by mechanical grinding of the coating to 0.05mm (0.002inches) in thickness and then ion beam thinning (11).

RESULTS

Metal Coatings

The two plasma-sprayed metal coatings exhibited similar morphological features. Particles which reached the substrate surface in the molten condition have spread out, Fig. 1. Some areas appear white due to charging of particles and show those particles which have reacted to produce oxides. The particles also exhibit pinholes which are characteristic of porosity due to evolution of absorbed gases.

Figure 2 shows X-ray element line scans and X-ray maps across the interface between the Ni-Al coating and mild steel substrate. Only a small amount of aluminum was detected in the coatings and this was unalloyed with the nickel (or iron) but present at the particle boundaries; most probably as an oxide. Some large areas high in aluminum were also detected at the interface and were found to be alumina grit which embedded into the substrate during surface preparation. The microstructure shows that some of the metallic particles have extensively deformed and were firmly attached to the substrate whereas other regions of the coating pulled away from the substrate. Oxides were presumably present at the regions of interfacial separation and support the theory that these are detrimental to coating adhesion. Dark areas were observed within the body of the coating and were shown to be either high or low in aluminum (Fig. 2a). The aluminum deficiency can result from the polishing operation when oxides pull out from

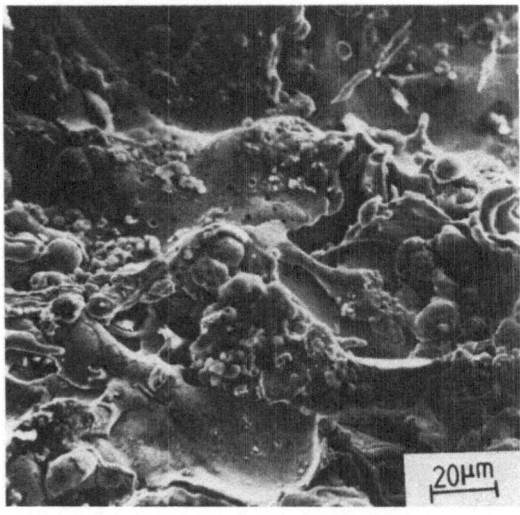

Fig. 1. Surface of Ni-Al composite coating.

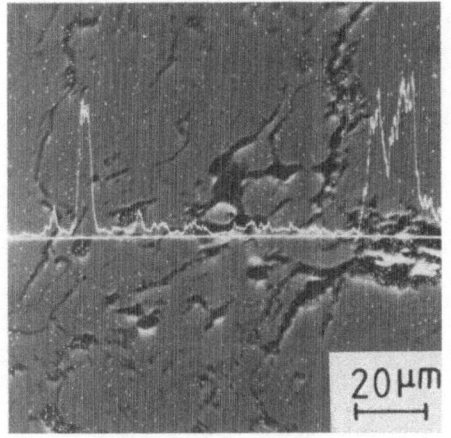

(a) Al

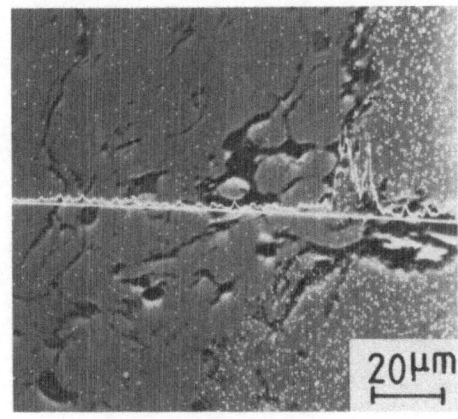

(b) Ni

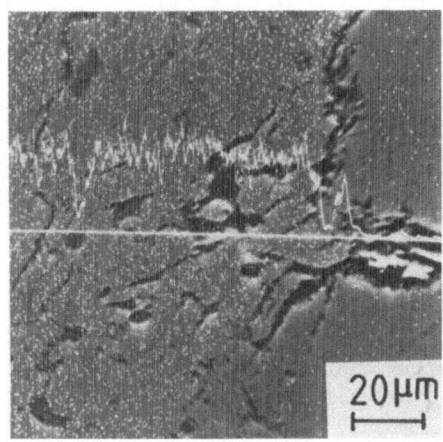

(c) Fe

Fig. 2. X-ray line scans and X-ray maps of nickel-aluminum coated onto a grit blasted mild steel substrate.

around the particle or it can be an area of porosity within the coating. Porosity can be present either between lamellar of particles (interlamellar porosity) or within particles (internal porosity). It is not clear what form is present in the examination of Fig. 2, but it should be noted that internal porosity has already been observed (Fig. 1) for Ni-Al coatings.

The fracture surfaces of mild steel coatings also exhibited particles which had flattened on impact with the substrate. Figure 3 shows the microstructure of particles which were in direct contact with the substrate. Dendrites were observed on the

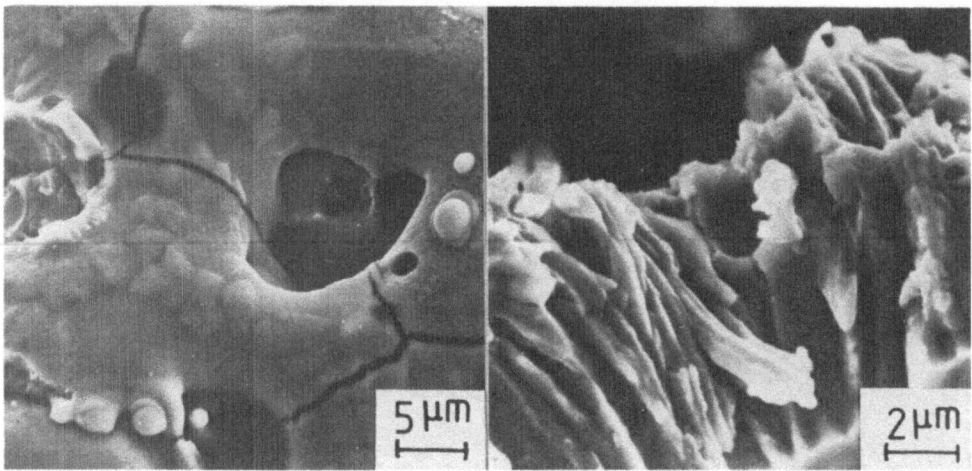

(a) Note microcracks and   (b) Note columnar grain
     dendrite artefacts.   growth in particle.

Fig. 3. Fracture surface of plasma-sprayed mild steel.

particle surface (Fig. 3a) and were associated with crystal
nucleation and growth which was driven by supercooling. The
particles do not exhibit any deformation arising from the tensile
test and microcracks are present. These may have arisen during
the formation of the coating (from shrinkage or impact by other
particles) or resulted from the tensile adhesion test.

Columnar grain growth approximately perpendicular to the
surface was distinguished (Fig. 3b) for other particles. This
particle has separated from the substrate and deformed
plastically. Thus good thermal contact of the particle with the
substrate (which resulted in columnar grain growth) may not
necessarily imply good mechanical keying and good adhesion. It is
also possible that the solidified particle may have separated from
the substrate after shrinkage or interaction with overlaying
particles.

Ceramic Coatings

The plasma-sprayed alumina coatings essentially exhibited the
same features as the metal coatings except that not all of the
particles were sufficiently molten to flow at the substrate
surface. Internal porosity was distinguished in some particles by
spot-like features on the surface (Fig. 4). The interlamellar
surfaces were revealed by fracturing the coatings. Generally they
were smooth with little evidence of the crystal growth pattern.
Some particle surfaces showed a dendritic pattern (Fig. 5) which
implies nucleation and oriented growth which is driven by

270

supercooling and a temperature inversion at the solid-liquid interface. On the other hand some fractured lamellar (Fig. 6a) revealed a columnar growth pattern consistent with nucleation and rapid growth from the region in contact with the underlying material. This would be expected if there was efficient heat transfer from the flattened droplet in intimate contact with the underlying solid.

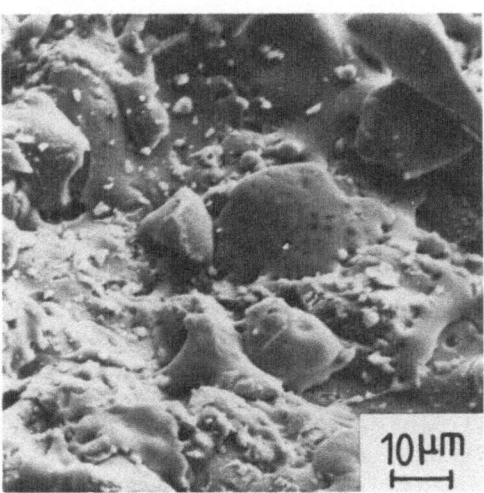

Fig. 4. Surface view of plasma-sprayed alumina.

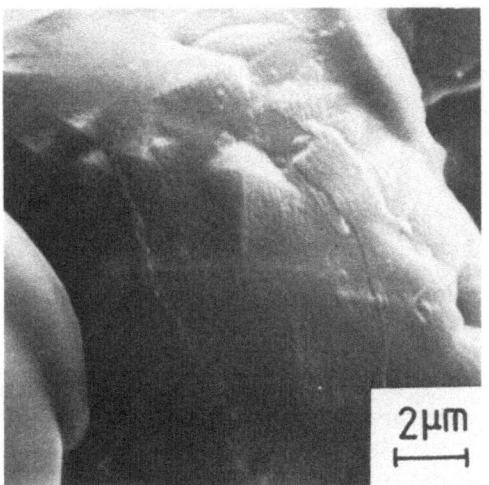

Fig. 5. Dendrites on the surface of a plasma-sprayed alumina particle.

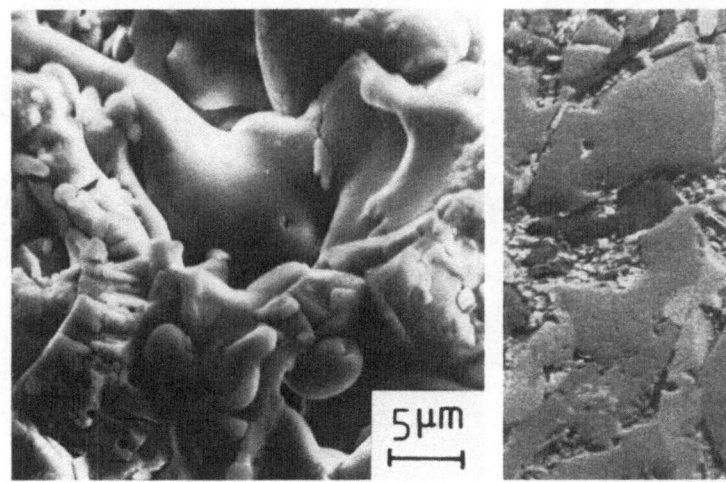

(a) Fracture surface.   (b) Polished section.

Fig. 6. Profile of plasma-sprayed alumina.

In general a large number of fine particles were observed on the surface of ceramic coatings, as seen in Fig. 4, and these were probably the lower size fraction of the powders. The coating profile (Fig. 6b) shows particles which have either the characteristic "saucer" shape or those which have not deformed appreciably on impact. Imperfect contact between adjacent lamellar would give rise to a continuous and very fine interlamellar porosity. This contrasts to voids which are formed within particles from gas evolution or solidification shrinkage.

The scanning electron micrographs of the alumina coating profile exhibited dark and light areas (Fig. 6b) which on X-ray analysis were distinguished as alumina and titania respectively (Fig. 7). The background of the X-ray maps is poorly resolved since the lens current had been adjusted for a large X-ray count rate. The alumina and titania particles can be differentiated from their contrast because the secondary electron emission is dependent on the atomic number. The white areas in Figs. 6b and 7 result from specimen "charging" in the electron microscope and show where epoxy adhesive has penetrated between the coating and substrate. Other regions of the coating have a fine granular structure around the particle boundaries and arise from the ceramic fracturing during the polishing operation. Transmission electron micrographs of yttria-zirconia coatings (Fig. 8) show either an equiaxed grain structure (Fig. 8a) or a fine dendritic phase growing in an amorphous region, Fig. 8b. The cooling rate may be estimated from the dendritic arm spacing (12) and has been

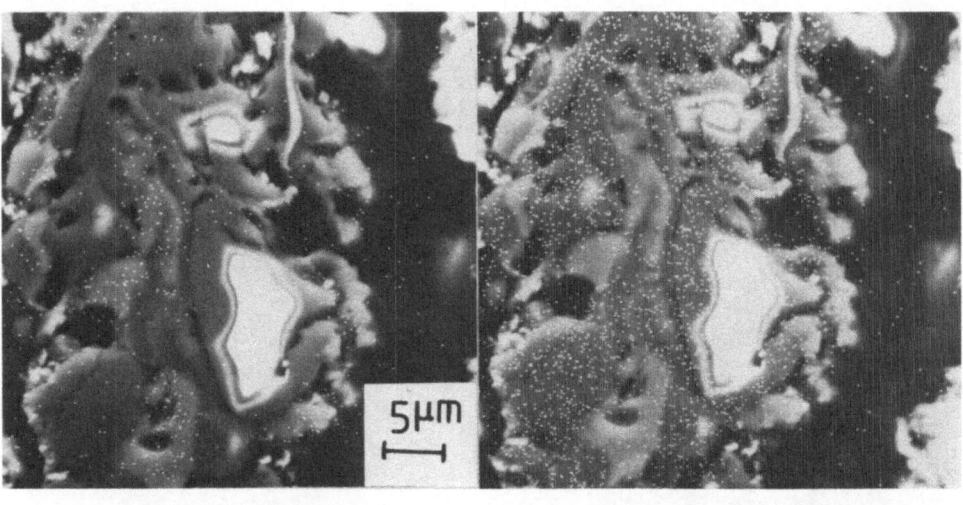

(a) Ti map   (b) Al map

Fig. 7. X-ray maps of a plasma-sprayed alumina coating. View is of the coating profile with the grit blasted mild steel substrate on the right-hand edge of the figure.

calculated as about 1.0×10^6 °C sec^{-1} during the solidification of this material. Other transmission electron studies on alumina coatings (13) have revealed imperfect contact between lamellar with voids of about 0.01 to 0.1μm present.

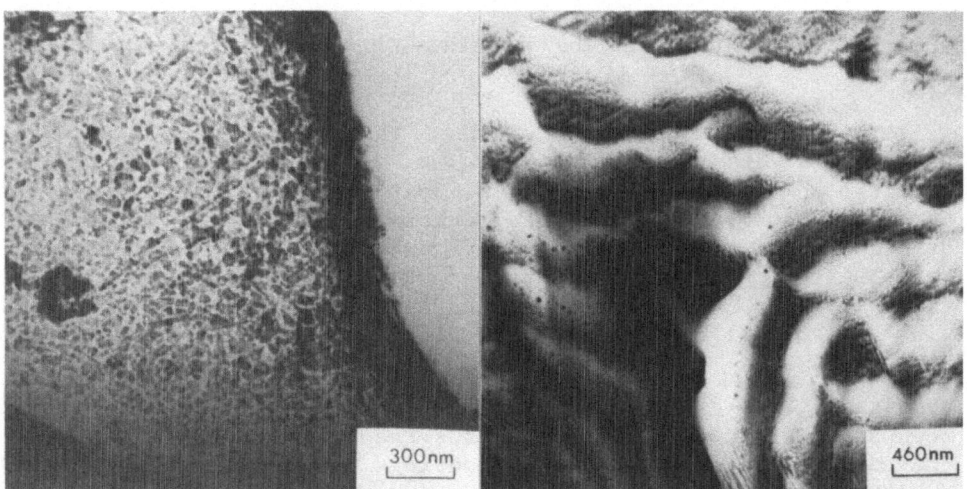

(a) Equiaxed grain structure.   (b) Dendrites within an
                                          amorphous structure.

Fig. 8. TEM of yttria-zirconia plasma-sprayed coatings.

DISCUSSION

The structure of nickel-aluminum composite coatings has been investigated extensively because they adhere well to a surface which may not have been grit blasted. It has also been established that ceramic coatings adhere more strongly to these composite coatings than to the grit blasted substrate alone, so it is common thermal spraying practice to use such a metal interlayer during the preparation of a coating. The interlayer is termed a 'bond coat' and the most suitable materials are based on Ni-Al, Ni-Cr and Mo. It is important to note that Ni-Al and Ni-Cr powders are not mixtures of the individual constituents but are composite powders consisting of a core material encased in the other component (14, 15).

There has been much discussion concerning the adhesion mechanism of bond coatings and it appears that some metallurgical bonding by chemical reaction occurs with the substrate. The high temperature prerequisite for alloying at the substrate surface may arise from chemical reaction of the powder constituents. This behavior has been verified (16) by observing an increase in the particle brightness during its time of flight. Gases are liberated during this reaction and may form the internal porosity that was observed within the coatings. This mechanism cannot be used to account for the high adhesion strength of molybdenum coatings; however it has been observed (17) that molybdenum reacts with the substrate. For example, with steel it forms a Mo-Fe intermetallic which can be distinguished by optical microscopy (18).

The alumina coatings exhibited the characteristic lamellar structure which revealed individual alumina and titania particles. It has been suggested that the first particles to be deposited cool most rapidly and form an amorphous layer (19) which has good mechanical and thermal contact with the substrate. Subsequent layers have a lower cooling rate because they are thermally insulated from the substrate by the amorphous layer. Also the thermal gradient is less since the coating surface may be heated to about 600 ^{o}C. Thick coatings (> 250 μm) exhibit columnar grain growth perpendicular to the substrate surface because of the slow cooling and thus the grain size increases with the coating thickness. Metal coatings on the other hand (20) exhibit a change in grain orientation which is dependent on the anisotropic heat exchange throughout the impinging particle. In the core region of the particle, which comes into initial contact with the substrate, heat is extracted through the substrate. In the peripheral area heat is extracted via the core region and the grains are thus radially oriented from the core.

The above characterization of plasma-sprayed ceramics does not detail the complete picture of the material which is in direct contact with the substrate; that is, the first deposited layer. The high rates of cooling, which occur when there is intimate physical contact between the substrate and deposit have produced amorphous structures which are usually identified as a "transparent rim" region in the ion beam thinned specimen. This model is different from one where the micrograins are oriented parallel to the substrate surface. In other words, oriented micrograins require imperfect contact of the particle to the substrate so that heat extraction occurs mainly through the central core of the particle. It would be expected then that coatings which reveal micrograins oriented in this manner do not adhere as well as coatings which exhibit an amorphous structure. It is important to further investigate the formation of these different structures - for example it is not known whether a structural change occurs across an individual particle or from particle to particle and how the material properties are influenced by these changes. These effects are only observed in the initially deposited layers. The microstructure of the coating distant from the substrate is modified by a less severe temperature gradient and therefore the grain structure is more equiaxed.

The material used in this work consists of a mixture of alumina and titania powders. Other Al_2O_3 - TiO_2 powders consist of either sub-micron titania particles which completely surround and adhere to the alumina particles (termed composite powders) or are produced as a powder which has been pre-reacted to form a mixture of α-Al_2O_3 and β-Al_2O_3.TiO_2 (termed fused powders). The purpose of adding a second phase to alumina is to increase the density and tensile bond strength of the coating (21), however the precise mechanism is not understood.

Scanning electron micrographs of the coating profile show that titania particles have deformed to a greater extent than the alumina. This would be expected from the lower melting point of TiO_2 (1830 to 1850 ^{O}C compared to that of Al_2O_3 which is 2015 $\pm$ 15 ^{O}C), even though the kinematic viscosity of alumina (22, 23) just above its melting point is similar to that of titania (24). Thus the kinematic viscosity of titania at its melting point is less than that of alumina at the same temperature and may extend the relative solidification time of the titania so that it flows around the prior solidified alumina. However it should be considered that equilibrium conditions are not attained during the plasma spraying process. Therefore it may be argued that relatively small changes in the temperature versus kinematic viscosity behavior of individual components will have little effect on the coating structure. What may be the more important parameter to examine is the wetting angle between the individual components of the powder. No data can be found on the wetting angle between alumina and titania. However data (25) on the contact angles of TiO_2 and Al_2O_3 on some high melting metals suggests that TiO_2 may flow around Al_2O_3 particles to produce a dense deposit.

Cavities and pores are formed between individual particles in the sprayed coatings because interparticle diffusion is limited and particle flow is hindered during the splat cooling process. Internal porosity can arise from (26, 27): (i) gases which are adsorbed by the particle in the plasma effluent may desorb when the particle solidifies; (ii) the particle may become superheated in the plasma and vaporize (2980 $\pm$ 60 ^{O}C for alumina and 2500 to 3000 ^{O}C for titania); or (iii) metal oxides may partially convert to nitrides in the nitrogen plasma and these may subsequently decompose and evolve nitrogen after contact with atmospheric oxygen. This work did not conclusively establish how the internal porosity was formed in the alumina coatings.

CONCLUSIONS

This basic study was aimed at describing some of the microstructures of plasma-sprayed coatings. The ceramic coatings exhibited similar morphological features to the metal coatings in the form of flattened particles, dendritic artefacts on the particle surface, a columnar crystal structure and porosity of various forms. The columnar structure that was observed in some flattened particles may be explained by assuming efficient heat extraction through the underlying structure (substrate or coating). In a similar manner the dendritic artefacts would be expected to preferentially nucleate on the surface of large particles where supercooling would be a maximum.

ACKNOWLEDGEMENTS

This work was initiated under contract No. 52 of the Australian Welding Research Association. It has also been supported by NASA - Lewis Research Center under grants NCC3-27 and NAG3-164.

(1) Fauchais, P., E. Boudrin, J.F.Coudert, and R. McPherson. "High Pressure Plasma and their Application to Ceramic Technology", Topics in Current Chemistry 107, (1983), 59-183.

(2) Safai, S., and H. Herman. "Plasma-Sprayed Materials", pp. 183-214 of Treatise on Materials Science and Technology, Vol. 20, Ed. H. Herman, Pub. Academic Press, 1981.

(3) Oberlander, B., and I. Kvernes. "Metallographic Studies of Plasma-Sprayed Duplex Coating System", Metallography, 16, (1983), 117-135.

(4) Andrews, C.W.D., and B.A.Fuller. "Effects of Substrate Materials and Powder Type on the Properties of Plasma Sprayed Ferrite", J. Mater. Sci., 10, (1975), 1771-1778.

(5) Wellner, P., E. Erdos, and G. Marxer. "Structure of Chromium-Carbide Nickel-Chromium Plasma Spray Coatings and their Heat Behaviour", Sulzer Technical Review, pp.56-66, Research No. 1978.

(6) Tucker, R.C.jr. "Structure Property Relationships in Deposits Produced by Plasma Spray and Detonation Gun Techniques", J. Vac. Sci. Tech., 11, (1974), 725-734.

(7) Apps, R.L. "The Influence of Surface Preparation on the Bond Strength of Flame-Sprayed Aluminium Coatings on Mild Steel", J. Vac. Sci. Tech., 11, (1974), 741-746.

(8) Kumar, K., and D. Das. "Equilibrium and Metastable Samarium-Cobalt Deposits Produced by Arc-Plasma Spraying", J. Thin Solid Films, 54, (1978), 263-269.

(9) Herpol, G.A,. and E. Tavernier. "Surface Preparation and its Relationship to Weakness of a Coating", Brit. Weld. J., 11, (1966), 683-389.

(10) B.S.4495:1969. British Standard titled "The Flame Spraying of Ceramic and Cermet Coatings".

(11) Wilms, V., and H. Herman. "Plasma Spraying of Al_2O_3 and Al_2O_3- Y_2O_3", J. Thin Solid Films, 39, (1976), 251-262.

(12) Wilms, V.H.S. Ph.D. Thesis "The Microstructure of Plasma Sprayed Ceramic Coatings", Ph.D. Thesis of SUNY at Stony Brook, June 1978, Avaiable from University Microfilm International No. 7819123.

(13) McPherson, R., and B. Shafer. "Interlamellar Contact Within Plasma Sprayed Coatings", Thin Solid Films, 97, (1982), 201-204.

(14) Dittrich, F.J., and A.P.Shepard. "Flame Spraying Exothermically Reacting Intermetallic Compound Forming Composites", US Patent 3,436,248, Apr. 1, 1969.

(15) Longo, F. N. "Metallurgy of Flame Sprayed Nickel Aluminide Coatings", Welding Research Supplement, 45, (1966), 66S-69S.

(16) Kudinov, V.V., F. I. Kitaev, and A. G. Tsidulko. "Strength Characteristics of Plasma Coating from Nickel-Aluminium Powder Mixtures", Soviet Powder Metallurgy and Metal Ceramics, 14, (1975), 637-642.

(17) Allsop, R.T., T.J. Pitt, and J.V. Hardy. "The Adhesion of Sprayed Molybdenum", Metallurgia, 63 (4), (1961), 125-131.

(18) Kampmann, C., and K. Kirner. "Metallographic Investigations on Molybdenum Spray Coatings", Pract. Metall., 9, (1972), 363-369.

(19) Safai, S., and H. Herman. "Plasma Sprayed Coatings: Their Ultra-microstructure". Paper 5 in Int. Conf. on Advances in Surface Coating Technology, London 13-15 Feb., 1978, Pub. The Welding Institute: London, 347pp, 1978.

(20) Safai, S., and H. Herman. "Microstructural Investigation of Plasma-Sprayed Aluminium Coatings", J. Thin Solid Films, 45, (1977), 295-307.

(21) Durmann, G., and F. N. Longo. "Plasma-Sprayed Alumina-Titania Composite", Ceramic Bull., 48, (1969), 221-224.

(22) Elyutin, V.P., V.I. Kostikov, B.S. Mitin, and Yu. A. Nagibin. "Viscosity of Alumina", Russian J. of Phys. Chem., 43, (1969), 316-319.

(23) Elyutin, V.P., B.S. Mitin, and Yu. A. Nagibin. "Properties of Liquid Al_2O_3", Inorganic Materials, 8, (1972), 416-418.

(24) Mitin, B.S., and Yu. A. Nagibin. "Properties of Liquid Titanium Dioxide", Inorganic Materials", 7, (1971), 709-711.

(25) Mitin, B. S., and V. Ya. Levin. "Interaction Between Liquid Oxides and High-Melting Metals", Inorganic Materials, 4, (1968), 1477-1481.

(26) Meyer, H. "Melting of Powders in a Plasma Jet", Deutsche Keramische Gesellschaft Berichte, 41, (1964), 112-119. Available under Translations Register Index Number TT-66-12343.

(27) Koubeck, F.J. "Microstructures of Melt Sprayed Oxide Ceramic Coatings as Observed by the Scanning Electron Microscope", pp. 393-400 of Proc. 3rd Ann. SEM Symp., 28-30 April, 1970, Pub. IIT Research Institute: Chicago, Illinois, 534pp, 1970.

EXPERIMENTAL AND THEORETICAL ASPECTS OF FREQUENCY-DOMAIN PHOTOPYRO-ELECTRIC SPECTROSCOPY OF CONDENSED PHASES (PPES), A NEW, SIMPLE AND POWERFUL SPECTROSCOPIC TECHNIQUE

Andreas Mandelis and John F. Zuccon

Photoacoustics Laboratory
Dept. of Mechanical Engineering, University of Toronto
Toronto, Ontario, M5S 1A4, Canada

INTRODUCTION

Photoacoustic Spectroscopy (PAS) has been used extensively in recent years with conventional microphones and piezoelectric detectors as signal transducers. The commonest piezoelectric transducers are usually made out of piezoelectric ceramic (e.g., lead zirconate titanate, PZT [1 - 3]). These ceramics exhibit a much higher frequency response than microphone transducers, and thus they became dominant in photoacoustic applications where fast transducer response is required, such as pulsed laser PAS [4]. Very recently Tam and Coufal [5] and Coufal [6] used polyvinylidene difluoride (PVDF) thin film piezoelectric transducers for pulsed and modulated photoacoustics. These films have distinct advantages over PZT transducers, the most important of which being their much higher, essentially flat frequency response between DC and tens of MHz [7]. In addition to piezoelectric behaviour, PVDF thin films (Pennwalt KYNAR^{TM}) exhibit strong pyroelectric properties [8] and, therefore, they can be used as detectors of thermal radiation. Coufal [6] utilized the pyroelectric character of 9 μm thin PVDF films to obtain pulsed and low modulation frequency spectra of Nd_2O_3 doped poly(methyl methacrylate) films. That application opened the way to the possibility of a new spectroscopic technique using pyroelectric films to detect optical absorption and non-radiative energy conversion processes in condensed phase matter. This paper is concerned with experimental and preliminary theoretical studies aimed at demonstrating the potential of Photopyroelectric Spectroscopy (PPES) as a new, simple and sensitive spectroscopic method for the in-situ study of physicochemical processes. Such processes may involve complex sample geometries which cannot be easily handled by conventional photoacoustic devices. Photopyroelectric data presented

279

in this work suggest that PPES has distinct advantages over PAS.

EXPERIMENTAL

Photopyroelectric Cell Characterization

Fig. 1 shows the photopyroelectric sample cell design which was used for the present studies. The teflon cell was enclosed in aluminum foil to minimize ambient electromagnetic interference. The output of the cell was directly connected to the input of a variable gain ITHACO preamplifier (Model 1201). The experimental assembly is shown in Fig. 2. The photopyroelectric cell was first tested as an optical radiation detector. A thin layer of Xerox Toner black powder (ca. 1 μm) was spread on the 28 μm thick Pennwalt KYNAR PVDF film, which was located at the bottom of the cell as part of the sample chamber. Strong PPES signals were thus obtained throughout the whole UV-VIS spectral region (260 nm – 920 nm) of the Xenon lamp.

The resulting spectrum (amplitude and Phase) is similar to the one obtained photoacoustically using an EG&G PAS cell model 6003.

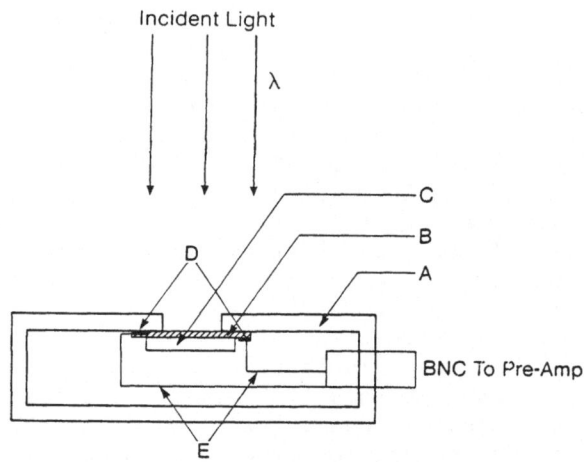

Figure 1: Photopyroelectric cell with a 0.5 in x 0.25 in PVDF sample holder. A: Teflon cell. B: 25 μm thick PVDF film (Pennwalt KYNAR 5380-11-8 Piezo Film). C: Teflon support. D: Solderable copper foil electrodes with conductive ad-adhesive. E: Silver wires.

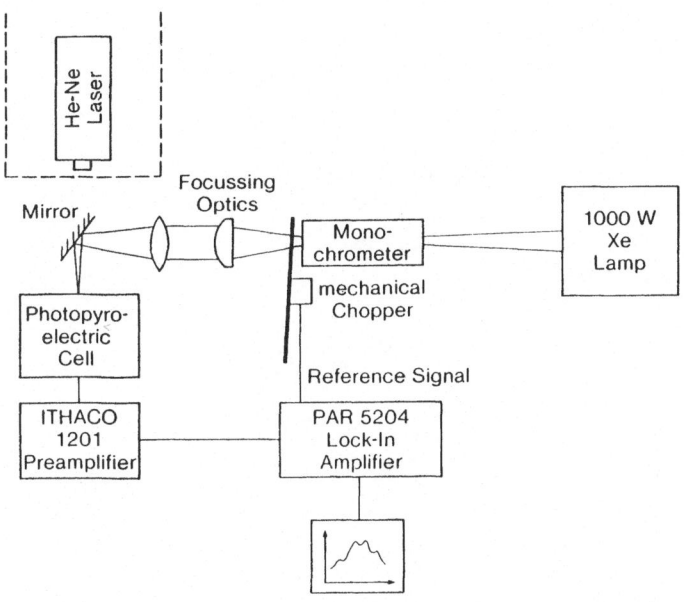

Figure 2: Photopyroelectric experimental apparatus. Both
broadband Xe lamp and He-Ne laser source are
shown.

The PPES phase, however, is essentially flat and constant throughout, whereas the PAS phase varies continuously by a few degrees
from 250 nm to 900 nm due to phase shifts contributed by the complex
microphone electronics. The frequency response of both cells is
shown in Fig. 3. The upper limit of the response is set by the
mechanical chopper. The frequency response of the PPES cell,
Fig. 3a, exhibits essentially two linear regions, one below 50 Hz
with a slope of -9.98 ± 0.03, and one above 50 Hz with a slope of
-1.5 ± 0.1. This behavior is to be compared with the EG&G PAS cell
frequency response curve Fig. 3b. The photoacoustic cell exhibits
non-linear Helmholtz – resonant behavior above 400 Hz. These re-
sonances are difficult to eliminate from any microphone PAS chamber
[9] and can seriously complicate frequency – domain studies of condensed matter specimens.

Spectroscopic, Chemical and Materials PPES Studies

Ho$_2$O$_3$ powder was used, mixed with tap water as a paste, so as
to achieve a good thermal contact with the PVDF substrate film.
Ho$_2$O$_3$ exhibits sharp absorption lines in the UV-VIS region and can
be used to test and calibrate the photopyroelectric cell. The

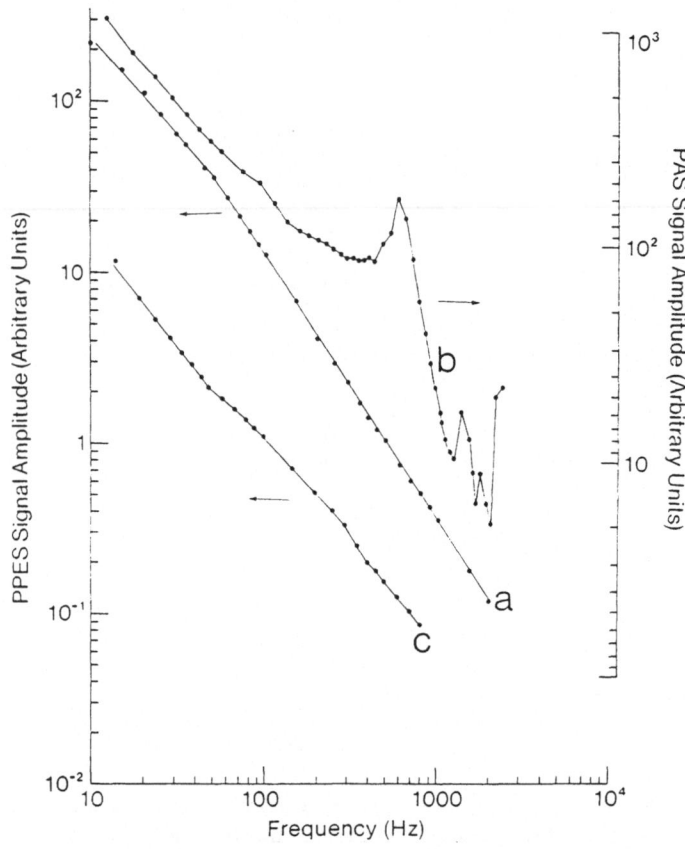

Figure 3: a) Frequency response of photopyroelectric cell using a 2 mW He-Ne laser light source. Pre-amplifier gain: 10. Pre-amplifier High Pass 3 Hz; Low Pass: 3 kHz.

b) Frequency response of EG&G PAS cell (model 6003) using a 2 mW He-Ne laser light source.

c) Frequency response of Cu plate after oxide removal. Light source: 2 mW He-Ne laser. Pre-amplifier Gain: 5000.

signal amplitude at 11 Hz, normalized by the Xe lamp spectrum, is shown in Fig. 4a. The spectrum is identical to published PAS spectra of Ho_2O_3 powders [10] within the spectral resolution limitations of the slitwidth used for this experiment (2 mm slitwidth; 8 nm resolution). Fig. 4b is the normalized PPES signal of the Ho_2O_3 paste taken at 50 Hz. The amplitude and the phase correlate with each other and, upon comparison with Fig. 4a, they are seen to anticorrelate with the features of the spectrum taken at 11 Hz.

The photopyroelectric probe was also used to monitor the kinetics of the chemical reaction:

$$CuO + 2H^+ \longrightarrow Cu^{2+} + H_2O \qquad\qquad (1)$$

A 0.34 mm thick Cu plate was oxidized thermally, so as to form a dark, thin oxide film on the surface. The sample was then mounted between the holder opening and the PVDF film (i.e., between A and B in Fig. 1). Physical contact of the pyroelectric with the back of the sample was achieved with the teflon backing plate C, Fig. 1,

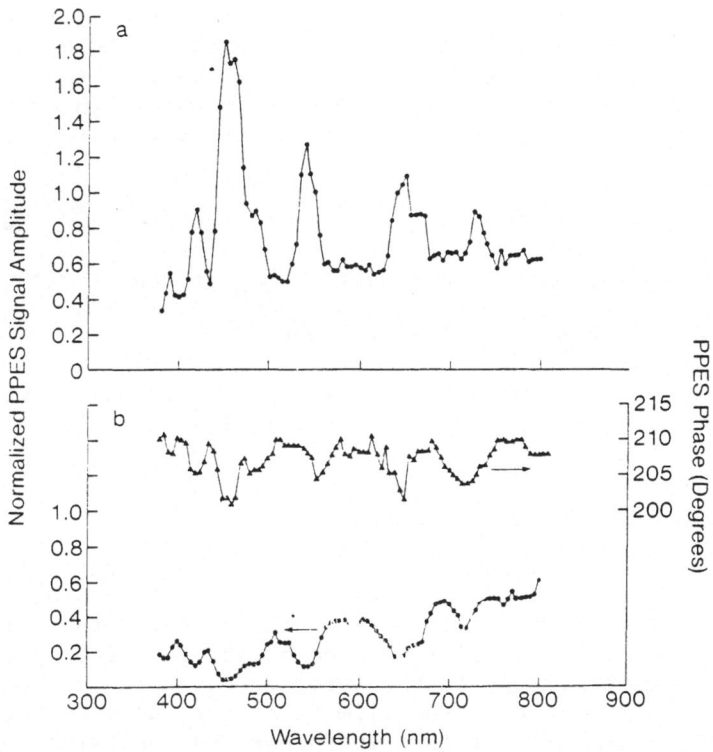

Figure 4: a) Photopyroelectric spectrum of Ho_2O_3 powder mixed with water. Chopping Frequency: 11 Hz. Pre-amplifier Gain: 500. Monochrometer Resolution: 8 nm.

b) Photopyroelectric spectrum (amplitude and phase) of Ho_2O_3 sample at 50 Hz. Pre-amplifier Gain: 1000. Monochrometer Resolution: 8 nm.

and plastic screws. The oxidized copper specimen was further illuminated with a He-Ne laser at 632.8 nm and the lock-in output at
10 Hz was recorded. Reaction (1) was initiated upon introducing
into the sample chamber a 0.1 M solution of HCℓ. Fig. 5 shows the
time evolution of the PPES signal amplitude from the start of the
reaction (t = 0) to its completion. Visual inspection showed metallic copper after ca. 4 minutes. The lock-in output was essentially constant beyond this time. The discontinuity observed in
Fig. 5 at t < 0 is due to the sudden loss of the PPES signal immediately after the introduction of the dilute HCℓ in the chamber.
After the recovery (ca. 30 sec) the earliest recordable signal was
already lower than the pre-mixing signal level by ca. 10%.

Fig. 3c shows the frequency response of the metallic copper
plate after the removal of the CuO. This curve exhibits a "knee"
at ca. 50 Hz, which was seen to be characteristic of the cell response. It also exhibits a clear break between 300 Hz and 350 Hz
with a subsequent change in the slope. By analogy to Photoacoustic
Spectroscopy, the thermal diffusion length μ_{Cu} in the copper can be
assumed equal to the plate thickness at the break frequencies, i.e.,
for 300 Hz $\leq f_b \leq$ 350 Hz [11]:

$$\mu_{Cu}(f_b) = (\alpha_{Cu}/\pi f_b)^{\frac{1}{2}} \cong L_{Cu} \qquad (2)$$

Eq. (2) gives the thermal diffusivity α_{Cu} of copper within the range:

$$1.09 \text{ cm}^2/\text{sec} \leq \alpha_{Cu} \leq 1.27 \text{ cm}^2/\text{sec}$$

This value range is in agreement with published values of α_{Cu} using
photoacoustic determination methods [11,12].

Discussion and Comparison to Theory

The experimental results demonstrate the essentially thermal
character of PPES. The PVDF Kynar films have been shown to be highly
sensitive spectroscopic detectors in accordance with preliminary
results obtained by Coufal [6]. The basic mechanism of operation
of this new spectroscopy is the pyroelectric effect: When the temperature of the sample - PVDF film interface increases by an amount
ΔT due to non-radiative relaxation processes following optical absorption, the polarized film aligns its polar constituent crystallites along the temperature gradient field, thus producing a net dipole moment which results in a pyroelectric voltage given by [8]

$$\Delta V(\omega) = \left(\frac{pL}{\varepsilon}\right) \Delta T(\omega) \qquad (3)$$

where p is the pyroelectric coefficient of the film, 3 x 10^{-9} Coul/
cm^2-°K, L is the film thickness, ε is the film dielectric constant,

284

$(1.06 - 1.13) \times 10^{-10}$ F/m, and ω is the angular frequency of modulation of the incident radiation.

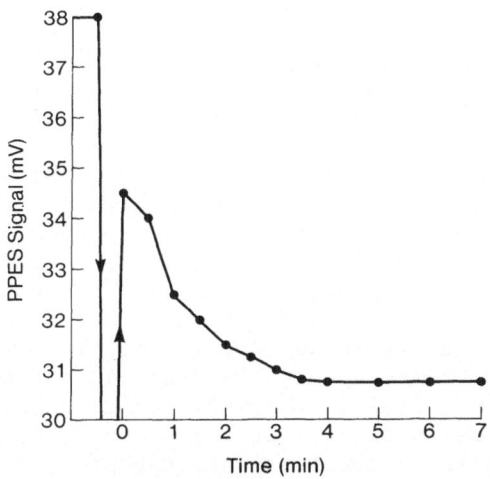

Figure 5: Kinetics of Eq. (1) monitored via photopyro-electric spectroscopy at 632.8 nm. Chopping Frequency: 10 Hz. Pre-amplifier Gain: 5000. Lock-in time constant: 1 sec.

The spectra obtained using this technique are of a quality similar to that of PAS. In fact PPES is related to Photoacoustic Spectroscopy in the sense that for a given sample geometry, PPES monitors the thermal energy transmitted to the substrate and gives a signal output proportional to the temperature of the interface between the sample and its backing. On the other hand, PAS monitors the thermal energy diffused to the sample surface and gives a signal output proportional to the spatial integral of the surface temperature [13]. It is, therefore, expected that PPES should be capable of providing more direct information about the sample than PAS which is an integrated and, therefore, an indirect technique. The phase profile obtained using a thin layer of Xerox Toner black powder is subject to optical absorption coefficient saturation considerations which are similar to those of PAS. The spectral feature reversal of Figs. 4a and 4b can be understood qualitatively in terms of the simple mechanism proposed above. At low chopping frequencies for which the sample is thermally thin ($\mu_s > L$), thermal energy diffuses into the backing PVDF film and gives rise to a pyroelectric signal. Opaque spectral regions, such as those between 435 nm – 500 nm of Ho_2O_3, produce large amounts of heat due to their high absorption coefficient values. Non-radiative deexcitation processes then generate a larger PPES signal than that obtained from optically transparent spectral regions. Upon a spectral scan this signal produces a replica of the absorption spectrum of the sample, such as the one shown in Fig. 4a. When the chopping frequency increases to levels such that the sample is thermally thick ($\mu_s > L$), only transparent regions can produce a pyroelectric signal through penetration across the sample thickness and direct absorption by the backing PVDF film. Such a transparent region is in Ho_2O_3 between ca. 560 nm – 630 nm. Strongly absorbing spectral regions retain the thermally converted optical energy close to the sample surface, as the condition $\mu_s < L$ prevents thermal communication with the backing. Such regions do not contribute to the generation of pyroelectric signals. Therefore, upon a spectral scan the pyroelectric effect produces a counter-spectrum, similar to a transmission spectrum, with strong signals at transparent regions and weak signals at opaque regions, such as shown in Fig. 4b. The exact peak-by-peak anti-correlation of the spectra in Figs. 4a and 4b demonstrates that at 11 Hz the Ho_2O_3 paste is thermally thin, whereas at 50 Hz it is essentially thermally thick. It may be concluded that PPES can be used as a conventional spectroscopic technique only at low chopping frequencies, or with very thin samples, so that the condition $\mu_s > L$ is valid. This is a restriction absent from Photoacoustic Spectroscopy, however, it is expected that an important contribution of PPES will arise in the study of thin film processes, such as the anodic growth of films on electrodes at the electrode – electrolyte interface.

A one-dimensional theoretical model has been developed [14], which describes the dependence of the photopyroelectric signal on the optical thermal, and geometric parameters of the sample/pyroelectric system, Fig. 6. Coupled heat diffusion equations were considered for the gas (air), sample, pyroelectric and backing regions, following absorption of the incident light by sample and pyroelectric, and non-radiative optical energy conversion to heat. The average pyroelectric voltage is given by

$$V = <Q>/C \qquad (4)$$

where C is the capacitance per unit area of the thin film, and $<Q>$ is the average charge. For two parallel charged plates of thickness L_p and dielectric constant K, Eq. (4) becomes

$$V(\omega_o) = [\frac{pL_p\theta_p(\omega_o)}{K\varepsilon_o}] \exp(i\omega_o t) \qquad (5)$$

286

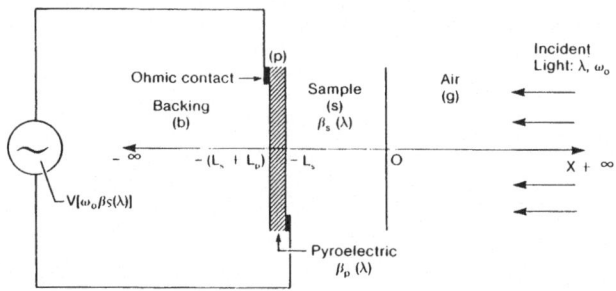

Figure 6:   One-dimensional geometry for the theory of
Photopyroelectric Spectroscopy of Condensed
Phases (PPES).

where

$$\theta_p(\omega_o) = \frac{1}{L_p} \int_{L_p} T_p(\omega_o, x)\,dx \qquad (6)$$

and ε_O is the permittivity constant of vacuum (8.85418 x 10^{-12} Cou-
lomb/Volt-m). $T_p(\omega_o, x)$ is the temperature field in the bulk of the
pyroelectric, a result of heat conduction processes through the radiation absorbing solid. For the geometry of Fig. 6 the field $T_p(\omega_o, x)$
was calculated from the solution of the coupled, one dimensional
thermal transport equations. The theory predicts both amplitude and
phase response saturation at large values of β_s (> 5 x 10^5 m^{-1}). Both
signal channels also become essentially independent of β_s for small
values of that parameter (< 10^3 m^{-1}). These can be considered to be
the useful upper and lower limits of β_s, between which the photopyroelectric effect can be utilized as a spectroscopic technique. Fig. 7
shows the spectral response to a Gaussian with the chopping frequency
as a parameter. A spectral inversion of the absorption amplitude
peak at higher frequencies is observed, due to the interplay in thermal energy contribution between the sample and the pyroelectric. No
inversion occurs in the phase lag, which anticorrelates with the amplitude for f < 20 Hz, and correlates with it for higher frequencies.
These trends are qualitatively entirely compatible with our experimental observations on Ho_2O_3 hydrated powders, Fig. 4. Physicochemical interface processes cannot be studied efficiently using conventional microphone photoacoustics due to the exponential damping
of thermal waves within the bulk of the upper medium [13]. Piezoelectric PAS can monitor such processes as a second order effect, because
the generation and launching of acoustic waves originating from thermal waves is required. PPES is sensitive to the thermal waves within
the lower medium, i.e, it has the advantages of a first order effect.
The CuO experiment, summarized in Fig. 5, substantiates the sensitivity of PPES to interfacial chemical processes and its power as a
tool for kinetic studies which are otherwise difficult to monitor

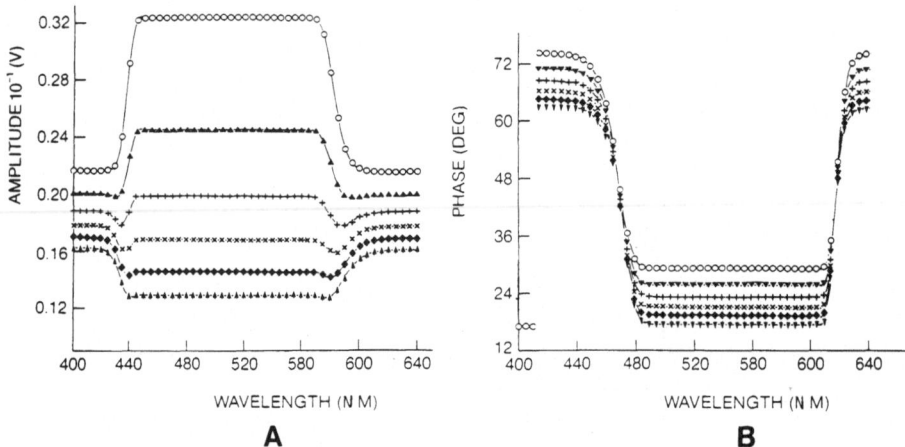

Figure 7: Photopyroelectric spectroscopic response to a Gaussian optical absorption band:

$$\beta_s(\lambda) = \beta_o \exp[-\frac{12400^2}{2}(\frac{1}{\lambda} - \frac{1}{500})^2]$$

as a function of the light modulation frequency: A) Amplitude, and B) Phase; $\beta_o = 1 \times 10^7$ m^{-1}

- • f = 10 Hz
- ▲ f = 15 Hz
- + f = 20 Hz
- x f = 25 Hz
- ♦ f = 30 Hz
- ↑ f = 35 Hz

due to the complex, in-situ requirements of the experimental configurations involved. The reaction rate constant of Eq. (1) can be calculated approximately from the kinetic information of Fig. 5, assuming a simple, first-order kinetic behavior of the form:

$$[CuO]_t = [CuO]_o \exp(-kt) \tag{7}$$

Eq. (7), when fitted to the data of Fig. 5, yeilds the following average value for k:

$$k = (1.55 \pm 0.6) \times 10^{-3} \text{ sec}^{-1}$$

Photopyroelectric Spectroscopy can also provide additional information concerning thermal parameters of materials, as shown in Fig. 3c. This capability appears to be of a degree similar to that of conventional Photoacoustic Spectroscopy. Perhaps the most important advantages of PPES over other photothermal spectroscopies are the

288

almost trivial ease of its implementation and the large amount of freedom allowed the experimenter regarding sample preparation, size, chemical and physical condition and pre-treatment.

In conclusion, a new spectroscopic technique, Photopyroelectric Spectroscopy, has been tested and characterized with respect to its spectral, chemical, and materials research capabilities. It was shown to be potentially more valuable than PAS (microphone and piezoelectric) for condensed phase samples, especially those involving complicated geometries. The main advantages of PPES are its extreme simplicity, sensitivity, in-situ non-destructive probing ability, direct (i.e., first-order) cause-effect signal interpretation, and adaptability to practical restrictions imposed by experimental system requirements. Preliminary theoretical calculations showed that the photopyroelectric voltage is governed by the interplay between optical absorption in the sample and in the pyroelectric transducer itself. They suggest that it is experimentally advantageous to work with optically opaque transducers, whose flat (i.e., photopyroelectrically saturated) spectral response does not interfere with spectral measurements on the sample. Opaqueness can be achieved through coating the pyroelectric surface with metallic thin layers (e.g., nickel), or black absorbing materials. The results of the theoretical considerations help establish photopyroelectric spectroscopy as a valid spectroscopic technique, to be added to the already rich arsenal of photothermal wave spectroscopies, with particularly high promise in the realm of flexible, in-situ nondestructive probing of samples with minimal preparation, and in transient applications in the ultrashort time domain as a piezoelectric or pyroelectric technique, due to the wide frequency response of pyroelectric thin films, such as PVF_2 [15].

References

1. A. Rosencwaig and G. Busse, Appl. Phys. Lett. 36 (1980) 725.
2. T. Sawada, H. Shimizu and S. Oda, Jpn. J. Appl. Phys. 20 (1981), L25.
3. A.C. Tam and C.K.N. Patel, Appl. Opt. 18 (1979), 3348.
4. C.E. Yeack, R.L. Melcher and H.E. Klauser, Appl. Phys. Lett. 41 (1982), 1043.
5. A.C. Tam and H. Coufal, Appl. Phys. Lett. 42 (1983), 33.
6. H. Coufal, IBM Research Report RJ 4023 (45181) (1983).
7. L. Bui, H.J. Shaw and L.T. Zitelli, Electron. Lett. 12 (1976), 393.
8. KYNAR^TM Piezo Film Technical Manual, Pennwalt Corp. (1983), 17.
9. N.C. Fernelius, Appl. Opt. 18 (1979), 1784
10. M.J. Adams, B.C. Beadle and G.F. Kirbright, Analyst 102 (1977), 569.
11. S.I. Yun and H.J. Seo, Technical Digest, 3rd. International

Meeting on Photoacoustic and Photothermal Spectroscopy, Paris, France (1983), Paper 6.2/2.

12. P. Charpentier, F. Lepoutre and L. Bertrand, J. Appl. Phys. 53 (1982), 608.

13. A. Mandelis, Y.C. Teng and B.S.H. Royce, J. Appl. Phys. 50 (1979), 7158.

14. A. Mandelis and M.M. Zver, J. Appl. Phys. (submitted for publication).

15. L. Bui, H.J. Shaw and L.T. Zitelli, Electron. Lett. 12, (1976), 393.

ULTRASONIC FLAW DETECTION IN MODEL CERAMIC SYSTEMS

Arthur J. Stockman and Patrick S. Nicholson

Ceramic Engineering Research Group, McMaster University

Hamilton, Ontario Canada

I   INTRODUCTION

The characterization of defects in high performance ceramics is important when the components are to be used in situations of high stress. The shape, size, and composition of these defects will determine whether or not the ceramic will fail. Defect characterization by scattered ultrasonic signal analysis assumes that information about defect material, size, and shape is contained in the scattered signals (1). However, backscattered signals must be correlated with optical observations of the scatterer which for opaque ceramics means that the samples must be carefully cut and polished to expose the defect. For this reason studies have been initiated of the ultrasonic signals reflected from defects in glass which can be nondestructively examined. The intention is to subsequently extend the work to opaque ceramics.

In previous work (2) on voids, a model was developed which showed that size information can be obtained from the backscattered signals of voids in the focal zone of a focussed ultrasonic transducer operating in the pulse-echo mode. The approach of Archer-Hall and Ali Bashter (3) has been utilized wherein the sound pattern of a continuous wave focussed transducer in the focal zone is approximated as a plane wave. Then the plane wave scattering theory of Ying and Truell (4) for small spherical defects is used to obtain the scattering amplitudes for a large range of frequencies. Finally, the pulse of sound is broken down into its component frequencies and this distribution is multiplied by the scattered amplitudes to obtain the total frequency spectrum expected for a backscattered signal. The frequency distribution is derived from the magnitude spectrum of the signal reflected from a flat surface in the focal plane of the transducer. This technique, which was used by Tittman et al. (5), compensates for the frequency response of the entire system. In the earlier work, the form of the pulse frequency distribution was approximated as Gaussian with good correlation to experimental results.

The accuracy of the model depends on how well this approximation for the frequency distribution matches the experimental distribution. When shock excitation is used to generate the sound from the transducer the frequency distribution depends upon fixed parameters such as transducer resonance, electrical impedance matching, and electrical and mechanical damping. However, in a "tone burst" system a short pulse of radio frequency (r.f.) voltage is applied to the transducer forcing oscillation at that frequency. The frequency and duration of this r.f. burst are controllable within limits of transducer and gating electronics, which means that the center frequency and width of the distribution can be predetermined. In the present work an attempt has been made to model the backscattered signal from voids using a tone burst rather than shock excitation system. The spectra generated are compared with those experimentally obtained by the signal from a tone-burst-excited transducer reflecting from voids in glass.

II THEORY

O'Neil has shown that the sound pressure profile for a monochromatic focussed transducer along the z-axis is (6):

$$p = \frac{-\omega^2 \rho\, s_o}{k}\ \frac{2}{1-A/z}\ \sin\left[\frac{k}{2}\,(\beta-z)\right] e^{\,i\left[\omega t - \frac{k}{2}(\beta+z)\right]}$$

where $\beta = \sqrt{(z-h)^2 + \dfrac{D^2}{4}}$ and $h = A - \sqrt{A^2 - \dfrac{D^2}{4}}$

Here p is the pressure at a distance z along the axis of the transducer. The quantity ρ is the density in the medium. The diameter and radius of curvature of the transducer are D and A respectively. The displacement of the transducer surface at z = 0 is s_0. The emitted sound wave has frequency, f, wavelength, λ, angular frequency $\omega = 2\pi f$ and wavenumber is $k = 2\pi/\lambda$. The angular frequency and wavenumber are related through the longitudinal speed of sound, c_L, where $\omega = c_L k$. One other material parameter needed for the calculation is the transverse speed of sound c_T.

Now the pressure can be approximated as:

$$p \sim \rho\omega^2 S_{z_f}\ \frac{e^{\,i(\omega t - k\Delta z)}}{k}$$

292

where s_{zf} is the displacement at the point of maximum pressure, z_f, and the distance along the z-axis from this point is Δz. In general, s_{zf} is not known but remains constant so that a relative pressure for the longitudinal backscattered wave from a void of radius, a, can be defined as:

$$P_r(k,a) = ka\ e^{-ika\left(\dfrac{z_f}{a}\right)} \sum_{m=0}^{\infty} (2m+1)A_m$$

where A_m are the scattering amplitudes determined from the stress-strain boundary conditions. This can be written as a function of frequency, f, since k is a function of frequency.

If R(f) is the net response of the system then the total frequency spectrum of the scattered wave is $G(f) = R(f)\ P_r(f,a)$. To approximate R(f) consider an ideal tone burst signal as depicted in Figure 1a. It is obtained by gating a continuous wave of frequency f_c = 30 MHz for a duration of t_0 = 161 ns. The magnitude of the Fourier transform of this signal is shown in Figure 1b. Mathematically the response function can be written as:

$$R(f) = \frac{\sin[\pi(f-f_c)t_0]}{\pi(f-f_c)t_0}$$

This form may also be described by specifying the center frequency and the full width at half maximum (FWHM) of the main peak, since the parameter FWHM can be related to the burst duration as:

$$FWHM = \frac{2\alpha}{\pi t_0}$$

where α is a constant determined numerically to be 1.8955. This gives the FWHM of Figure 1b as 7.5 MHz. By using this approximation for R(f), model calculations for G(f) can be performed so that magnitude frequency spectra and time base spectra can be obtained and compared with experiment.

III EXPERIMENTAL

The use of glass provided an easy means of optically identifying and characterising the voids. Glass samples were taken from sections of plate glass in which the voids were either spherical or slightly ellipsoidal. In the analysis only the minor or smaller dimensions have been considered.

The equipment used was a 1-320 MHz signal generator of a continuous r.f. signal and a broad band gated amplifier was used to amplify and gate the signal before sending it to a 25 MHz f/4

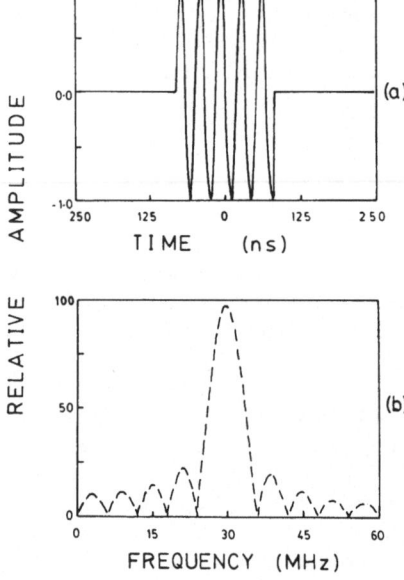

Figure 1. Ideal tone burst pulse a) 30 MHz signal of 161 ns duration and b) magnitude frequency spectrum FWHM = 7.5 MHz.

focussed transducer. The receiving section consisted of a broad band (50 Hz - 400 MHz) preamplifier and a broad band receiver (2 - 200 MHz). The r.f. waveforms from the receiver were converted to digital form by a transient digitizer with a 200 MHz sampling rate. Signals so captured were sent to a minicomputer where they were stored and subsequently windowed to obtain the waveform section of interest. The windowed data were then fast Fourier transformed to obtain the frequency spectra. Quadratic interpolation was then performed on the magnitude spectra to determine the the peak and full width at half maximum. Waveforms presented as figures are reproductions from digitally stored or generated data which has been fitted by a graphic terminal's built-in fitting routine so the peak amplitudes are not necessarily as they would appear on an oscilloscope trace.

The sound was coupled between the transducer and the glass sample using water as the couplant. Voids were located by scanning

294

in the X-Y plane. The height and X-Y position of the transducer were adjusted to obtain a maximum reflected signal intensity and the duration of the r.f. tone burst was adjusted on the gated amplifier and the resulting signal recorded. This final procedure could be repeated as many times as possible to obtain signals of varying pulse duration. Generally, only three settings were useful as pulses of long duration were interfered with by the top and bottom surface reflections. After the defect signals were recorded the transducer was raised to yield the maximum signal intensity for the reflection from the top surface of the glass. Again the r.f. burst duration was adjusted and the signal recorded. It was assumed that the flat reference surface or void was resident in the focal plane of the transducer when the signal was at maximum intensity.

IV RESULTS AND DISCUSSION

Electronic gating seldom produces sharp transitions from on to off or vice versa. Furthermore, the mechanical damping of the transducer does not permit all the frequencies shown in Figure 1b to be present in the sound beam. To accomodate this the approximate spectrum, as shown in Figure 1b, was cut off after the second side lobe for use in the calculations. Figures 2a and 2c show examples of signals reflected from a flat surface in the focal plane of the transducer. The signals do not show the sharp on-off characteristics of Figure 1a. The signal depicted in Figure 2a was generated by setting the signal generator to 30 MHz and burst duration to a minimum. The signal of Figure 2c was generated by a longer burst duration with the same input signal frequency intensity Figure 2b is the magnitude spectrum of Figure 2a and the FWHM of this spectrum is 13 MHz. Similarily, Figure 2d is the magnitude spectrum of Figure 2c. The FWHM of this spectrum is 7.5 MHz, the same as that of Figure 1b even though the number of peaks in Figure 2c is greater than that of Figure 1a. Furthermore, the side lobes of Figure 2d are greatly reduced with respect to Figure 1b. However, the sharpness of the main peak increases as the burst duration is increased, so the main lobe of the spectrum shows some resemblance to that of Figure 1b.

How the differences between the theoretical response function of Figure 1b and the actual response functions of Figures 2b and d can be seen by comparing observed and calculated signals and spectra. Figures 3a and 3c are signals reflected from a void 130 μm by 163 μm. The signal in Figure 3a was obtained for the minimum burst duration as used for Figure 2a, Figure 3c was obtained using the same setting for the burst duration as used for Figure 2c. The spectra of the signals from the void are shown in Figures 3b and 3d respectively. For a spherical void in glass the model predicts the spectra of Figures 4b and 4d and the signals

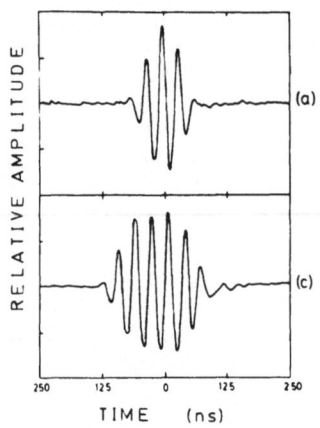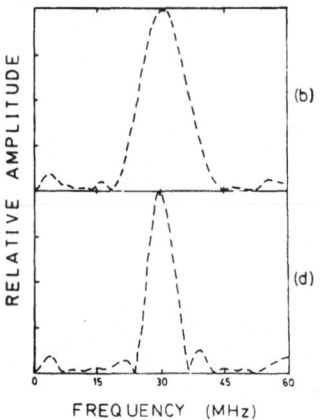

Figure 2. Response function as measured from signals reflected from sample surface with input signal frequency = 30 MHz. a) Digitized signal for minimum pulse duration and b) corresponding magnitude spectrum (FWHM = 13 MHz). c) Digitized signal for longer pulse duration and b) corresponding frequency spectrum (FWHM = 7.5 MHz)

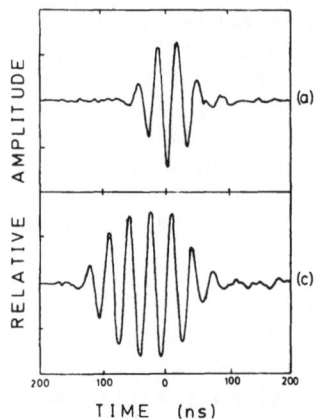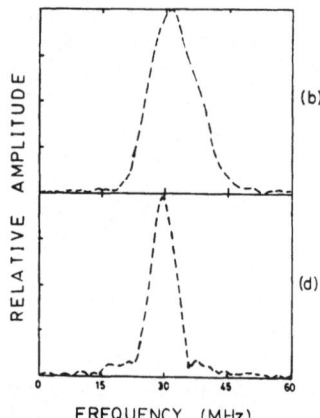

Figure 3. Backscattered signals and their corresponding spectra received from a void of 130 μm x 163 μm. a) Signal for minimum pulse duration and b) its spectrum. c) Signal for longer pulse duration and d) its spectrum.

296

shown in Figures 4a and 4c. The parameters used for the approximate response function were center frequency of 30 MHz and FWHM of 13 MHz for Figures 4a and 4b and 7.5 MHz for Figures 4c and 4d. As expected, the calculated time base signals tended to rise sharply to their maximum value whereas the observed signals did not. The result was that the spectra of the observed signals were broader at the half maximum level than those of the model. However, both model and signal spectra did show asymmetries which reflect the response of the void to different frequency components of the incoming signal.

By performing calculations for voids over a range of sizes it was possible to plot the variation of Frequency at Maximum Amplitude and FWHM of the expected spectra. These are shown in Figures 5 and 6 as dashed lines while the input parameter appropriate to the Figure is represented as a solid line. Figures 5a and 6a were calculated for an input FWHM of 13 MHz, Figures 5b and 6b were calculated for an input FWHM of 7.5 MHz, and Figures 5c and 6c were calculated for an input FWHM of 3.9 MHz. In all cases the input r.f. signal was 30 MHz. The model curve shapes are unusual especially for the short duration bursts. The strong frequency dependence of the scattering amplitudes for wavelengths in the order of 2π times the void radius are responsible for the changes in spectrum shape which are reflected in the measured values. At 30 MHz the wavelength of sound for the longitudinal wave in glass is about 190 µm and the corresponding void diameters are approximately 100 µm.

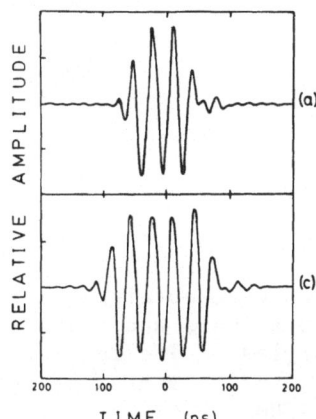 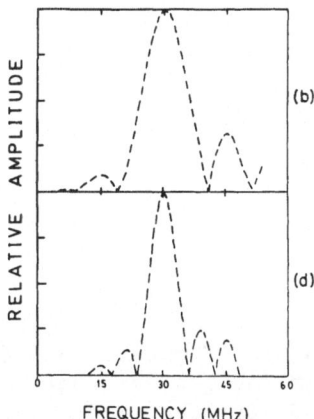

Figure 4. Calculated backscattered signals and spectra from a spherical void of 130 µm diameter. Input FWHM = 13 MHz a) signal and b) magnitude spectrum. Input FWHM = 7.5 MHz c) signal and d) magnitude spectrum.

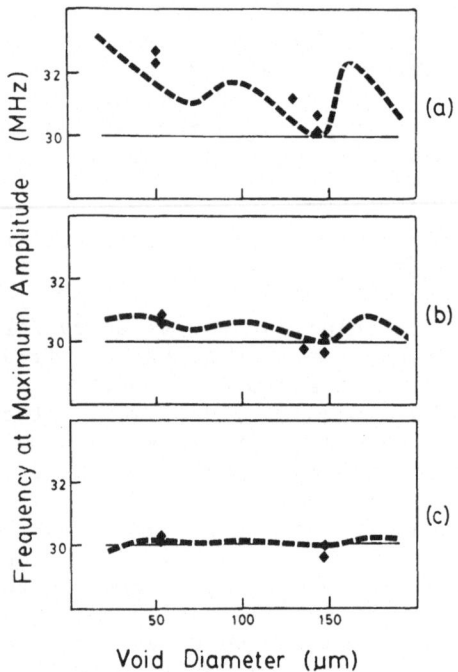

Figure 5. Variation of Frequency at Maximum Amplitude of magnitude spectra with void diameter. Dashed line represents values calculated from the model and triangles represent values determined from experimental data. Applied signal frequency = 30 MHz, input FWHM a) 13 MHz, b) 7.5 MHz, and c) 3.9 MHz.

The same parameter measurements were obtained from the fast Fourier transforms of the received signals and these points are included in Figures 5 and 6 as triangles. The results show that the FWHM and center frequency are shifted to a greater amount than predicted by the model for larger FWHM. This is to be expected because of the differences between the theoretical and observed response functions. What is important in comparing the model and the experimental values is that the points tend to follow the non-linear trend indicated by the model. Furthermore as the burst duration increased, the correspondence of data points with the

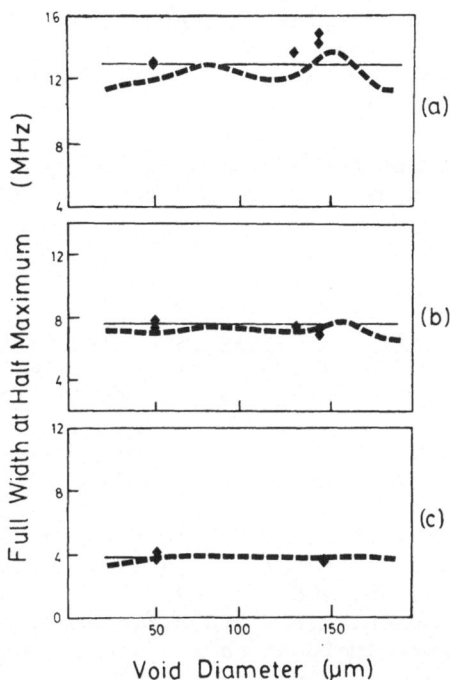

Figure 6. Variation of Full Width at Half Maximum (FWHM) of magnitude spectra with void diameter. Dashed line represents values calculated from the model and triangles represent values determined from experimental data. Applied signal frequency = 30 MHz, input FWHM a) 13 MHz, b) 7.5 MHz, and c) 3.9 MHz.

predicted curves increased. This indicates that the effects of the electronics and mechanical damping are less important. It is possible that the changes in the parameters with varying pulse duration will provide additional information about the character of the defect.

V   CONCLUSIONS

Voids in glass having dimensions as small as 50 μm were examined by ultrasound of principal frequency 30 MHz, which corresponds to a wavelength of about 190 μm in glass. Furthermore, the model developed, based upon scattering of ultrasound by spherical voids in a planar continuous wave field, predicts reasonably well the response of such small voids to a pulse of ultrasound produced by tone burst excitation in the focal zone of a focussed transducer. With modifications to the theoretical response function, the tone burst system which permits varying the frequency content of the signal seen by the void can be used with the model to examine defects and predict their characteristics.

REFERENCES

1 Ultrasonic Spectral Analysis for Nondestructive Evaluation, Fitting, D.W. and Adler, L. (Plenum Press, New York, 1981).
2 Stockman, A and Nicholson, P.S., submitted for publication to J. Am. Cer. Soc.
3 Archer-Hall, J.A. and Ali Bashter, A.I., "The Diffraction Pattern of Large Aperture Bowl Transducers", NDT International, April 1980, 51.
4 Ying, C.F. and Truell, R., "Scattering of a Plane Longitudinal Wave by a Spherical Obstacle in an Isotropically Elastic Solid", J. Appl. Phys., 36, 1086 (1956).
5 Tittman, B.R., Cohen, R.E. and Richardson, J.M., "Scattering of Longitudinal Waves Incident on a Spherical Cavity in a Solid", J. Acoust. Soc. Am., 63, 68 (1978).
6 O'Neil, H.T., "Theory of Focusing Radiators", J. Acoust. Soc. Am., 21, 516 (1949).

THE HYGROSCOPIC BEHAVIOR OF SiO_2 AND $Si_xO_yCl_z$ THIN FILMS ON Si USING ATR FTIR SPECTRA

R. Koba, J. R. Monkowski and R. E. Tressler

226 Steidle Building
The Pennsylvania State University
University Park, PA 16802

ABSTRACT

Attenuated total reflection (ATR) has been used with Fourier transform infrared spectroscopy (FTIR) to study the structure and composition of SiO_2 and $Si_xO_yCl_z$ thin films (thicknesses between 15 and 160 nm). The conjunction of ATR with FTIR resulted in excellent sensitivity in the spectral region between 3800 cm^{-1} and 1500 cm^{-1}. The films were thermally grown on a silicon internal reflection element avoiding any prism-sample discontinuities. The hygroscopic behavior of $Si_xO_yCl_z$ films was determined by monitoring the –OH stretch band at 3650 cm^{-1} as a function of exposure time in a 75°C + 93% RH ambient. An HCl absorption band was identified at 2800 cm^{-1} which increased in intensity proportional to exposure time. Both water-created bands saturated with time. The identity of the bands was verified by deuteration shift and by comparison with the spectrum of SiO_2. After water saturation, spectra were collected at various etch depths which proved that water absorption occured over the entire oxide thickness. ATR-FTIR spectra were compared with secondary ion mass spectrometry (SIMS) depth profiles of similar oxy-chlorides to give a more quantitative idea of the total H content as well as another view of the H and Cl profiles.

INTRODUCTION

Chlorinated silicon oxide films are routinely used in the production of microelectronic devices. These films are usually thermally grown in HCl/O_2 or Cl_2/O_2 ambients at temperatures between 1000° and 1100°C. They have been studied extensively,

301

especially with regard to the distribution and effects of chlorine in the SiO_2 matrix [1,2]. The properties of these films were found to be a function of the ratio of oxygen activity to chlorine activity inside the film. Monkowski et al. [3] reported that chlorine tended to accumulate at the oxide/silicon interface in the form of a chlorosiloxane phase when the local $a(O_2)/a(Cl_2)$ ratio inside the film was reduced to a value of 16.3 or lower. They published calibrated secondary ion mass spectrometry (SIMS) depth profiles of chlorine in various oxides and found that the Cl distribution became uniform as the $a(O_2)/a(Cl_2)$ ratio in the furnace ambient approached 16.3. In the same study, hydrogen depth profiles were determined by nuclear reaction analysis and SIMS. The hydrogen profiles paralleled the chlorine profiles, and mass spectroscopy indicated evolution of HCl in chlorinated films exposed to high humidity. The hygroscopic nature of the films reinforced the theory that the chlorine-rich regions were a form of chlorosiloxane.

The objective of this study was to examine the hygroscopic behavior of a uniformly chlorinated thermal oxide. Two complimentary techniques were employed: secondary ion mass spectroscopy (SIMS) and attenuated total reflectance (ATR) as used in a Fourier transform infrared spectrometer (FTIR). SIMS provided good quantitative analysis for hydrogen, while ATR permitted the qualitative identification of the hydrogen-containing species. In addition, the SIMS was used to independently calibrate the ATR for semi-quantitative analysis. Previous ATR research on SiO_2 thin films has not included any independent calibrations [4,5]. Instead, published molar extinction coefficients have been used to calculate the concentrations of the water-related modes. This practice is inappropropriate because 1) infrared extinction coefficients are usually empirical, i.e. instrument and operator dependent [6], and 2) most molar extinction coefficients are based upon gravimetric measurements of water loss [7]. To examine oxide films less than 200 nm thick, ATR needed to be used in conjunction with a quantitative surface analysis technique in order to more accurately calculate the concentration of hydrogen-related species [8].

PROCEDURE

All films were thermally grown at 1100°C in a polysilicon tube furnace to minimize uncontrolled water contamination. Dry oxides were grown using "zero grade" O_2 at 1 SLPM. Chlorinated oxides were grown in an atmosphere of 0.06 SLPM of Cl_2 and 0.94 SLPM of O_2. Silicon substrates were cleaned according to the standard RCA procedure prior to oxidation. Film thicknesses were measured to within ± 1 nm with a Geartner Ellipsometer assuming a film index of 1.46. All films were stored in a vacuum dessicator until use.

Films were exposed to a high humiditiy ambient to study the absorption of water as a function of time. Oxides grown in 6% Cl_2/O_2 were placed in a Blue M controlled temperature and humidity chamber at 75 $\pm$ 3°C and 93 $\pm$ 5% relative humidity for up to 155 hours. Samples were removed at various times for analysis by SIMS or ATR. To compare the ATR spectrum of "wet" SiO_2 to "wet" $Si_xO_yCl_z$, a 100% O_2 oxide was also exposed to the same ambient for greater than 100 hours, but the resulting water-related modes were barely discernible. In order to incorporate more water, a SiO_2 film was exposed to steam at 450°C for 3 hours. To identify the hydrogen-related IR absorption modes, a chlorinated film was placed above a bath of D_2O at ~100° C and ~90% R.H. for 30 hours. This deuteration process shifted the absorption freqencies by a factor of 0.737.

ATR-FTIR Analysis

Oxides were directly grown onto an intrinsic ($>$ 500 Ω-cm) silicon internal reflection element (IRE). The IRE (Harrick Scientific EE2321) was shaped as a trapezoid with θ = 30°, a thickness of 0.215 cm and an aperture-aperture distance of 5.10 cm, resulting in a total of N = 41 internal reflections. After oxidation, the film thickness was measured at several locations on both sides of the IRE. The oxide grown on entrance and exit apertures was carefully stripped with HF.

Spectra were collected on a Digilab FTS-15E Fourier transform infrared spectrometer fitted with a Foxboro Wilks Model 9 ATR attachment. The spectral range was between 3800 and 1500 cm^{-1} at a resolution of 8 cm^{-1}. Both non-polarized and polarized infrared radiation were used to collect spectra. A mid-IR wire grid polarizer (Perkin-Elmer model 186-0212) was placed in the exit aperture of the ATR attachment with the E field directed perpendicular to the plane of incidence. ATR power spectra were digitized and stored on magnetic tape. This procedure permitted great flexibility since the power spectrum of a "wet" sample could be ratioed to that of the bare Si prism or, (to increase sensitivity to water related modes) to the power spectrum of the same oxide before humidity exposure.

To perform quantitative analysis with ATR spectra, polarized IR radiation must be employed. Absorption mode "valleys" on the reflection spectra must be analyzed according to Fresnel's equations since Lambert's law does not apply to reflection spectroscopy [9]. The appropriate Fresnel equations for perpendicularly polarized reflected intensity in a three media system [10] were coded into a computer program. This program was written to convert percent attenuation below the baseline into the appropriate absorption coefficient α by iterative solution of Fresnel's equations. Once α was known, then Beer's law could be applied:

$$\alpha = \sigma (C - C_o) \tag{1}$$

where σ is the absorption cross section in units of cm^2 (it may be thought of as an atomic extinction coefficient), C is the concentration of the absorbing species of interest in units of $\#/cm^3$, C_o is the instrumental detection limit ($\#/cm^3$), and α is in units of cm^{-1}.

Empirical polarized spectra were always ratioed to the "dry" oxide to increase sensitivity. (No water modes were detected in the "dry" oxide when ratioed to the Si IRE.) Two approaches were used to determine the ratio I/I_o at an absorption mode: 1) the traditional method of measuring the distance of the valley minimum beneath the interpolated baseline, and 2) dividing the integrated area beneath the valley by the total area defined by the endpoints of the interpolated baseline. The latter method was essentially the averaging of peak depth ratios at all points along the absorption valley. Each method resulted in different values of α and σ [8], but yielded approximately the same values of concentration C (because of separate calibrations).

SIMS

A secondary ion mass spectrometer was calibrated for quantitative analysis of hydrogen using a SiO_2 film implanted with H^+ ions. A detailed discussion of the calibration procedure and operation of the SIMS appears elsewhere [11]. Only the 6% Cl_2 /O_2 samples exposed to the high humidity were analyzed by SIMS. Samples were immediately transfered from the humidity chamber into the SIMS analysis chamber. The base pressure of the SIMS was ~5 x 10^{-9} torr. The time interval between sample insertion and analysis was approximately 3 hours. Quadropole $H^+/^{30}Si^+$ signal ratios were collected at every 3 nm of sputter depth. Appropriate calibration relations were used to convert the signal ratios into hydrogen concentrations or hydrogen areal densities. Integral-mean concentrations were defined as the ratio of areal density to film thickness.

RESULTS & DISCUSSION

Detailed Spectra and Total Hydrogen Concentrations

Non-polarized ATR spectra of SiO_2 and $Si_xO_yCl_z$ after water absorption appear in Fig. 1. The SiO_2 film was exposed to steam

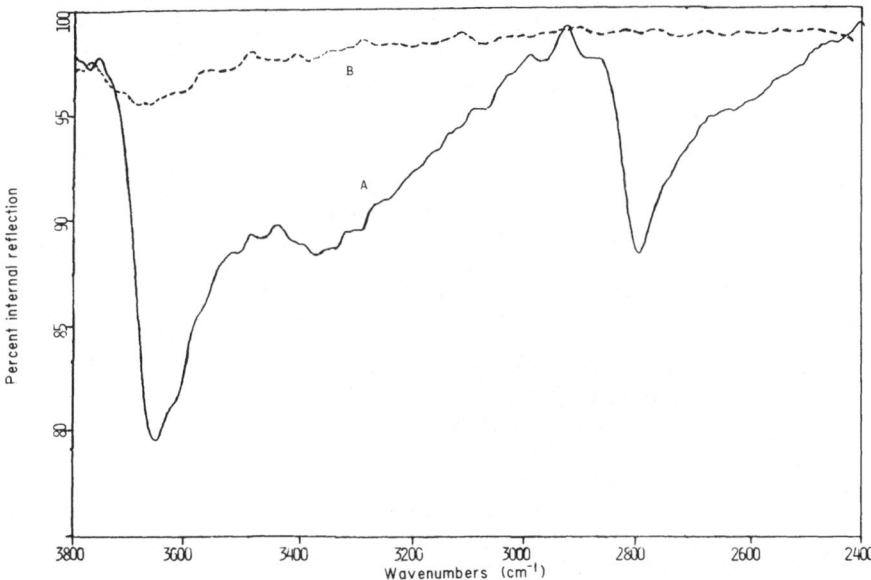

Fig. 1.   ATR infrared spectra: a) 152 nm thick 6% Cl_2/O_2
oxide exposed to 75°C and 93% relative humidity for
4530 min. b) 156 nm thick 100% O_2 oxide exposed to
steam at 450°C for 3 hours.

at 450°C for 3 hours and the 6% Cl_2/O_2 film was exposed to 75°C &
93% R.H. for 75 hours. Note the absence of the 2800 cm^{-1} HCl
mode in SiO_2. The -OH stretch of hydroxyls bonded to silicon
occurs at 3650 cm^{-1}, while the broad band centered at 3350 cm^{-1}
is attributed to hydrogen-bonded silanol hydroxyls plus modes of
molecular water (viz. the HOH symmetric stretch, the
antisymmetric stretch and the overtone of the symmetric bend.)
It is also thought that the 3650 cm^{-1} mode may be partially due
to molecular water [12], but this issue will be discussed later.
By ratioing spectra to the dry film before humidity exposure, the
SiO_2 multiphonon lattice absorption at 1620 cm^{-1} could be
subtracted out, leaving behind the fundamental HOH bend mode at
1620 cm^{-1}. This mode provides unambiguous proof of the presence
of molecular water in silica [13]. Unfortunately, due to its
small intensity, the measurement of its I/I_0 had the greatest
amount of uncertainty (as compared to the other modes.) An Si-H
stretch at 2250 cm^{-1} was not observed in the spectrum of any film
("wet" or "dry") examined in this study.

The kinetics of water vapor absorption by a 6% Cl_2/O_2 oxide
at 75°C and 93% R.H. were tracked by both SIMS and ATR. (The
thickness of the films studied by SIMS was 120 nm while the film
grown on the Si IRE was 152nm.) These two techniques were

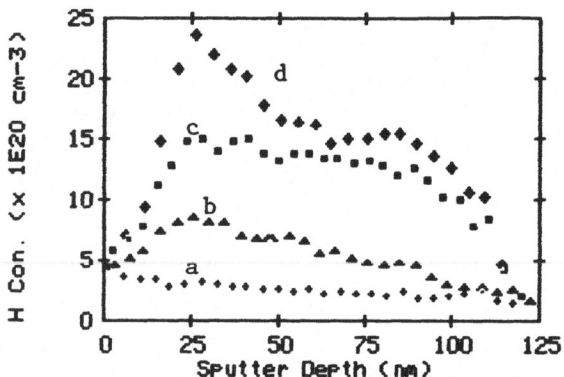

Fig. 2. SIMS hydrogen concentration profiles in 6% Cl_2/O_2 oxide
exposed to 75°C and 93% R.H. for a) 0, b) 100, c)
1020 and d) 2820 minutes.

complimentary since SIMS could measure the total H content and
ATR could discriminate the three major chemical forms of
hydrogen. Fig. 2 contains hydrogen concentration profiles of
films after various exposure times. Fig. 3 lists the
integral-mean hydrogen concentrations versus time and includes
the least squares regression equation:

$$C_H(t) = 1.84 \times 10^{20} + 3.73 \times 10^{19} t^{.46} \qquad (2)$$

where C_H is in atoms/cm^3 and t is in minutes. The SIMS profiles
of chlorine were uniform over depth and did not change in shape
as water was absorbed. Quantitative analysis of the HCl, Si-OH
and H_2O modes was performed using perpendicularly polarized
spectra ratioed to the dry oxide. Table I lists the absorption
coefficients calculated from valley depths while Table II lists
the α values calculated from absorption areas.

Forms of Hydrogen Incorporation

In order to apportionate the hydrogen concentrations among
Si-OH, HCl and H_2O, we were forced to make a priori assumptions
about the mechanism of water incorporation. According to M.D.
Monkowski [14] chlorinated oxides are hygroscopic due to the
presence of unstable Si-Cl groups which are readily displaced by
water. The net reaction is:

$$H_2O + \geqq Si-Cl \rightarrow \; \geqq Si-OH + HCl \qquad (3)$$

(The Si-Cl species could not be tracked by infrared spectroscopy
since its absorption freqency overlaps the fundamental modes of
silica. The silica modes are extremely intense and make

306

Table I Absorption Coefficients (cm^{-1}) Calculated using Valley Depths Beneath the Baseline (* refers to interpolated values)

Time (min)	C_H ($\times 10^{20}/cm^3$) ($\pm 1 \times 10^{20}$)	α(Si-OH) ($\pm 15\%$)	α(HCl) ($\pm 15\%$)	α(H_2O) ($\pm 50\%$)
0	1.85	0	0	0
35	4.34	0	0	0
100	5.33	0	0	0
120	5.96	0	0	0
420	8.94*	50.8	36.4	0
600	10.1*	77.0	50.8	0
1020	12.8	116*	61.0*	29*
1440	14.5	146*	71.4*	42*
1530	14.3*	208	73.9	43
2820	17.2	224*	95.9*	43*
3420	19.5*	244	106	43
4530	21.8*	256	114	52

Table II Absorption Coefficients (cm^{-1}) Calculated using Integrated Areas (* refers to interpolated values)

Time (min)	C ($\times 10^{20}/cm^3$) ($\pm 1 \times 10^{20}$)	α(Si-OH) ($\pm 15\%$)	α(HCl) ($\pm 15\%$)	α(H_2O) ($\pm 50\%$)
0	1.85	0	0	0
35	4.34	0	0	0
100	5.33	0	0	0
120	5.96	0	0	0
420	8.94*	27.6	15.1	0
600	10.1*	41.7	25.8	0
1020	12.8	63.3*	28.4*	4.7*
1440	14.5	70.6*	33.6*	7.7*
1530	14.3*	85.2	35.2	8.7
2820	17.2	86.5*	45.9*	16*
3420	19.5*	99.2	50.4	17
4530	21.8*	112	54.6	27

background subtraction impracticable.) It was assumed that the ATR detection limit of H_2O in the film was zero, so that up until H_2O was observed in the IR, all hydrogen in the films was equally divided between Si-OH and HCl. Since IR spectra collected at 420 and 620 minutes included significant Si-OH and HCl bands, but lacked the 1620 cm^{-1} H_2O band, these two spectra were used to calculate the absorption cross section for Si-OH (3650 cm^{-1}) and

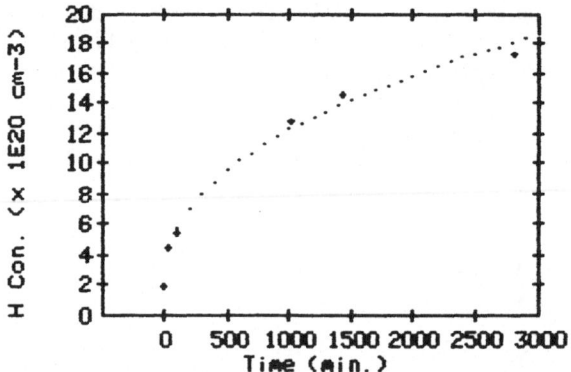

Fig. 3. SIMS integral−mean hydrogen concentrations after
exposure to 75°C and 93% R.H. as a function of
time.

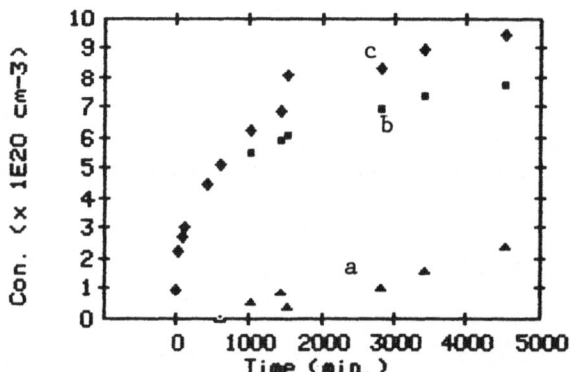

Fig. 4. Concentrations of various species as determined by ATR:
a) H_2O, b) HCl and c) SiOH.

HCl (2800 cm^{-1}).   Assuming a detection limit of 3 x 10^{20}/cm^3 for
HCl and Si−OH, we find σ(Si−OH) = 3.6 (± 1.5) x 10^{-19} cm^2 and σ
(HCl) = 2.5 (± 1.0) x 10^{-19} cm^2 when using the valley depth
method. Using integrated areas, σ(Si−OH) = 2.0 (± 0.8) x 10^{-19}
cm^2 and σ(HCl) = 1.1 (± 0.6) x 10^{-19} cm^2.

Once molecular water was known to be in the oxide, a
decision had to be made as to how to analyze the 3650 cm^{-1} band.
The appearance of H_2O definitely affected this band since the
(3650 cm^{-1})/ α(2800 cm^{-1}) ratio significantly increased at the
appearance of H_2O. Many authors state that the 3650 cm^{-1} mode is
the sum of the −OH of silanols and molecular water [15]. Others
feel the 3650 cm^{-1} is exclusively the result of silanols [5]. In

308

Table III The Three Fundamental Vibrational Mode
Frequencies of H_2O (in cm^{-1})

Phase	Symmetric Stretch $A_1 \nu_1$	Symmetric Bend $A_1 \nu_1$	Antisymmetric Stretch $B_1 \nu_1$
Vapor	3657	1595	3756
Liquid	3219	1627	3445

this study, it was thought best to attribute the 3650 cm^{-1}
exlusively to silanol groups, since the 3650 cm^{-1} mode is more
characteristic of water vapor, not liquid water (see Table III.)
Since the HOH bend occurs at 1620 cm^{-1}, the molecular water in
the oxide is in the condensed state (polymerized either with
itself or with silanols.) Therefore, the symmetric and
antisymmetric stretch modes lie in the band between 3500 and 3000
cm^{-1} (Fig. 1.) (Historically, the attribution of 3650 cm^{-1} to
molecular water in silica may have been due to the low resolution
of the grating spectrometers. The ATR-FTIR instrument provides
superior resolution of these fundamental modes.) The creation of
additional silanol groups by molecular water is a well known
phenomenon in SiO [16]:

$$H_2O + (\equiv Si-O-Si\equiv) \rightarrow 2 (\equiv Si-OH) \tag{4}$$

The concentrations of HCl and Si-OH were calculated from the
aforementioed absorption cross-sections at various exposure
times. The H_2O concentration was calculated as half the
concentration of hydrogen remaining after subtraction of the
Si-OH and HCl concentrations. The results appear in Fig. 4,
based upon the averaged results of Tables I & II. Admittedly,
the conclusion about the identity of the 3650 cm^{-1} mode was based
upon qualitative data. If the experimental uncertainty in the
values of the absorption coefficient at 1620 cm^{-1} were less, then
a statistically definitive conclusion about the 3650 cm^{-1} mode
could be reached. The shape of the concentration vs time plots in
Fig.4 seem to indicate that water absorption is approximately
diffusion limited.

The nonpolarized ATR spectrum of 155 hours was similar to
the 75.5 hour spectrum indicating that the 6% Cl_2/O_2 film was
almost saturated with water. The film was then etched to various
thicknesses and spectra were collected. (By necessity, the
resulting reflection spectra were ratioed to the bare IRE, so the
molecular water bend mode at 1620 cm^{-1} could no longer be
isolated.) The HCl mode at 2800 cm^{-1} is shown after various
times and at various thicknesses in Fig. 5. The -OH modes are

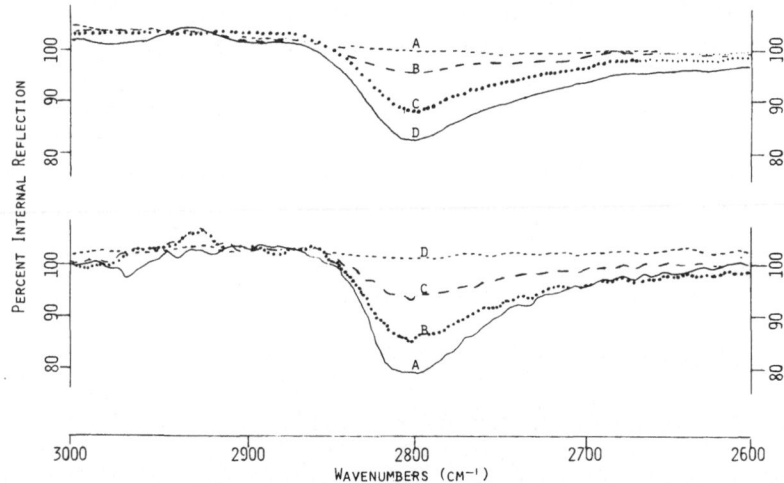

Fig. 5. HCl absorption band in 6% Cl_2/O_2 oxide. Top: exposure
times at 75°C, 93% R.H. a) 0, b) 420, c) 1530 and d)
4530 minutes. Bottom: after 9300 minutes, various
oxide thicknesses: a) 151.5 nm, b) 77.5 nm, c) 41.3 nm
and d) 5.5 nm.

shown at various thicknesses in Fig. 6. Assuming the evanescent
wave extended outward to the film surface, it is concluded that
the molecular water was concentrated at the outer half, but the
HCl was distributed uniformly over the bulk. SIMS profiles of
chlorine in "wet" and "dry" films were both uniform, indicating
no HCl outflux. By extrapolation to 155 hours, the "saturated"
HCl content is estimated at ~8 x 10^{20}/cm^3. In the SIMS Cl
profiles published by Monkowski et.al. [3], the chlorine levels
in 5% Cl_2/O_2 oxides were reported as ~2 x 10^{21}/cm^3. (This level
may be slightly lower than found in the films studied here.)
Either there was an absolute drop in Cl concentration which the
SIMS was too imprecise to detect or the film saturated at 75 C
93% R.H. with only ~40% of the available Si-Cl sites reacted.
The former hypothesis is less plausible since the out diffusion
of chlorine would probably be indicated by a significant gradient
in the Cl depth profile. Incomplete reaction of Si-Cl groups was
probably limited by thermodynamics since the Si-OH concentration
level may have risen high enough to drive reaction (3) back to
the left.

Model for Water Incorporation

From these results, the following model for water
incorportion in a uniformly chlorinated oxide (at 75°C) is
proposed. The atomic structure of an originally "dry" film is

310

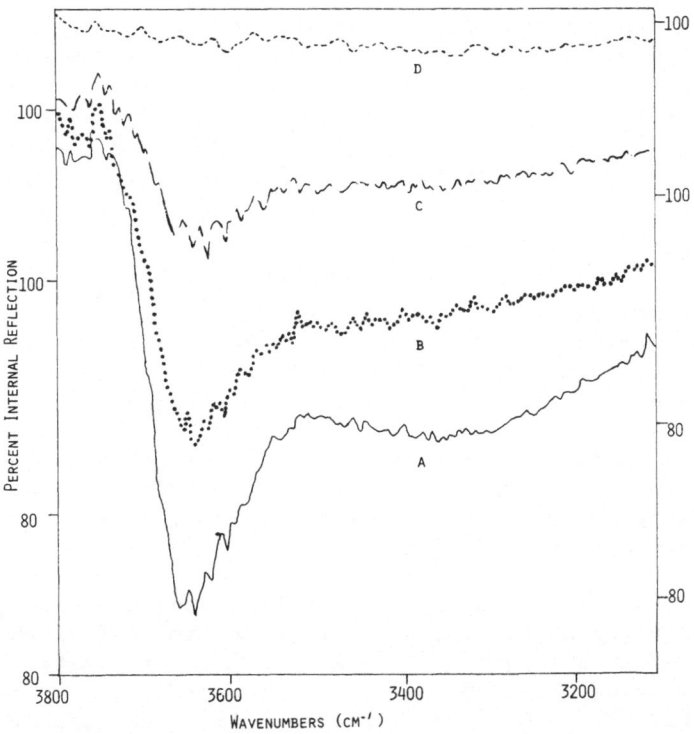

Fig. 6.   6% Cl_2/O_2 oxide after 9300 minutes exposure, various oxide thicknesses: a) 151.5 nm, b) 77.5 nm, c) 41.3 nm and d) 5.5 nm.

quite open, due to numerous non-bridging silica tetrahedra created by terminal Si-Cl groups. This discontinuous network probably has a higher solubility and diffusivity for water molecules than does pure silica. Si-Cl groups are weakly bonded and are thermodynamically unstable in the presence of water; they constitute a uniform distribution of hydrogen traps. The outer surface is covered by a carpet of silanol groups and water vapor molecules are adsorbed onto this carpet by hydrogen bonding [16,17]. Water molecules diffuse into the bulk until dissociating at Si-Cl sites. Because of the openness of the network, some fraction of H_2O is always able to reach the oxide/Si interface. There are approximately two stages of water incorportion. In the initial stage, there are numerous Si-Cl groups which limit the "residence time" of molecular water. Water reaches all points of the film containing Si-Cl bonds. In the second stage, depletion of Si-Cl sites near the surface permits a local build-up of H_2O. The increase in H_2O concentration makes feasible the breaking of many strongly bonded

311

Si-O-Si bridges (reaction 4). One may imagine a region of concentrated molecular water which advances from the surface of the film into the bulk. This region has achieved a steady state condition with regard to reactions 3 and 4. Therefore, the overall kinetics of water incorporation seem to be limited by the diffusion of molecular water.

CONCLUSIONS

Uniformly chlorinated silicon oxide films are hygroscopic due to the presence of Si-Cl bonds. Molecular water diffuses in and reacts to form Si-OH and HCl. There is sufficient molecular water incorporated to dissociate Si-O-Si bridges and create additional silanol groups. After 155 hours in 75°C at 93% R.H., the film is approximately saturated with less than half of the total chlorine atoms reacted to form HCl. Hydrogen absorption is approximately parabolic with respect to time, indicating that the kinetics are diffusion limited.

Perpendicularly polarized ATR spectra have been independently calibrated using SIMS. The ATR detection limit of Si-OH and HCl is estimated as $3 \times 10^{20}/cm^3$. When the intensity ratio I/I_0 is calculated using the traditional baseline method, the absorption cross-section of Si-OH is $3.6 (\pm 1.5) \times 10^{-19}$ cm^2 and $\sigma(HCl)$ is $2.5 (\pm 1.0) \times 10^{-19}$ cm^2.

ACKNOWLEDGMENTS

The authors thank Mr. George Arnold of Sandia National Labratories for hydrogen implantation of the SiO_2, Dr. Cheryl Houser for technical assistance with the SIMS, and Mr. Gary Walters for technical assistance with the FTIR.

REFERENCES

1. J. Monkowski, Solid State Technology, 58 (July 1979), 113 (August 1979).
2. H. L. Tsai, S. R. Butler, D. B. Williams, H. W. Kraner and K. W. Jones, J. Electrochem. Soc. 131,2:411 (1984).
3. J. R. Monkowski, M. D. Monkowski, I. S. T. Tsong and J. Stach, in: "Silicon Processing, ASTM STP 804," D.C. Gupta, ed., American Society for Testing and Materials (1983)
4. K. H. Beckmann and N. J. Harrick, J. Electrochem. Soc. 118, 4:614 (1971).
5. A. Harstein and D. R. Young, Appl. Phys. Lett. 38, 8:631 (1981).

6. H. H. Willard, L. L. Merritt, J.A. Dean and F.A. Settle, "Instrumental Methods of Analysis," D. Van Nostrand, New York (1981) 209.

7. W. A. Pliskin, J. Vac. Sci. Technol. 14,5:1064 (1977).

8. W. A. Lanford and M. J. Rand, J. Appl. Phys. 49, 4:2473 (1978)

9. N. J. Harrick, "Internal Reflection Spectroscopy," Wiley Interscience, New York (1967).

10. W. N. Hansen, J. Opt. Soc. Am. 58, 3:380 (1968).

11. R. Koba, C. G. Pantano, R. E. Tressler and J. R. Monkowski, to be published in Appl. of Surface Science (1985).

12. W. A. Pliskin and H. S. Lehman, J. Electrochem. Soc. 112, 10:1013 (1965).

13. F. M. Ernsberger, J. Am. Ceramic Soc. 60, 1-2:91 (1977).

14. M. D. Monkowski, personal communication (1982).

15. R. F. Bartholomew, J. Non-Crystalline Solids 56:331 (1983).

16. R. H. Doremus, in: "Reactivity of Solids," J. W. Mitchell, R.C. DeVries, R.W. Roberts and P. Cannon, eds., Wiley, New York, (1969).

17. V. A. Bershtein and V. V. Nikitin, Izvestiya Akademii Nauk SSSR, Neorganicheskie Materialy (USSR), 10, 2:316 (1974).

A NEW FT-IR TECHNIQUE FOR THE SITU STUDY OF THE MECHANISM OF SOLID STATE MATERIALS PREPARATIONS

William R. Moser, John C. Chaing and Jack E. Cnossen

Dept. of Chem. Eng., Worcester Polytechnic Institute
Worcester, Massachesetts   01609

INTRODUCTION

The study of the mechanisms of hydrogel preparations of solid state inorganics by vibrational and electronic absorption spectroscopic techniques such as infrared and UV/VIS methods is severely limited due to the fact that solids are formed during the course of the investigation preventing the passage of the optical beam through the solution. We wish to report the development of a general technique [1,2] for the study of such systems based on a recently developed Cylindrical Internal Reflectance (CIR) spectroscopic method [3,4]. This technique utilizes directly stirred, high pressure autoclaves which were fabricated containing embedded cylindrical internal reflectance crystals. This configuration ensured that the reacting solution of inorganic ions or suspended solids was continuously, well mixed and in contact with the analyzing crystal at all times.

This new technique reported here permits the study of the preparation of solid state inorganics and materials such as zeolites [5], and high surface area materials resulting from hydrogel synthesis [6,7]. The method, in principle, has the capability for the study of solid state inorganic synthesis once very high temperature reactors are fabricated.

SPECTROSCOPIC METHOD

The cylindrical internal reflectance (CIR) spectroscopic method was invented by Wilks and Sting [3,4], and an analytical system for the infrared examination of solutions is commercially available [8]. We have fabricated several high pressure reactors of different configurations and designs with CIR crystals built directly into their pressure walls, and all of these use the spectroscopic bench developed earlier [8]. An advanced Nicolet 60SX FTIR spectrometer was used in these studies. A typical reactor which was fabricated and used for several hundred high pressure studies is shown in Figure 1. This autoclave uses the direct drive high pressure head of a Parr Mini Reactor [9], and a bottom which was fabricated [10] to contain a CIR crystal with a high pressure closure. The internal volume of the reactor may be adjusted between 40 to 150 ml through the use of teflon inserts. A wide range of catalytic and chemical studies [11] have demonstrated that the CIR reactors offer several important advantages for the in situ study of reactions at high pressures and temperatures. The configuration in Figure 1 was tested at 126 MPa (125

atmoshperes) at 150°C. They have been routinely used at 10.3
MPa (102 atm.) at 150°C for catalysis studies.

Figure 1   Stirred high pressure reactor equipped with an
embedded cylindrical internal reflection crystal

The principle of the CIR spectroscopic technique is
fundamentally an Internal Reflection Spectroscopic Method (IRS)
or Attenuated Total Reflectance (ATR) [12]. The embedded
crystal in Figure 1 is a cylindrical rod pointed on each end at
usually a 45° angle. The crystal rod used in this case was
3.25 in. long and 0.25 in. in diameter. The crystal may be
fabricated from several high performance ceramics like ZnSe,
ZnS, Si, saphire or others with desirable strengths and optical
transmission properties. The reactors are mounted on an

optical bench where the infrared beam is focused perpendicular onto the 45° conical end. The beam makes 10 reflections before passing out the other end of the crystal where it is directed to the detector. At the point of each internal reflection in the crystal a 1.0 to 1.2 um penetration occurs into the solution or material to be analyzed on the outside of the crystal. Thus, the effective pathlength of the crystal used in Figure 1 was 10-12 um which may be increased through the use of other crystal geometries.

REACTOR CAPABILITIES

The outstanding capabilities of the high pressure reactors with embedded CIR crystals was previously described [11] and will be summarized here. They are especially important to inorganic hydrogel syntheses since this CIR method is the only infrared or UV/VIS technique which permits the infrared analysis of a well stirred solution containing a suspended solid. The pathlength does not change when high pressures and modest temperatures are utilized, which is a severe deficiency of transmission infrared methods. Quantification of the solution components is easy and highly reliable. Since it is a short pathlength method, the infrared spectra of aqueous solutions are easily measured. Subtraction of the water spectrum permits the ready identification of the solution and gel intermediates. It is important to recognize that the suspended crystalline solids are not seen by the infrared beam.

318

However, crystalline solids have been packed into a low volume reactor similar to the device in Figure 1 which yielded high quality spectra of the solid. The most important advantage for mechanistic studies is that the technique is a truly in situ method where the reacting solution, and even a dissolved gas is at constant equilibrium with respect to the analyzing crystal.

HIGH PRESSURE IN SITU STUDIES

To illustrate the capabilities of the CIR reactors, as seen in Figure 1, the mid-range infrared spectrum of an in situ catalytic study is shown in Figure 2. The infrared spectrum shown in the figure was taken under autogeneous conditions of 1200 psi (8.27 MPa) at $150^{\circ}C$. The reaction studied in Curve A of Figure 2 is the cobalt iodide catalyzed reaction of methanol and carbon monoxide to produce methyl acetate. Curve B in Figure 2 used the same components except that ruthenium trichloride trihydrate was added as a promoter.

The important features of this reaction are to note that the solvent is pure methanol which is strongly absorbing in the IR and does not totally absorb at any part of this spectra; quantitative IR data in methanol solvents using traditional transmission spectroscopic techniques is normally quite difficult to obtain due to its strong absorbance. The CIR method permits the easy and reliable quantification of the rate of methyl acetate formation by integration of the band at

1735 cm^{-1} without solvent band interference. Furthermore, the
infrared bands for the catalytically important intermediate
[Ru(CO)$_3$(I)$_3$]$^{-1}$ were easily identified at 2035 cm^{-1} and 2104
cm^{-1} while the reaction mixture was stirred at 150°C under 1200
psi of carbon monoxide pressure.

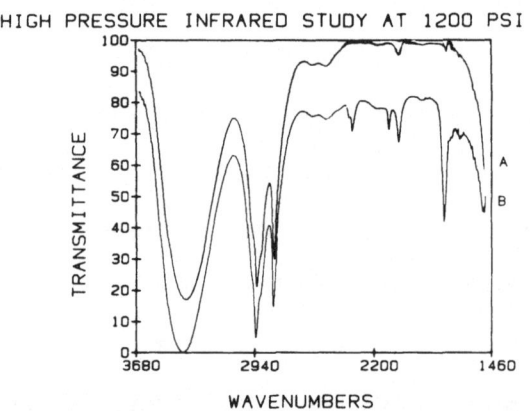

Figure 2   Reaction of methanol and carbon monoxide at 1200
(8.27 MPa) and 150°C after 250 minutes reaction
time producing methyl acetate (1735 cm^{-1}) using
cobalt iodide catalyst (Curve A) and cobalt
iodideruthenium trichloride catalyst (Curve B).

A variety of such studies [1,2,11] showed that reactions
at high pressures and temperatures could be readily studied
where the equipment required was rapidly and reliably assembled
and required only 5 to 20 minutes to record high quality

spectra. Depending on the absorbance of the infrared bands of interest, the time of high quality spectra measurement may be reduced to as little as 30 seconds for the stronger bands.

AUTOGENEOUS STUDIES IN SOLID STATE INORGANIC SYNTHESIS

The high pressure reactor equipped with a crystal of zinc sulfide embedded in the reactor shown in Figure 1 was used for the synthesis of two classes of solid state inorganics.

The synthesis of aluminophosphates was studied in a well stirred aqueous solution at a reaction temperature of 154°C reaching an autogeneous pressure of 85 psi (0.591 KPa). The synthesis of this class of materials with the trivial name of ALPO were described previously in the literature [13]. Our studies utilized a suspension of solid aluminum hydroxide in an aqueous solution of phosphoric acid and the organic template, which was reported [13] to form ALPO-5, triethanol amine. The time lapse infrared spectra recorded in Figures 3 and 4 show the mid infrared region of the autogeneous reaction using 500 scans per spectra (ca. 4 min total measurement time) at 4 cm^{-1} spectral resolution. Although the details of our investigation are reported elsewhere a few important aspects of the data will be described here as a means to illustrate only a few aspects of the CIR technique.

An examination of the 1375 cm^{-1} region of the spectra in
Curves A to E shows that no phosphoric acid ester derivative
with triethanol amine were formed in this reaction until the
reaction temperature reached 152°C. Curves F-I demonstrate a
band at 1377 cm^{-1} due to an organic phosphoric acid ester [15].
At the end of this reaction, the solid aluminophosphate was
isolated and calcined at 600°C in air. This reaction was

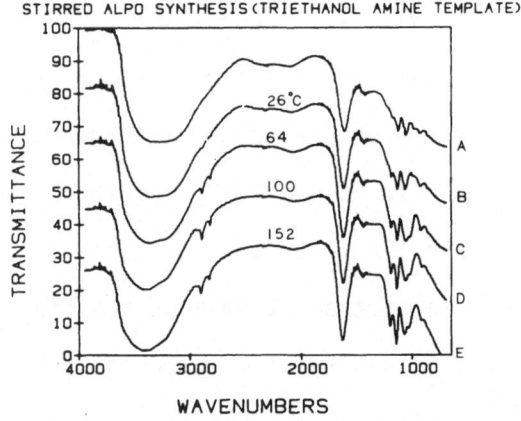

STIRRED ALPO SYNTHESIS(TRIETHANOL AMINE TEMPLATE)

Figure 3   Stirred situ infrared study of ALPO-5 synthesis
           from aqueous solution at 152oC and 85 psi (0.59KPa)
           Curve A shows phosphoric acid and triethanol amine in
           water. Curves B to E resulted after the aluminum
           hydroxide was added and the mixture heated in the
           sealed CIR cell.

compared to an identical reaction which was not stirred at any time during the course of the reaction. The latter data are not illustrated here, but this reaction showed no evidence for the formation of an organic phosphoric acid ester even after two days at 155°C. The X-ray diffraction of the calcined

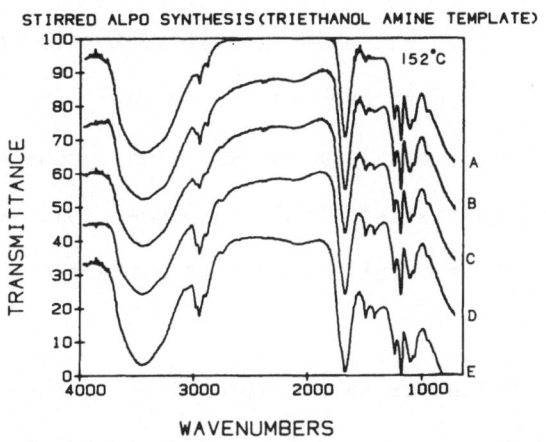

STIRRED ALPO SYNTHESIS (TRIETHANOL AMINE TEMPLATE)

Figure 4 Strirred ALPO-5 synthesis at constant temperatures of 152°C and 85 psi. Infrared Curve A to E taken at reaction times of 1.8, 2.3, 3.3, 4.3 and 5.8 hours respectively.

323

solids for the unstirred reaction showed that it was mainly ALPO-5 [13] while the calcined solid isolated from the stirred reaction was a mixture of a small amount of ALPO-5 and a second crystalline aluminophosphate. We suggest that the difference in these reactions is that the stirred reaction led to the formation of a phosphoric acid ester template which leads through steric control to a different product as compared to the product found when the triethanol amine is not esterified. There data also suggest a dynamic crystallization re-solution and recrystrallization process is taking place during the reaction continuously.

The other general aspect that one should appreciate from the above data is that these reactions, either stirred or unstirred, were carried out under normal synthesis conditions where the infrared spectra even in the strongly absorbing aqueous solutions were easily measured. The second advantage is that the reaction started with a suspended solid and ended with a slurry of the crystalline product without interfering with the IR analysis.

Several other studies have examined the synthesis of synthetic zeolites like ZSM-5 under a variety of pH conditions [14]. In these cases one could observe a concentration of the organic template in the gel.

INFRARED ANALYSIS OF POWDERED SOLID STATE MATERIALS

By embedding a CIR crystal in a low volume reactor of

about 1 ml, one may obtain high quality infrared spectra of powdered solids. To accomplish this, the powdered sample is filled into one end of the reactor with the crystal already mounted. A vibrator is used to pack the solid into the reactor so that it has intimate surface contact with the CIR crystal.

A few of the specrtra of solid state inorganics obtained in this way are shown in Figures 5,6, and 7 for a ZSM-5 of Si/Al ratio of 1000/1, a sample of silica gel, and a sample of gamma-alumina respectively. Although these spectra are fundamentally ATR spectra, the advantage of this type of analysis is that the CIRCLE [8] bench and optics may be used without the need of a traditional ATR optical bench.

The data shown here and in other reports [1,2,11,14] have illustrated the unique capabilities of high pressure reactors equipped with embedded cylindrical internal reflectance crystals to perform in situ reaction studies under autogeneous conditions. For solid state applications, synthesis from solutions or hydrogels at high temperatures and pressures are easily accomplished even where a suspended solid is used or produced in the reaction. Solids may be analyzed as undiluted powders using the same CIRCLE [8] optical bench which the high pressure reactors employ. We are currently examining the possibility of using the CIR technique for high temperature solid state synthesis.

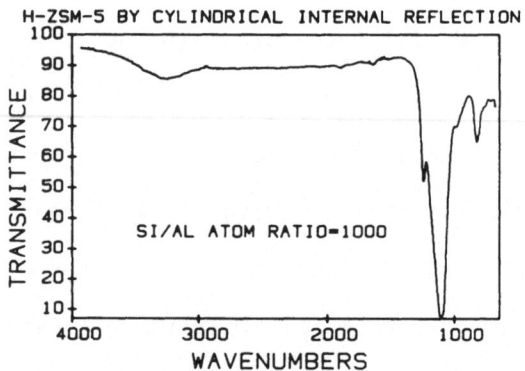

Figure 5   Infrared spectra of a solid sample of ZSM-5
            (Si/Al=1,000) packed into CIR reactor.

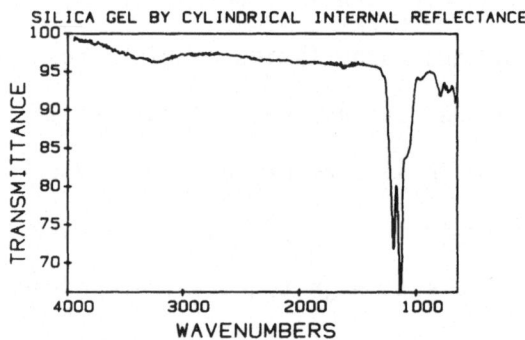

Figure 6   Solid sample of silica gel packed into a CIR
            reactor.

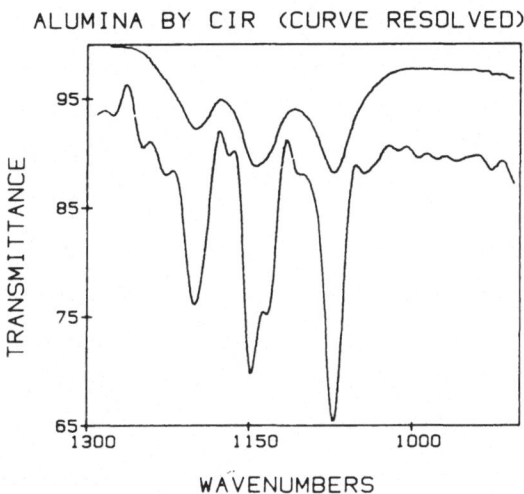

Figure 7 Curve resolved IR spectrum of alumina in a CIR
reactor.

REFERENCES

[1] W.R. Moser, J.E. Cnossen, and S.A. Krouse, 187th National
     ACS Meeting, St. Louis, MO April 8-13 (1983).

[2] W.R. Moser, Scan Time, January/February 1984 Issue,
     Barnes Analytical-Spectra Tech Publication, Stamford,
     CT, USA.

[3] A. Rein and P.A. wilks, American Laboratory, Oct. 1982.

MERCURY POROSIMETRY OF DRY SILICA GELS

E. M. Rabinovich and S. D. Poulsen

AT&T Bell Laboratories
Murray Hill, New Jersey 07974

INTRODUCTION

A recent explosion in research of silica gels has been stimulated by prospects of glass preparation at temperatures significantly lower than temperatures of conventional melting and by opportunities to prepare unusual glasses which cannot be produced by melting (1,2). Dried gels represent highly porous structures which can be sintered to monolithic glasses at temperatures below the liquidus. Two types of sol-gel glass preparation have emerged: 1) hydrolysis and polymerization of alkoxides (1,2), and 2) gelation of colloidal sols, which includes dispersion of a powder in water, drying, redispersion, gelation, drying and sintering (2-5). In both cases the size and mutual distribution of pores are very important for processing after gelation: drying and sintering. Mercury porosimetry (6-8) provides information about pore size distribution. This paper deals mainly with methodological problems of gel study by this method.

METHOD AND EQUIPMENT

Mercury porosimetry is based on the property that mercury does not wet most solid surfaces. Therefore mercury must be forced into pores under pressure. As the pressure P increases, mercury will intrude into smaller pores, in correspondence with Washburn's equation:

$$d = \frac{-4\gamma \cos\theta}{P} ,$$

(1)

where d is a pore diameter, γ is the surface energy and θ is the contact angle between mercury and a material under study (6).

Experiments were conducted using an automatic porosimeter Auto-Pore 9200* (9), where intrusion of mercury into a specimen enclosed in glass penetrometers was monitored by capacitive measurements. There are 4 low-pressure and 2 high-pressure stations in the instrument. The low-pressure stations are used for evacuation of specimens in penetrometers, filling by mercury and pressurization up to 0.21 MPa (30 psi). The high-pressure stations provide pressure up to 414 MPa (60,000 psi). In our study we accepted $\gamma = 485 \times 10^{-3}$ J/m^2 and $\theta = 130°$ (other workers (7) use either 130° or 140° for θ). Using these values,

$$d \text{ (in nm)} = \frac{1247}{P \text{ (in MPa)}} \qquad (1a)$$

or

$$d \text{ (in nm)} = \frac{180725}{P \text{ (in psi)}} . \qquad (1b)$$

The minimal pore diameter which can be detected by this porosimeter is 3 nm. Calibrated penetrometers with a total volume 5 cm^3 and intrusion volumes of 1.1 and 1.8 cm^3 were used. Calibration was done to determine the precise penetrometer volume; in this case the penetrometer could be used as a pycnometer for measurement of the bulk density and true density if all pores are open and larger than 3 nm. Because it is not always the case, below this last parameter is referred to as the apparent (skeletal) density.

REPRESENTATION OF RESULTS AND MODIFICATIONS

The porosimeter includes a computer and a printer. The intrusion-extrusion takes place according to a desired program which may include as many as 85 points (9). The equilibrium time at each point is also fixed by the user. Upon completion of the test, the instrument prints a report on 6-7 pages, including the summary of results, the table of original data and a graphical interpretation. The latter comprises cumulative and incremental intrusion volume and surface area, all vs. pore diameter presented in logarithmic scale. This graphical presentation, although useful, has several shortcomings.

First, although the instrument allows an arbitrary 85 point program, the graphs contain only 56 points fixed according to the logarithmic scale. Therefore the plots show not real experimental points but interpolated values. Hence, programming of additional points in specific pore size ranges is not reflected on these plots, and for better resolution the plots should be manually redrawn from the table. If the program excludes large pores and begins the experiment from, e.g., 20 μm pores, the instrument extrapolates data and gives meaningless curves for the range of 300-20 μm.

Second, the plots of incremental volume dv (or surface ds) vs. diameter d

---

* Micromeretics.

does not represent a real distribution which should be given as differential volume dv/dd and surface ds/dd, or $dv, s/d \log d$. However, because the distance between points on the diameter scale $\Delta \log d =$ const, the plots still provide the correct shape of the curves, corresponding to the distribution.

Third, the numbers on the ordinates are per cents from maximum values but not real values and this makes comparison of the plots for different specimens difficult.

Using a separate microcomputer we have overcome these shortcomings. The microcomputer used is a LNW 80;* Micro-Soft Basic, LNW Basic, NEWDOS-80, Auto Plot and Micro-Systems software were used with modifications. Transferring and replotting the data from the Auto-Pore 9200 computer was done in four steps:

1) The porosimeter was connected with the microcomputer through 300 baud modems and a telephone line.

2) A program referred to as a Bulletin Board System was modified to be compatible with the porosimeter computer.

3) After the data file was uploaded to the micro-computer, the test editor removed data which do not represent information to be plotted; the remained data were saved in a format compatible with the graphic plotting program.

4) These data were plotted using the program's auto-scale option. The data could be plotted in many forms, scales and paper sizes. In general, to represent the full curve, from 300 to 0.003 μm in pore diameter, the logarithmic scale was used, but the most interesting parts of the curve could be replotted enlarged, using either linear, or logarithmic scale for diameter. Ordinate scales are always linear and gave true values and not percentage from maximum values; thus it is convenient to compare different curves if the same scale has been used.

The graphs plotted were: cumulative volume (v) and surface (s) vs. diameter d (Fig. 1a), differential volume (Fig. 1b) and differential surface vs. diameter. The last 2 curves, representing pore size distribution, could be plotted in two ways. The normal representation (for volume) should be dv/dd vs. d or $\log d$ (curve 1, Fig. 1b), but in this case some features of the curve could be lost. Plotting $dv/d \log d$ vs. $\log d$ (curve 2, Fig. 1b) allowed the resolution of an additional maximum which had remained unresolved on the dv/dd curve. To resolve this peak on the dv/dd curve it would be necessary to plot separately the maximum near 0.5 μm in an enlarged scale.

REPRODUCIBILITY OF POROSIMETRIC DATA AND PORE DISTRIBUTION

To check reproducibility of data, three measurements of the same twice dispersed colloidal silica were done. The results for different pore parameters are given in Table 1. As seen, the bulk density, intrusion volume (total pore volume),

---

* LNW Research Corp.

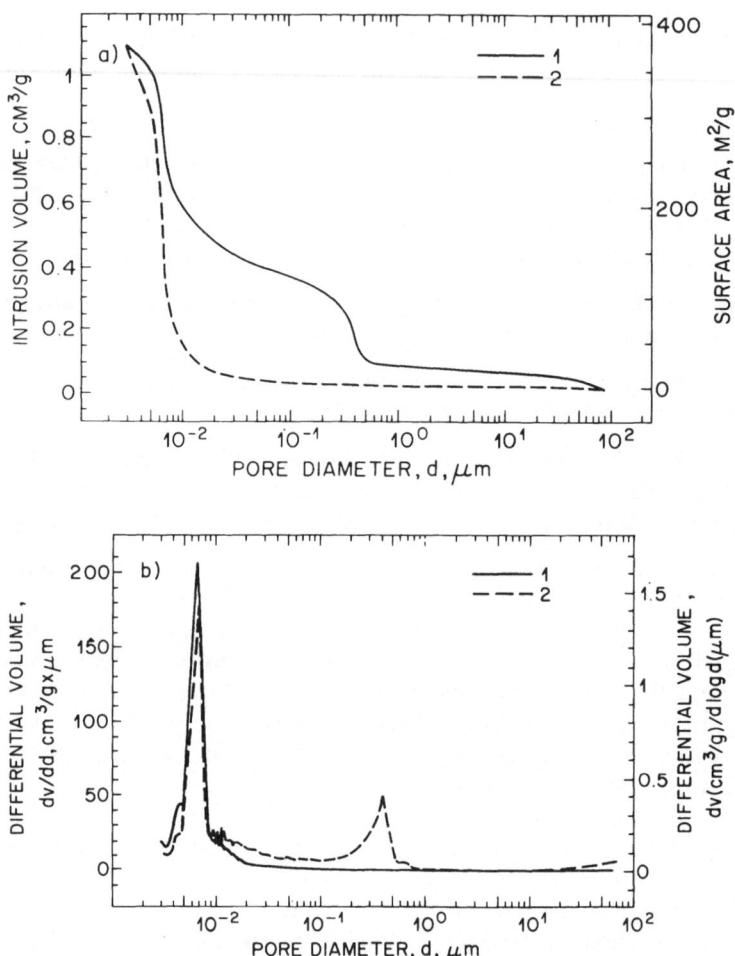

Fig. 1. Examples of porosimetric curves (twice dispersed colloidal silica powder) after their replotting with microcomputer LNW 80: a) — cumulative volume (1) and surface (2) vs. pore diameter; b) — differential volume vs. pore diameter plotted by 2 ways: curve 1 — dv/dd vs. $(\log)d$, curve 2 — $dv/d \log d$ vs. $\log d$.

Table 1. Reproducibility of Porosimetric Data of Twice Dispersed Colloidal Silica

PARAMETER	DATA			AVERAGE			AVERAGE DEVIATION ±, %
Intrusion volume, cm^3/g	1.07	0.95	1.08	1.03	±	0.06	5.8
Surface area, m^2/g	380	373	383	379	±	4	1.1
Average pore diameter, nm	11.2	10.1	11.2	10.8	±	0.5	4.6
Median pore diameter (volume), nm	18.6	12.4	18.7	16.6	±	2.8	16.9
Median pore diameter (surface), nm	5.4	5.4	5.4	5.4			0
D_{max1}, nm	4.9	4.9	4.9	4.9			0
d_{max2}, nm	402	315	402	373	±	39	10.4
Bulk density, g/cm^3	0.631	0.641	0.641	0.637	±	0.005	0.8
Apparent density, g/cm^3	1.93	1.63	2.06	1.87	±	0.16	8.6

surface area, average pore diameter and position of the 1st maximum (in the nanometric range) can be determined with error 0-6%; position of the 2nd maximum and the related median pore diameter (volume) are determined with greater error.

The apparent (skeletal) density is determined from the total pore volume and bulk density. If the theoretical density (TD) of the material is known (for amorphous silica it is 2.2 g/cm^3), the value of the apparent density allows one to judge which portion of the total pore volume remained undetected. If the apparent density is close to TD, this means that practically all porosity was detected; if it is smaller than TD, the volume of undetected pores (<3 nm) can be calculated from the difference. Unfortunately, this parameter is determined with a significant error for the twice dispersed material.

The reproducibility of pore parameters for different batches of colloidal silica prepared by hydrolysis of tetraethyl orthosilicate (TEOS) $Si(OC_2H_5)_4$ is shown in Table 2. As seen, the reproducibility of preparation is rather good and does not increase significantly the method's error. Absence of the 2nd dispersion and the 2nd maximum reduces noticeably the error for the median pore diameter (volume) and for the apparent density.

Table 2. Reproducibility of Pore Parameters for 5 Different Batches of TEOS-Derived Silicas

PARAMETER	DATA					AVERAGE	AVERAGE DEVIATION ABSOLUTE, ±	AVERAGE DEVIATION IN %	AVERAGE DEVIATION OF METHOD, %
Intrusion volume, cm^3/g	0.74	0.80	0.77	0.83	0.96	0.82	0.06	7.3	5.8
Total surface area, m^2/g	432	418	404	428	464	429	15	3.5	1.1
Average pore diameter, nm	6.8	7.7	7.6	7.8	8.3	7.6	0.3	3.9	4.6
Median pore diameter (volume), nm	6.0	6.6	6.8	6.7	7.2	6.7	0.3	4.5	16.9
Median pore diameter (surface), nm	5.8	6.3	6.4	6.6	6.7	6.4	0.2	3.1	0
d_{max}, nm	4.9	6.3	6.3	6.3	6.3	6.0	0.5	8.3	0
Bulk density, g/cm^3	0.81	0.79	0.81	0.74	0.68	0.77	0.04	5.2	0.8
Apparent (skeletal) density, g/cm^3	2.01	2.16	2.12	1.96	1.97	2.04	0.08	3.9	8.6

Table 3. Comparison of Surface Areas Obtained by BET Method and HG Porosimetry

MATERIAL	SURFACE AREA, m^2/g	
	BET	Hg
Powder from TEOS	113	109
Once dispersed colloidal powder No. 8	311	483
The same powder after 2nd dispersion	311	417
TEOS Gel A	292	518
TEOS Gel B	220	543

SURFACE AREA FROM MERCURY POROSIMETRY

The surface area from the mercury intrusion curves is calculated according to:

$$S = \frac{1}{\gamma \cos\theta} \int_{v_{\min}}^{v_{\max}} P dv , \qquad (2)$$

where v is Hg intrusion volume (6). Table 3 gives a comparison of the surface areas obtained from mercury porosimetry curves ("Hg surface area") with the BET surface areas. All specimens presented in Table 3 had the apparent density close to TD, which means that practically all pores were $\geqslant 3$ nm, otherwise the comparison would be senseless. As seen, only the first sample with a relatively low surface area shows about the same value for the both methods; in most cases the Hg area is significantly higher than the BET area. Three reasons can account for this difference: 1) Hg area is calculated on the assumption that pores have cylindrical shape, while the BET area is indifferent to the shape; 2) our choice of $\theta = 130°$ can be responsible for a 19% increase in the area values compared with $\theta = 140°$; 3) hydroxyl groups on the surface of the pores (10) can significantly alter the values of θ and γ, especially at high pressures, compared with those values for oxides (or metals). However, as seen from Tables 1 and 2, the Hg surface area is a well-reproducible parameter and can be used for comparison of different gels.

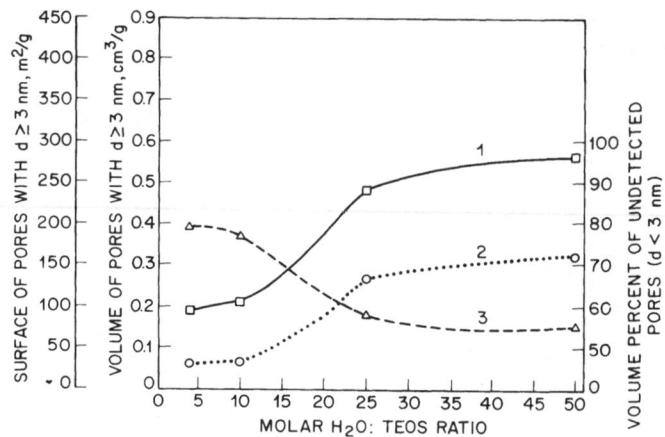

Fig. 2. Dependence of volume (curve 1) and surface area (curve 2) of pores ≥3 nm and the volume of undetected pores ($d < 3$ nm) (curve 3) of dry silica gels on the $H_2O:Si(OC_2H_5)_4$ ratio.

It is interesting that while the BET surface area shows no difference between once and twice dispersed colloidal silicas, the Hg area shows a steady decrease (4-14%) after the 2nd dispersion and the apparent density is also reduced on 5-9%. As we pointed out in (3), during dispersion and gelation the neck growth can occur between particles in contact. The necks lead to formation of "ink-bottle" pores where part of pore entries can be less than 3 nm; these pores will be closed for mercury, thus reducing both the surface area and apparent density. The overall intrusion volume increases after the 2nd dispersion by as much as 30%, but this increase takes place due to large pores between the aggregates, which have little effect on the surface area. The bulk density decreases correspondingly.

GELS WITH PORE SIZES BELOW 3 NM

All TEOS-derived gels considered above were hydrolyzed in presence of NH_4OH (pH = 11). This results in formation of colloidal powders. Hyldrolysis in presence of HCl yields three-dimensional silicon-oxygen-hydroxyl network with a BET surface area of 500-1000 m^2/g and with a significant fraction of pores <3 nm in diameter. Mercury porosimetry with pressures up to 414 MPa can give only limited information about these materials. Fig. 2 shows the dependence of surface area and volume of pores $\geqslant 3$ nm on a H_2O:TEOS ratio when the pH of water was adjusted to 1 with HCl (4 moles C_2H_5OH per 1 mole TEOS were also added). The volume percentage of undetected pores X was calculated from the Hg intrusion volume v and total pore volume per 1g of gel V as determined from the bulk density ρ and real density of glassy silica (2.2 g/cm^3):

$$V = \frac{1}{\rho} - \frac{1}{2.2} \qquad (3)$$

$$X = \frac{V - v}{V} \times 100 \qquad (4)$$

As seen from Fig. 2, an increase in the amount of water taken for hydrolysis results in a reduction of the volume of undetected pores and, correspondingly, results in an increase in both the volume and surface area of pores with diameter $\geqslant 3$ nm.

SUMMARY

Mercury porosimetry provides comprehensive and reproducible information about pore size distribution of colloidal and alkoxide-derived silica gels when all the pores are larger than 3 nm in diameter. Only limited information can be received for acid-catalyzed polymerized gels with pores <3 nm, but the volume fraction of these pores can be also estimated from the mercury intrusion data. The absolute values of the surface area measured by mercury porosimetry are larger than those by BET, but the Hg intrustion values are well reproducible and can be used for description of gels. Twice dispersed colloidal gels exhibit 2 peaks on the pore size distribution curve.

ACKNOWLEDGEMENT

We appreciate informative discussions with D. W. Johnson, Jr.

REFERENCES

1. S. Sakka, Gel method for making glass, in: "Treatise on Materials Science and Technology, vol. 22, Glass III," M. Tomozawa and R. H. Doremus, ed., Academic Press, NY-Lnd. (1982).

2. Glasses and Glass Ceramics from Gels, Proc. 2nd Intern. Workshop on Glasses and Glass Ceramics from Gels, *J. Non-Crystalline Solids*, 63 (1984).

3. E. M. Rabinovich, D. W. Johnson, Jr., J. B. MacChesney, and E. M. Vogel, Preparation of high-silica glasses from colloidal gels: I, Preparation for sintering and properties of sintered glasses, *J. Amer. Ceram Soc.*, 66:683 (1983).

4. D. W. Johnson, Jr., E. M. Rabinovich, J. B. MacChesney, and E. M. Vogel, Preparation of high-silica glasses from colloidal gels: II, Sintering, ibid., 66:688 (1983).

5. D. L. Wood, E. M. Rabinovich, D. W. Johnson, Jr., J. B. MacChesney, and E. M. Vogel, Preparation of high-silica glasses from colloidal gels: III, Infrared spectrophotometric studies, ibid., 66:693 (1983).

6. S. Lowell, "Introduction to powder surface area," Wiley, New York (1979).

7. H. M. Rootare, A review of mercury porosimetry, in: "Perspectives in Powder Metallurgy, vol. 5, Advanced Experimental Techniques in Powder Metallurgy," J. S. Hirschhorn and K. H. Roll, ed., Plenum Press, N.Y.-Lnd. (1970).

8. L. K. Frevel and L. J. Kressley, Modifications in mercury porosimetry, *Anal. Chem.*, 35:1492 (1963).

9. Instruction Manual, Autopore 9200 (version 2.02 Software) Micromeretics, MIC No. 921/42801/00 (1982).

10. E. M. Rabinovich, Alkoxide and colloidal silica gels and glasses, a comparison, *J. Non-Crystalline Solids* (1985) (in publication).

CHARACTERIZATION OF MICROSTRUCTURAL

EVOLUTION BY MERCURY POROSIMETRY

C. Han, I. A. Aksay, and O. J. Whittemore

Department of Materials Science and Engineering
College of Engineering
University of Washington, Seattle, WA 98195

ABSTRACT

Evolution of microstructure during the sintering of α-Al_2O_3 compacts formed by colloidal filtration was studied by mercury porosimetry. It is shown that the decay of the flow channels created during filtration plays an important role in the densification process. The retention of open channels until the final stages of sintering was essential in achieving high sintered densities.

INTRODUCTION

In the processing of ceramics by powder consolidation techniques, steps taken in the presintering stages play an important role in the densification behavior during sintering. For example, one essential requirement is to form the green compacts with uniform pore size distribution in order to minimize the variations in local densification rates during sintering. Experimental[1-5] and theoretical[6,7] studies have clearly demonstrated that nonuniformities in a green compact are often amplified during the subsequent sintering stages as a result of differential sintering rates.

A recent trend in ceramics processing, particularly when working with submicron size powders, has been the use of colloidal dispersion techniques to attain uniform microstructures.[8] In colloidal techniques, submicron size powders that otherwise spontaneously agglomerate due to van der Waals attractive forces are

339

kept dispersed in a fluid medium by controlling the repulsive interparticle interactions.[9] Once a uniform dispersion in the colloidal suspension stage is attained, the next goal is then to retain this uniformity while the fluid medium is being eliminated during consolidation.

Studies have shown that transitions from a dispersed colloidal suspension to a consolidated state resemble fluid to solid phase transitions observed in atomic systems.[10] For a given interparticle interaction potential, this fluid to solid phase transition takes place at a critical particle concentration with the nucleation of densely packed particle clusters as a multiple site nucleation process. A direct outcome of this transition by a multiple site nucleation process is that the resultant microstructures always contain microagglomerates which we will refer to as domains hereafter. Due to the formation of these domains, at least two types of pores result: (1) small intra-domain pores, and (2) larger inter-domain pores. The most important implication of the domain formation is that even the colloidal suspension routes cannot provide an easy solution to the problem of packing density inhomogeneities. Furthermore, if the suspensions are consolidated by filtration, the inter-domain pores are modified by the flow of the dispersion medium which goes into the filter medium through the inter-domain voids, while the intra-domain voids remain unchanged.[11] The bimodality of the pore size distribution then becomes more distinct.

In sintering of these compacts, domains sinter faster than the inter-domain regions due to their higher packing density and smaller pore size (Fig. 1). Because of this differential sintering rate, uniform sintering cannot be expected. In this paper we present data on the sintering behavior of $\alpha-Al_2O_3$ compacts formed by colloidal filtration and illustrate that the pores associated with the inter-domain regions, i.e. filtration channels, control the densification behavior during the final stages of sintering. In addition, we provide data on the sintering behavior of monosize and bimodal particle size systems and illustrate that sintering rates can be enhanced with bimodal systems. Finally, we illustrate the importance of mercury porosimetry, especially in the characterization of the decay of filtration channels during sintering.

EXPERIMENTAL PROCEDURE

Nearly monosize $\alpha-Al_2O_3$ powders* with median particle diameters of 0.78 and 0.21 μm were used as the coarse and fine particle

---

*Sumitomo Chemical America, Inc., New York, NY, AKP-15 (0.78 μm) and AKP-50 (0.21 μm), with >99.99% purity.

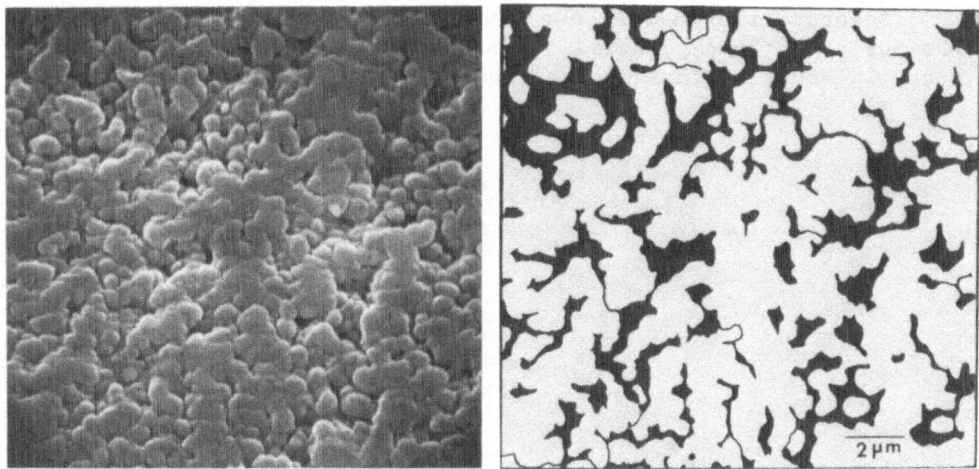

Fig. 1. Microstructure (SEM) of 0.78 μm powder compact after partial sintering at 1200°C for 1 hr. Inter-domain regions
are highlighted in the negative image.

systems. In addition, these powders were used to prepare four
binary mixtures at fine particle fractions of 0.14, 0.17, 0.20, and
0.25. Aqueous colloidal suspensions of the monosize and binary
systems at particle concentrations of 55 % by volume were prepared
at pH=2.5 with HCl additions. Ultrasonic vibration was applied
occasionally to facilitate the breakup of the soft agglomerates.
Suspensions were then cast on gypsum molds. Samples were prepared
from the filtered cakes and sintered in air at temperatures between
1200°C and 1500°C for 1 hour. Mercury porosimetry and scanning
electron microscopy were used for characterization.

RESULTS AND DISCUSSION

The variation of packing density with the fraction of fine
particles is shown in Fig. 2. The packing densities of compacts
initially increased with the addition of fine particles as expected
based on previous theoretical and experimental studies.[12-17] When
fine particles are added to a system of much larger coarse particles, fine particles first fill the voids between the coarse particles. Therefore, until all the voids are filled, the partial molar
volume of fine particles is zero, and the packing density of the
mixture increases with the fraction of the fine particles. After
all the voids are filled, additional fine particles expand the
volume of the mixture compact, and then the packing density decreases with the fraction of fine particles.

A theoretical model on the variation of packing density in homogeneously packed bimodal systems was first developed by Furnas.[12] However, in subsequent experimental studies,[13,15] it has

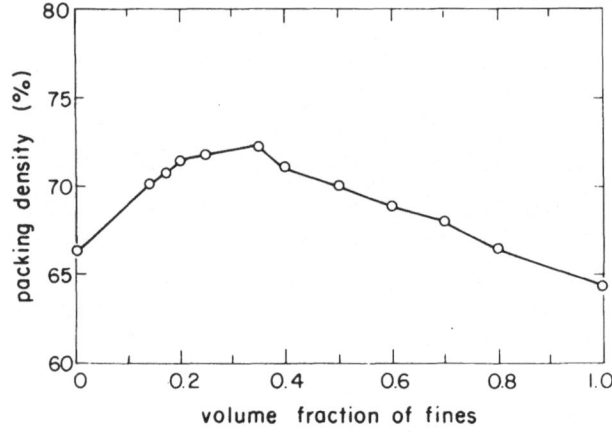

Fig. 2. Packing density variation in the bimodal system with end members of 0.78 and 0.21 μm size particles.

been shown that experimental data always displayed lower densities than the ones predicted by the model. Our data as presented in Fig. 2 similarly display deviation from Furnas' idealized model. In previous studies,[16-17] inhomogeneities in the mixing of coarse and fine particles were proposed as the cause of this deviation. When the mixing of bimodal particles is not uniform, the partial molar volume of the fine particles does not follow the value predicted by the idealized model but deviates from ideality as the inhomogeneity of the mixing increases.

When working with colloidal systems, homogeneity in mixing can be achieved in the suspension stage prior to consolidation. However, problems in long-range particle segregation (on a dimensional scale larger than the particle domains) may arise during the consolidation stage due to differential sedimentation. In the present study, with the use of suspensions that contained >50 % by volume of particles, long-range segregation of particles during the consolidation stage was prevented as confirmed through electron microscopic examination and hydraulic resistance measurements.[18] In spite of this precaution, the fact that our density data still

342

displayed deviation from Furnas' model would suggest short-range inhomogeneities. The implication is that, consistent with the formation of particle domains and the formation of filtration channels around these domains, fine particles do not decrease the inter-domain porosity as effectively as the intra-domain porosity.

Channel sizes as measured by mercury intrusion continuously decreased with increasing fraction of fine particles (Fig. 3). This behavior is consistent with the concept that, in uniformly mixed bimodal compacts, fine particles act as space filling units and thus the intrusion neck size decreases as fine particles occupy space between the coarse ones. However, it is important to note that, during the intrusion of mercury into pores with many openings, mercury penetrates into the

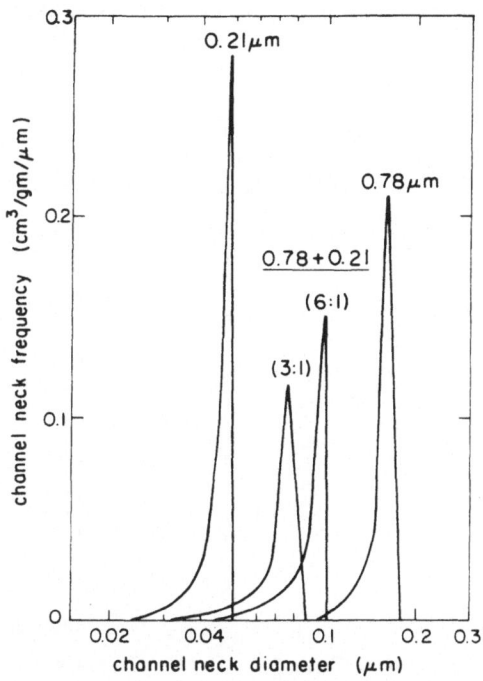

Fig. 3   Channel neck size frequency of monosize and bimodal compacts obtained by mercury intrusion.

pores through the largest opening, one which is still smaller than the largest dimension of the channels connected by the necks. Therefore, mercury intrusion data interpret pores as being cylindrical in shape with a diameter equal to the neck diameter. In fact, a comparison of the data presented in Fig. 3 with the microstructure of Fig. 1 reveals that, while the channel necks determined by porosimetry are all smaller than 0.2 µm, the microstructure clearly possesses many pores that are 10 times larger than the necks.

During sintering of powder compacts, smaller pores disappear faster than larger pores. Similarly, the channel necks decay faster than the main body of the channel and, at a certain stage of densification, this results in the formation of isolated pores. At this stage, whether these isolated pores will be in the shrinkage regime or not is determined by the number of grains surrounding a pore and the dihedral angle at the pore/grain boundary junctions.[19,20]   Therefore, it becomes necessary to control the size of the isolated pores and thus the number of grains surrounding a pore through variations in channel morphology that in turn affects their decaying behavior.

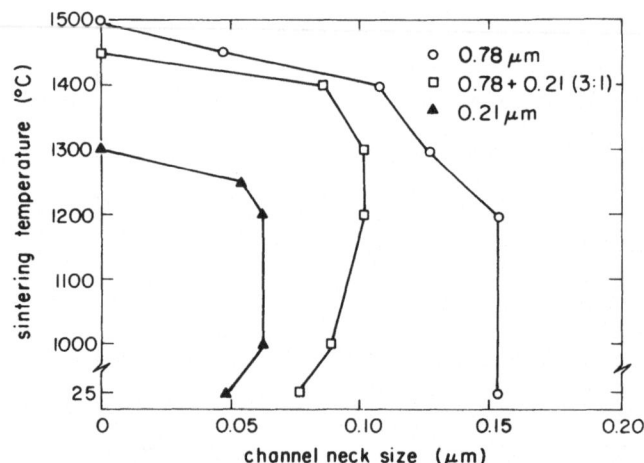

Fig. 4.   Change of the largest neck size during the sintering of monosize (0.78 and 0.21 μm) and bimodal (3:1) systems.

Variations in the channel neck size during the densification of the fine, coarse, and bimodal systems are compared in Fig. 4. During the initial stage of sintering, channel neck growth was observed in the mixture and the fine particle compacts, whereas no neck growth was observed in the coarse particle compacts. On further sintering, necks shrank in all cases and eventually closed at approximately around the temperature of the last data point in each corresponding curve. The temperatures at which the neck closure occurred decreased with decreasing initial neck size. However, it is important to note that, at the point of neck closure, the compact with 100% fine particles still contained 5% porosity (now as closed pores) while both the binary mixture as well as the coarse particle compacts had less than 2% porosity.

The detrimental effect of this early neck closure upon the densification behavior of fine particle compacts in the final stage of sintering is illustrated in Figs. 5 and 6. Fine particle compacts initially displayed the fastest sintering up to 1300°C

344

(Fig. 5). At this point, due to the isolation of channels as closed pores that appear to be in the growth rather than the shrinkage regime, further densification practically stopped. On the other hand, coarse particle compacts and the mixtures showed lower densification rates initially; however, final sintered densities were higher than that of the fine particle compacts. In comparing the mixture compacts and the coarse particle compacts, it is apparent that the mixtures always sintered faster than the coarse particle compacts (Fig. 5).

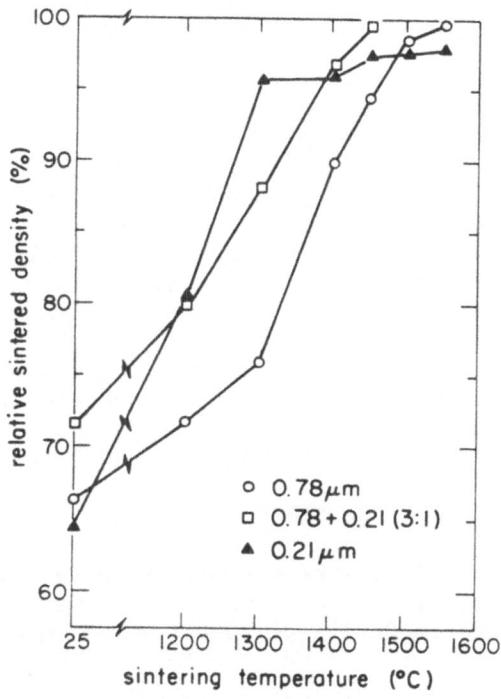

Fig. 5. Sintered density of mono-size (0.78 and 0.21 μm) and bimodal (3:1) systems.

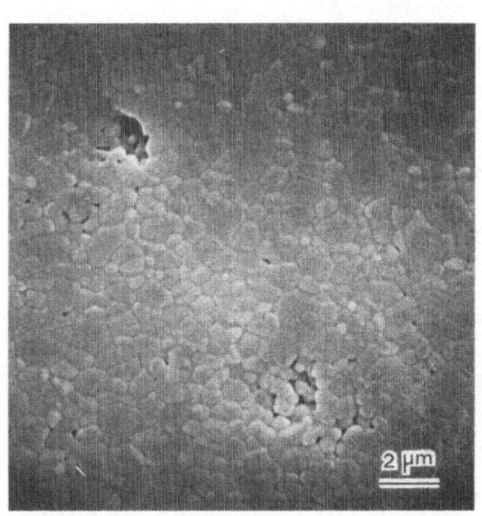

Fig. 6. Microstructure (SEM) of 0.21 μm powder compact after reaching an end point density of 98.5 % TD at 1300°C for 3 hr.

CONCLUSIONS

In this study of microstructural evolution, where mercury porosimetry was used as the major characterization tool, the following conclusions can be made:

(1) In the densification of alumina compacts formed by colloidal filtration of submicron size particles, the decay of channels, which are formed during the filtration stage, affected the overall microstructural evolution. Channel closure at lower densities resulted in lower end-point densities.

(2) The addition of finer particles was shown not to be detrimental to sintering. On the contrary, the sintering temperature was reduced in bimodal mixture compacts.

ACKNOWLEDGEMENTS

This work was supported by the Advanced Research Projects Agency of the Department of Defense and was monitored by the Air Force Office of Scientific Research under Grant No. AFOSR-83-0375.

REFERENCES

1. K. D. Reeve, Am. Ceram. Soc. Bull., 42 [8] 452 (1963).
2. T. Vasilos and W. Rhodes, in "Ultrafine-Grain Ceramics," J. J. Burke, N. L. Reed, and V. Weiss, eds., Syracuse University Press, New York (1970), p. 137.
3. W. D. Kingery, in "Ceramic Processing Before Firing," G. Y. Onoda, Jr. and L. L. Hench, eds., John Wiley & Sons, New York (1978), p. 291.
4. F. F. Lange and M. Metcalf, J. Am. Ceram. Soc., 66 [6] 398 (1983).
5. F. F. Lange, B. I. Davis, and I. A. Aksay, J. Am. Ceram. Soc., 66 [6] 407 (1983).
6. A. G. Evans, J. Am. Ceram. Soc., 65 [10] 497 (1982).
7. R. Raj and R. K. Bordia, Acta Met., 32 [7] 1003 (1984).
8. Chapters in Part 4 of "Ultrastructure Processing of Ceramics, Glasses, and Composites," L. L. Hench and D. R. Ulrich, eds., John Wiley & Sons, New York (1984) provide reviews on recent developments in colloidal processing techniques.
9. J. Th. G. Overbeek, J. Colloid Interface Sci., 58 [2] 408 (1977).
10. I. A. Aksay, in "Advances in Ceramics," J. A. Mangels and G. L. Messing, eds., Am. Ceram. Soc., Columbus, OH, (1984), vol. 9, p.94.

11. I. A. Aksay and C. H. Schilling, in "Ultrastructure Processing of Ceramics, Glasses, and Composites," L. L. Hench and D. R. Ulrich, eds., John Wiley & Sons, New York (1984), p.439.

12. C. C. Furnas, Relations between specific volume, voids, and size composition in systems of broken solids of mixed sizes, U.S. Bur. of Mines Res. Invest., No. 2894 (1928).

13. A. E. R. Westman and H. R. Hugill, J. Am. Ceram. Soc., 13, [10] 767 (1930).

14. R. K. McGeary, J. Am. Ceram. Soc., 44 [10] 513 (1961).

15. F. N. Rhines, in "Ceramic Processing Before Firing," G. Y. Onoda, Jr. and L. L. Hench, eds., Wiley-Interscience, New York (1978), p.321.

16. G. L. Messing and G. Y. Onoda, Jr., J. Am. Ceram. Soc., 61 [1-2] 1 (1978) and 61 [7-8] 363 (1978).

17. J. P. Smith and G. L. Messing, J. Am. Ceram. Soc., 67 [4] 238 (1984).

18. C. Han, "Sintering of Bimodal Powder Compacts," M. Sc. Thesis, University of California, Los Angeles, CA (1985).

19. R. Cannon, in "Oxidation of Non-Oxide Ceramics," Case-Western Reserve University report to AFOSR, Contract #F49620-78-C-0053, June 1981.

20. F. F. Lange, J. Am. Ceram. Soc., 67, [2] 83 (1984).

ELECTRICAL CHARACTERIZATION OF CERAMICS AND POLYMERS

Larry C. Burton

Departments of Electrical Engineering and Materials
Engineering
Virginia Polytechnic Institute & State University
Blacksburg, VA 24061

INTRODUCTION

Electrical characterization of ceramic and polymeric materials can be of both fundamental and diagnostic value, even for materials not used directly in electronic applications. We would like to review some selective DC and AC measurements to illustrate this point. Electrical conduction is of particular interest in certain ceramics, capacitors and insulators being examples. DC conduction and thermoelectric measurements are discussed below specifically for the case of ferroelectric ceramic used in capacitors.

For the case of conductive polymers, resistance and resistance relaxation measurements can serve several functions. Such measurements are discussed for carbon black filled styrene butadiene and natural rubber. An AC model for metal-polymer contact resistance is also presented.

CERAMIC MEASUREMENTS

The main electrical parameters related to ceramics are dielectric and piezoelectric constants and electrical resistivity. High dielectric constant materials (K $\gtrless$ 2,000) are used chiefly in multilayer ceramic capacitors. Insulation resistance is a key parameter for applications as diverse as integrated circuits and high voltage power lines.[1]

The principal form of degradation in ceramic capacitors is

insulation resistance, not dielectric constant.[2] Therefore, a
knowledge of leakage current mechanisms is of great interest. We
have studied several ceramic capacitor types, and capacitor ceramic
The two major types of current we have found are ohmic and
space charge limited. These currents depend on voltage (V) as
follows:[3]

Ohmic: $I = q A n \mu V/T$ (1)

Space Charge: $I = 9 A \varepsilon \mu V^2/8T^3$ (2)

where q = electronic charge, A = area, n = carrier condentration,
μ = mobility, ε = permittivity and T = thickness. (Equation (2)
represents only one of many forms of space charge currents
possible. The reader is referred to reference 3 for greater
detail.)

It can be easily deduced that space charge currents should
be present for thin layers (several mils or less) of high dielectric constant ceramics (K $\gtrsim$ 2000). For resistivity of 10^{14} Ω-cm,
an ohmic current density (I/A) of about 4×10^{-10} A/cm^2 is expected for T = 1 mil and V = 100 V. From equation (2), using
$\varepsilon = 2000 \varepsilon_o$ and $\mu = 10^{-4}$ $cm^2/Vsec$ we obtain a space charge limited
current density of about 10^{-2} A/cm^2. Thus, space charge currents
should be dominant at much lower voltages, for ceramic with the
above parameters.

This is illustrated in Figure 1, where current-voltage
dependence is shown for a somewhat degraded ceramic capacitor
(previously lifetested at 125°, twice rated voltage[4]).

The slopes of unity (ohmic) and two (space charge) are
clearly distinguishable. The transition current and voltage are
indicated as I_x and V_x. Equations (1) and (2) can be equated and
solved for the carrier concentration and mobility in terms of
these. Results for this device are n ~ 5×10^{13} cm^{-3} and
μ ~ 10^{-8} $cm^2/Vsec$.

These estimates of carrier concentration and mobility
deduced from I-V characteristic for a leaky ceramic capacitor are
still quite small. In order to ascertain more definitively the
roles of n and μ, thermoelectric voltages were measured on several
pieces of reduced ferroelectric ceramic of the same type used in
the capacitor represented in Figure 1. Concentration and mobility
were estimated assuming small polaron hopping transport.[4] It
was found in these studies that the reduction in resistivity is
due as much to increased mobility as to increased carrier concentration. In fact, the resistivity temperature dependence (expressed as an activation energy) is almost entirely due to mobility, since the donors in these ceramics are almost completely

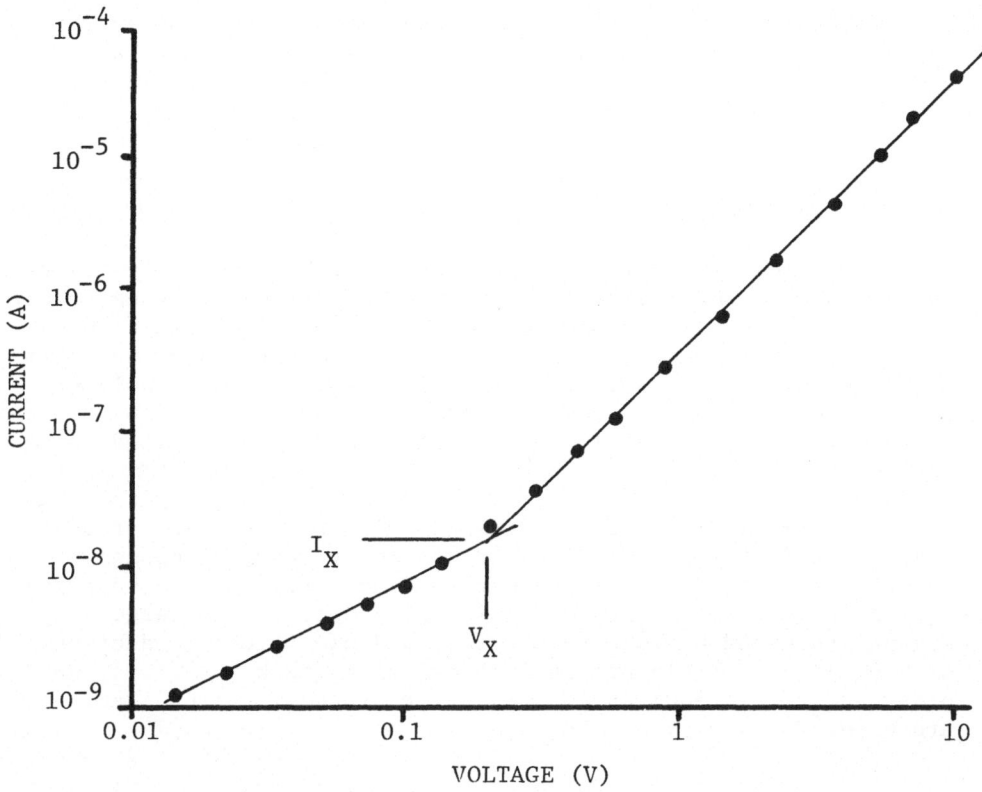

Figure 1. Current-voltage dependence for leaky ceramic capacitor (144pF, X7R).

ionized even at room temperature. Thermal activation energies for conductivity is separated into the concentration and mobility contributions in Table 1, for two samples of X7R type $BaTiO_3$ based ceramic.

Table 1

Thermal Activation Energies (eV)

Sample	Conductivity	Carrier Concentration	Mobility
As made	1.22	0.19	1.03
Reduced	0.34	~ 0	0.34

The decrease in activation energy has also been noted for degraded capacitors[4], thus the similarity between the latter and the reduced ceramic respresented in Table 1 is apparent.

POLYMER MEASUREMENTS

Electrical properties of polymers have been recently reviewed in the book by Seanor.[5] Electrical conduction processes are of interest for antistatic applications, and for novel electronic applications of conducting polymers. We are interested in applying electrical measurements to certain types of filled polymers for diagnostic purposes and to correlate electrical and mechanical characteristics. Some of these results have been reported in an earlier paper.[6]

Electrical properties of filled polymers (such as carbon black filled styrene butadiene or natural rubber, discussed below) are largely due to the carbon black network. These properties include the increased dielectric constant (as large as several thousand at low frequencies) and electrical conduction.[7] Carbon black particle and agglomerate size are key factors, but other additives required in the curing process (sulfur, ZnO, etc.) are expected to play a role.

It is also highly probable that similar bonds determine both electrical and mechanical properties of the rubber. These include carbon black particle to particle bonds, carbon to polymer bonds, polymer cross links, polymer chain bonds and vander Waals forces between carbon black aggregates and between the carbon black and the matrix. Electrical conduction transients can therefore potentially be used to monitor disruptions suffered by these bonds when subjected to mechanical stress, as indicated below.

An interesting type of electrical resistance transient that occurs due to disruption and subsequent healing of the links mentioned above is shown in Figure 2, for two samples using different polymer types. Both contained 45% by weight carbon black, and had brass plates vulcanized directly to them to minimize contact resistance.[8] (Additional fabrication details were described elsewhere[6]).

In figure 2, resistance increases drastically for compressional stress of 60 lb/in^2 ("loaded"), and subsequently relaxes upon removal of stress ("released"). Both the resistance and its increase/relaxation depend on sample composition and history. The relaxation characteristics in fact provide a "fingerprint" for a given sample type, with seven different compositions showing unique characteristics.

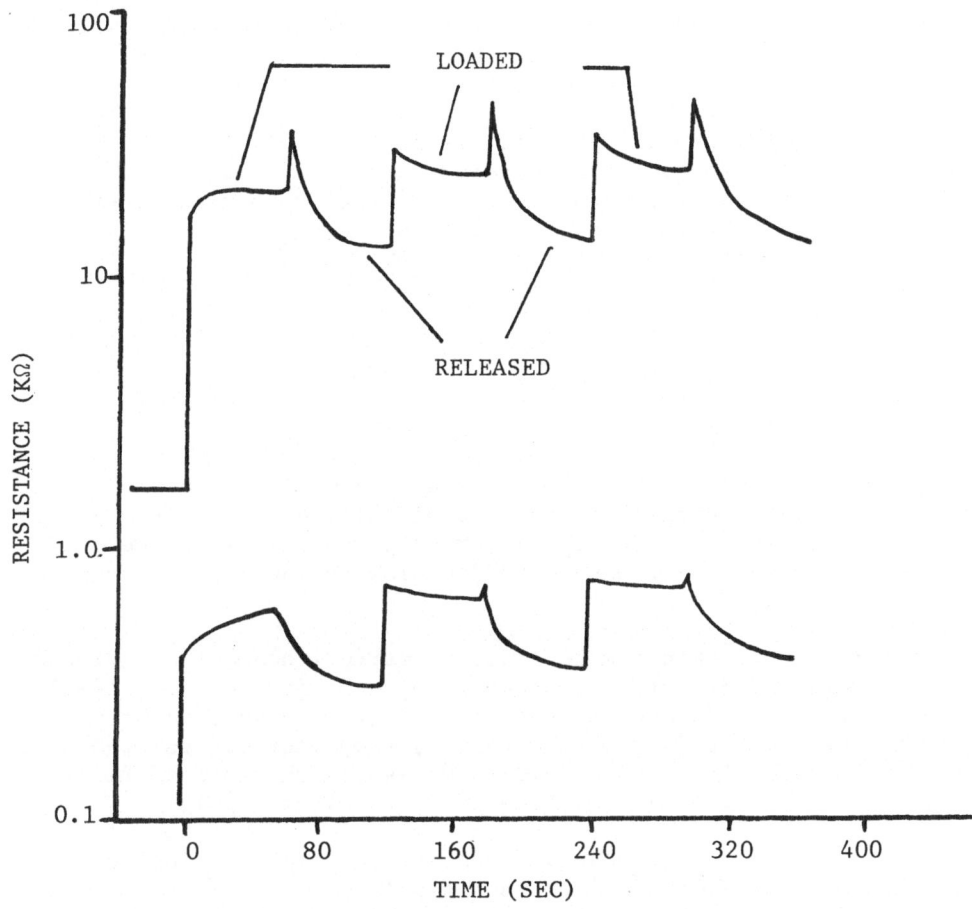

Figure 2. Electrical resistance relaxation characteristics for
two rubber types.

As noted above, electron motion through a given sample is
degraded by the disruption of certain mechanical links. The
amount of disruption depends very strongly on the pressure, the
resistance increase varying nearly exponentially on applied
pressure. While still under stress, fractured current paths can
be partially repaired. Hence, the resistance tends to relax back
toward its unstressed value. However, when the stress is removed,
these recently repaired bonds are re-fractured, since the matrix
is attempting to return to its original state. One possibility
is that the same bonds and chains may be rupturing on stress
application, and again on its removal.

The apparent increase in resistance following stress, which
does not decay to zero over a time period of one month, is
analogous to the mechanical "set" of vulcanized rubber. Thus it
appears that a certain fraction if the mechanical links that give

353

rise to conductivity are permanently or near-permanently disrupted during stress. These can be restored by heating for several hours at 70°C.

Such resistance increase-relaxation characteristics are unique for each sample type measured, and can be used to define a signature for each type. We anticipate that such measurements can find direct application to carbon black filled products for diagnostic and perhaps even screening purposes.

CONTACT RESISTANCE

In the previous experiment, brass plates were cured directly onto the rubber samples in order to minimize contact resistance. We are including this section in order to show a technique for measuring contact resistance, and to discern under what conditions it can be neglected.

For a long, thin sample (i.e. a "strip") contact resistance can be measured directly by probing the voltage along the sample when a DC current passes through it.[7] If the voltage drop across the contact is V_c at current I, then contact resistance is V_c/I. However, for thin polymer samples (in the form of films, sheets or plates) this technique is difficult to apply.

We have therefore modelled contact resistance using AC impedance measurements. A contacted polymer plate or sheet can be represented as shown in Figure 3a. The parallel combination that is read out by the measuring circuit (in this case an HP 4192A Impedance Analyzer) is shown in Figure 3 b.

In this model, G_c and C_c are the contact conductance and capacitance respectively, and G_B and C_B are the bulk rubber parameters. The Impedance Analyzer displays only net values G and C (Figure 3b.) The network transformation equations at frequency ω are

$$C = \frac{(C_c G_B + C_B G_c)(G_B + G_c) - (G_c G_B - \omega^2 C_c C_B)(C_B + C_c)}{(G_B + G_c)^2 + \omega^2 (C_B + C_c)^2} \tag{3}$$

$$G = \frac{(G_c G_B - \omega^2 C_c C_B)(G_B + G_c) + \omega^2 (C_c G_B + C_B G_c)(C_B + C_c)}{(G_B + G_c)^2 + \omega^2 (C_B C_c)^2} \tag{4}$$

354

At low frequencies these can be approximated by

$$C = \frac{C_B + (G_B/G_C)^2 C_c}{(1 + G_B/G_c)^2} \qquad G = \frac{G_c \, G_B}{G_c + G_B} \qquad (5)$$

and at high frequencies by

$$C = \frac{C_c \, C_B}{C_c + C_B} \qquad G = \frac{C_c^2 G_B + C_B^2 G_c}{(C_B + C_c)^2} \qquad (6)$$

These limiting cases indicate that if contact resistance is large (G_c is small) the net (measured) capacitance will be higher at low frequencies than with zero contact resistance. At high frequencies, measured capacitance is independent of contact resistance because the capacitances shunt the resistors.

This model was verified quantitatively by measuring the capacitance values for two samples with different contacts, and two with different thickness. One sample (C1) had silver epoxy contacts (high contact resistance). Another sample (C2) had aquadag contacts (low contact resistance). Capacitances versus frequency for these two samples are seen in Figure 4a. Two additional samples of different thickness (sample A1 at 1mm and A2 at 6 mm) were also prepared using silver epoxy contacts. Their capacitance values are shown in Figure 4b.

In Figure 4, the trends predicted by the model are verified. In Figure 4a, sample C2 (with lower contact resistance) has lower capacitance at low frequencies. Above about 100KHz, the capacitance values become nearly identical, as predicted.
In Figure 4b, for samples of different thickness but with the same type of (high resistance) contacts, the capacitance values are nearly identical at lower frequencies, where the capacitance is governed almost entirely by the contact. At higher frequencies, the thinner sample has significantly higher capacitance as expected, since capacitance is inversely proportional to thickness.

From the limiting low and high frequency values of Figure 4a (and from conductance values measured over the same frequency range) sample C1 can be represented by the following values:

$$R_c = 1/G_c = 2.1 \times 10^5 \Omega, \quad C_c = 2.1 \text{nF}, \quad R_B = 1/G_B = 2.7 \times 10^5 \Omega,$$

$$C_B = 0.3 \text{nF}.$$

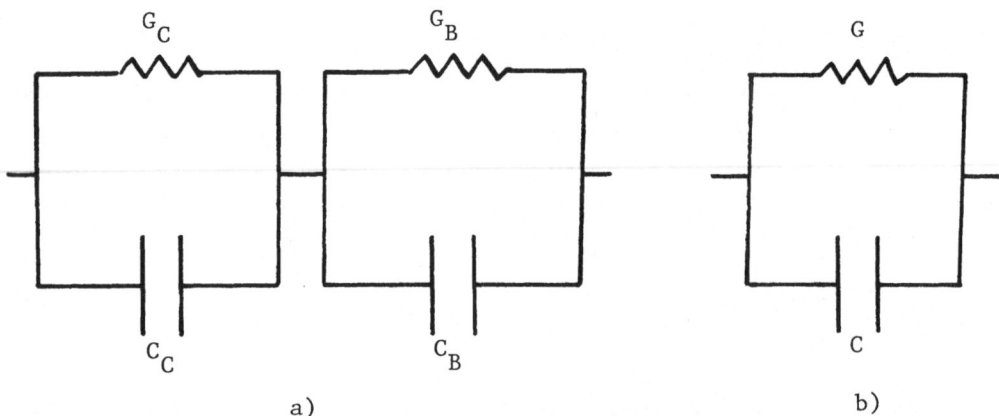

Figure 3.a) Lumped model for rubber and contacts; b) Parallel
combination read out by Impedance Analyzer.

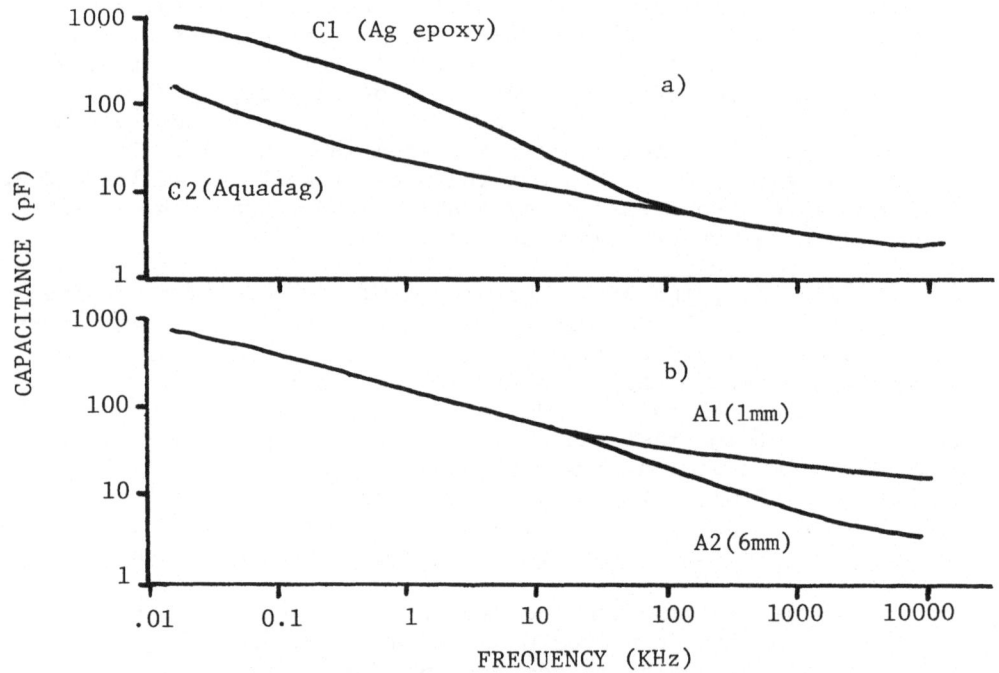

Figure 4. Capacitance versus frequency for samples with
a) different contacts (same thickness);
b) different thickness (same contacts)

It is seen that, for this sample, the contact resistance is nearly as large as the bulk rubber resistance. The contact capacitance is much larger than that of the bulk rubber, which is oh so detrimental as it appears, since

$$C = \frac{C_c C_B}{C_c + C_B} = C_B \text{ if } C_c \gg C_B.$$

Thus, for samples/contacts such as these, contact resistance is not significant at frequencies above about 100 KHz. However, for lower frequency measurements (including DC) it can be detrimental and must be taken into account.

Acknowledgements

We would like to acknowledge the Office of Naval Research and U.S. Army-TACOM for supporting this work.

REFERENCES

1. Leslie E. Cross, Bull. Amer. Ceram. Soc. 63 [4], 586 (1984).
2. William J. Minford, IEEE CHMT Proc., CHMT-S, 297 (1982).
3. M. A. Lampert and P. Mark, Current Injection in Solids, Academic (1970).
4. L. C. Burton, Proc. 1984 Electronic Components Conference, p. 177 (IEEE Catalog No. 84CH2030-5) To appear in IEEE Trans. CHMT.
5. D. A. Seanor (Ed.), Electrical Properties of Polymers, Academic (1982).
6. L. C. Burton, 125th Meeting Rubber Division, Amer. Chem. Soc. 1984 (paper No. 80) To appear in Rubber Chem. & Technol.

7. E. K. Sichel (Ed.), Carbon Black-Polymer Composites, Dekker (1982).

8. R. H. Norman, Conductive Rubbers and Plastics, Elsevier (1970).

THERMAL ANALYSIS OF ORGANIC BINDERS FOR CERAMIC PROCESSING

J. P. Pollinger and G. L. Messing

Pennsylvania State University
Department of Materials Science and Engineering
University Park, PA 16802

INTRODUCTION

Organic binders are used extensively in ceramic fabrication processes such as dry and isostatic pressing, tape casting, injection molding, slip casting and extrusion. Depending on the specific application, the organic may have to provide plasticity during forming,[1,2] strength to the green part after forming,[3] and be easily removed by either oxidation and/or pyrolysis without adversely affecting the green microstructure.[4,5] While a wide range of organics exist, only a limited number of polymer systems meet all of these requirements as well as the rheological needs for powder processing prior to forming.[6-8] For a given polymer class properties vary as a function of polymer molecular weight, composition, and the amount and type of plasticizer. While an extensive literature exists for most of the important polymer classes, the properties that are important for ceramic processing have often not been evaluated. Consequently, selection of organics for ceramic processing is based on experience in the industry and not on the known properties of the organic. While in many cases this approach results in adequate formulations, many times the engineer is stymied when alternative binder systems must be considered. Polymer properties influencing binder selection for ceramic processing include the glass transition temperature (Tg), crystallization temperature (Tc), melting temperature (Tm), viscoelastic response, the initial and final decomposition temperatures, and amount of residue after binder decomposition. The glass transition temperature is of particular importance for ceramic processing and is defined as the temperature or range of temperatures below which an amorphous polymer is glassy and rigid and above which it is flexible and deformable.[9] Clearly, a polymer's Tg must be less than the forming temperature for forming processes requiring binder deformability such as in dry powder compaction or tape lamination for fabrication of multilayer structures. For multilayer fabrication it is also important that the tape remain flexible for room temperature handling; thus, the binder Tg must be less than room temperature. It is clear that present state of the art thermal analysis equipment can significantly assist in the selection of organic additives by providing detailed fundamental information about the thermal response of the organic.

This paper addresses organic binder characterization by thermal analysis as a means of establishing a more fundamental basis for binder selection and formulation. Specific thermal analysis techniques that will be discussed include differential scanning calorimetry (DSC), thermomechanical analysis (TMA), dynamic mechanical analysis (DMA), and thermogravimetric analysis (TGA). Using polyvinyl butyral as an example, each technique will be discussed in terms of the fundamental mechanism of data accumulation, sample preparation, data interpretation and relative merit as a characterization tool.

ORGANIC SYSTEM

Polyvinyl butyral (PVBu) is a widely used organic binder for organic solvent-based tape casting. It is a terpolymer of vinyl butyral, vinyl alcohol, and vinyl acetate and is available in grades ranging from 80-88% polyvinyl butyral, 9-21% polyvinyl alcohol (PVOH), and 0-2.5% polyvinyl acetate (PVAc). The hydroxyl and acetate content effect solubility, water absorption, and other film properties. BVBu also exists in a range of weight averaged molecular weights from 30,000 to 225,000 which has a major effect on solution viscosity, binder glass transition temperature, and mechanical properties.[10]

The PVBu used in these examples has an average composition of 17.5-21.)% PVOH, 0-2.5% PVAc and 80% PVBu with an average molecular weight of 125,000.[a] Polyoxyethylene aryl ether (POAE),[b] a plasticizer, was also added in various amounts to PVBu to examine its effect on binder system thermal response.

Experimental preparation of organic binder samples for subsequent thermal analysis consisted of casting thin films of plasticized and unplasticized PVBu to simulate actual processing

---

[a] Butvar B-74, Monsanto Co., Springfield, MA.
[b] Pycal 94, ICI Americas, Inc., Wilmington, DE.

conditions encountered by the polymer during ceramic forming. Films were formed by slowly adding a 5 wt% binder-plasticizer mixture to isopropyl alcohol. The solutions were placed in glass petri dishes and dried at 70°C for 48 hours to evaporate the solvent and then stored in a dessicator at room temperature until thermal analysis.

THERMAL ANALYSIS TECHNIQUES

A heating rate of 10°C/min was used for thermal analysis. Purge gas flow rates were 0.2 SCFH for DSC, TMA, and DMA and 0.5 SCFH for TGA. The extrapolated onset temperature of a transition was measured by the intersection of tangents method described in ASTM Standard D3418.[11]

Differential Thermal Analysis (DTA) is a technique where the temperature difference between a substance and a reference material is measured as a function of temperature and/or time whereas DSC is a technique where the energy necessary to establish a zero temperature difference between a substance and a reference material is determined as a function of temperature and/or time. Typical DTA and DSC instruments available commercially are shown in Figures 1 and 2, respectively. Both techniques measure the same thermal properties although DSC is a more sensitive technique since it measures energy flow instead of differences in temperature. Commercial DSC instruments are limited to a maximium temperature of 725°C, whereas DTA cells are available up to 1600°C and higher.

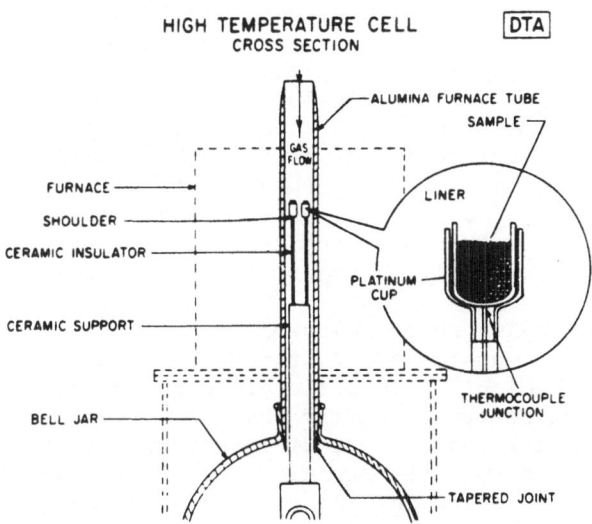

Fig. 1. A schematic of a typical DTA cell (Du Pont Co.).

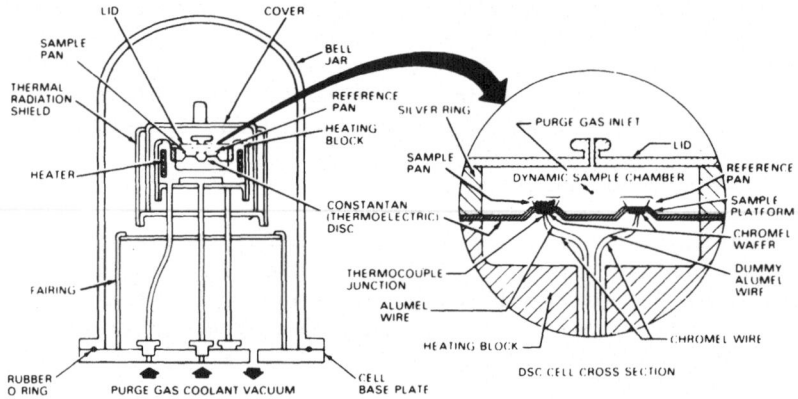

Fig. 2. A schematic of a typical DSC cell (Du Pont Co.).

Because of its greater sensitivity and the lower temperature needs
for thermal analysis of organic binders, DSC is the preferred
instrument for organic binder thermal analysis. For DSC either a
film or powder is placed in a 0.5cm diameter sample pan, typically
aluminum, but which also may be platinum or gold for use at higher
analysis temperatures (e.g>600°C) or when reactive samples or
atmospheres are used.

Thermal properties usually measured by DSC include transition
temperatures such as melting, crystallization, and decomposition, as
well as the glass transition temperature. The enthalpy of a
transition can be directly measured from the area underneath the
transition peak. In DSC[c] of unplasticized PVBu, 10 mg samples of
0.1 cm thick films were heated in open aluminum pans in air and
nitrogen (Figure 3). In air PVBu decomposes through a complex series
of exothermic oxidation reactions between 300 and 500°C while in
nitrogen PVBu decomposes by a broad endothermic pyrolysis reaction
between 350 and 450°C. Because of its large molecular weight and
physical structure PVBu does not crystallize and thus
crystallization and melting transitions are not observed by DSC.

A material's Tg can be measured by DSC since its heat capacity
changes discontinuously at this transition[12] resulting in a
baseline shift which is proportional to the heat capacity difference
of the material before and after the transition. In some systems
this shift is small and may be difficult to detect. Therefore, to
optimize its detection the organic sample is encapsulated between
aluminum sample pans to increase its physical contact with the pans
and to isolate it from reaction with the atmosphere. To measure the
Tg when it was less than room temperature the chamber and sample
were rapidly cooled with liquid nitrogen-cooled nitrogen gas. Shifts

[c]Model 910, Analytical Instruments Div., Du Pont Co., Wilmington, DE.

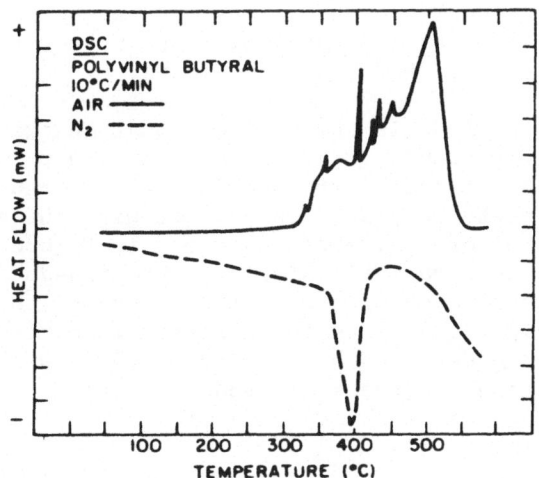

Fig. 3. DSC of PVBu in air and in nitrogen.

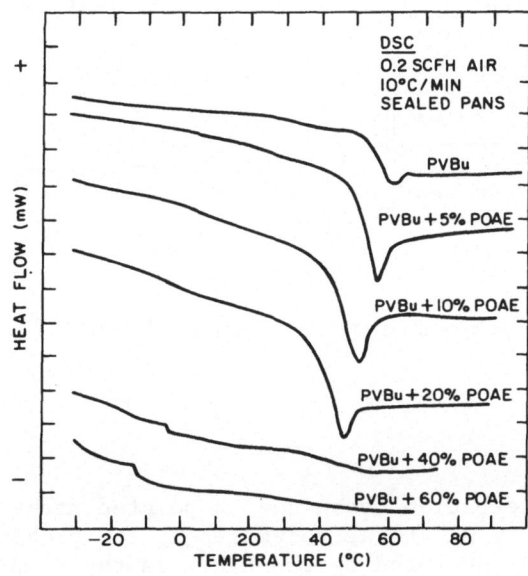

Fig. 4. Effect of plasticizer content on the Tg of PVBu determined
by DSC.

in the Tg of PVBu as a function of POAE content are shown in Figure 4. Plasticizer additions to 20% reduce Tg from 54 to 37°C and 40% POAE reduces the Tg to -3°C. For fabrication processes requiring a flexible polymer at room temperature, the 40% plasticized PVBu would be a good candidate.

TMA is a technique in which dimensional changes are measured as a function of temperature and/or time with either small compressive or tensile loads applied to the sample during measurement. Typically a fused quartz probe is placed in contact with the surface of a sample film of uniform thickness as shown in Figure 5. The probe is attached to a linear variable displacement transducer (LVDT) to measure sample expansion or contraction. Because the thermal expansion coefficient of a material changes discontinuously at its Tg[12] the Tg can be detected by any one of three different TMA techniques- penetration, compression, and tension[13].

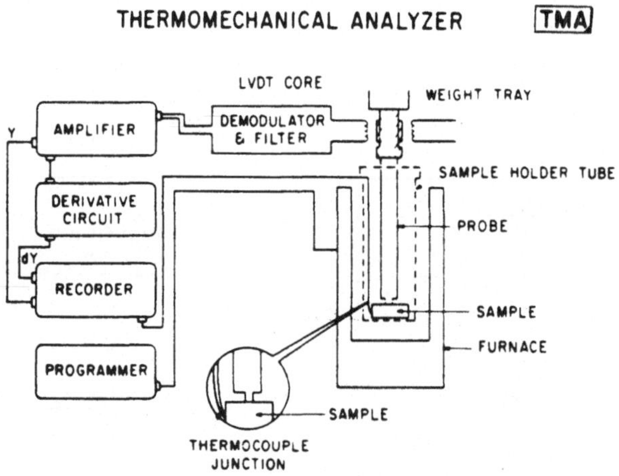

THERMOMECHANICAL ANALYZER   TMA

Fig. 5. A schematic of a typical TMA cell (Du Pont Co.).

In the penetration technique, a pointed probe is placed in contact with a film. At the polymer's Tg the probe penetrates into the smple and is detected by the LVDT. In the compression technique a compressive load is applied to a flat faced fused quartz probe that contacts the sample surface. The sample must have parallel flat surfaces so that the probe can squarely contact it. Either an increase in the thermal expansion coefficient or a sharp contraction of the sample is generally observed for polymers at the Tg. In the tensile technique a tensile load is applied to opposite ends of a film sample in the TMA and a marked elongation is observed as it is heated through Tg.

The effect of plasticizer content on the Tg of 0.5cm by 0.5cm by 0.1cm thick PVBu films was measured by the compression method. From Figure 6 it is seen that increasing the plasticizer content decreases the Tg from 53°C for unplasticized PVBu to -20°C for 60 wt% POAE which is in agreement with the DSC results.

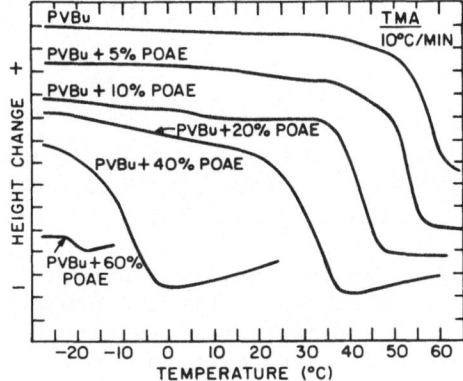

Fig. 6. Effect of plasticizer content on the Tg of PVBu determined by TMA

An advantage of TMA is that sample preparation and setup is simpler than for DTA and DSC where more sample handling is required to insure accurate measurements. The TMA also clearly exhibits the actual binder response to a compressive or tensile load at a specific temperature simulating binder mechanical response in some ceramic forming operations.

In DMA a material of known dimensions is oscillated at constant amplitude and the resonant frequency and damping are measured as a function of temperature and/or time (Figure 7). From this data viscoelastic properties of the polymer can be calculated such as tensile modulus, shear modulus, and mechanical loss.[14] Since the

d Model 942, Analytical Instruments Div., Du Pont Co., Wilmington, DE.

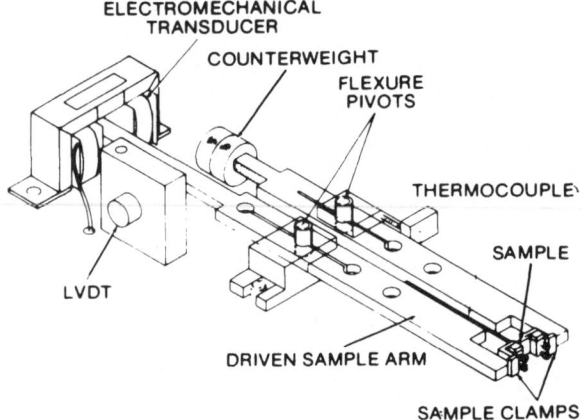

Fig. 7. A schematic of a typical DMA cell (Du PontCo.).

tensile modulus decreases as a material is heated through its Tg and
mechanical loss is maximized, the Tg can be accurately detected. In
fact, DMA is one of the most sensitive thermal analysis techniques
for the detection of second order transitions and is particularly[15]
useful when DSC or TMA are not sensitive enough. Its
disadvantage is that sample preparation and sample mounting are much
more involved than for DSC or TMA. Rectangular sections of known
dimensions of polymer films are used as the DMA samples. The samples
must have sufficient stiffness to support a resonant frequency
measurement. Low stiffness materials may be analyzed by increasing
sample thickness, placing the sample between two metal shims and
measuring the difference in response between it and the metal shims
alone, mounting the sample in the DMA so that it is resonated across
its width rather than its thickness, or by impregnating a fiberglass
weave with the polymer. A typical DMA[e] analysis of unplasticized
PVBu film of dimensions 1cm by 4cm by 0.1cm shows a sharp decrease
in resonant frequency and a corresponding maximum in damping at
PVBu's Tg of 54°C (Figure 8).

In TGA the weight of a substance is measured as a function of
temperature and/or time in an environment heated or cooled at a
constant rate. Weight changes due to solvent evolution,
decomposition, and the amount of residue after decomposition or

[e]Model 982, Analytical Instruments Div., Du Pont Co., Wilmington, DE.

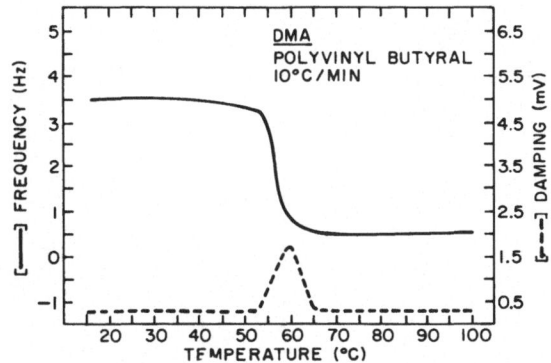

Fig. 8. Detection of the Tg of PVBu by DMA.

pyrolysis can be analyzed and the effect of environment (e.g. reactive versus inert), and heating rate on decomposition can be determined. Samples are placed in either aluminum or platinum boats connected to a microbalance[f] as shown in Figure 9. Figure 10 shows the effect of PVBu sample preparation technique upon decomposition.

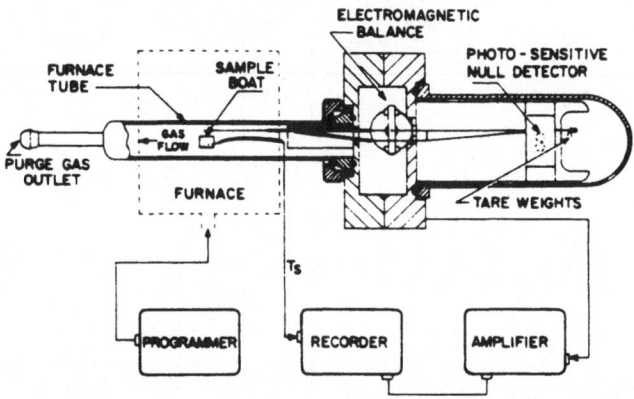

Fig. 9. A schematic of a typical TGA cell (Du Pont Co.).

[f]Model 951, Analytical Instruments Div., Du Pont Co., Wilmington, DE.

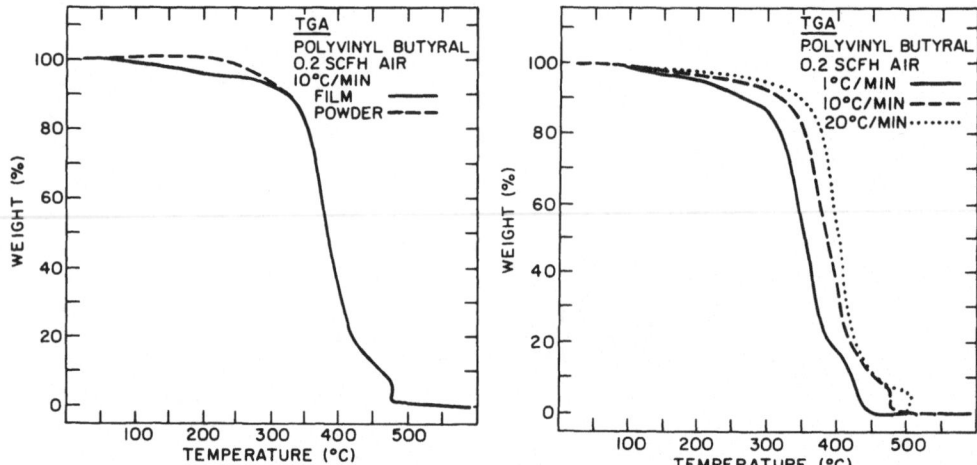

Fig. 10. Effect of PVBu sample preparation on PVBu decomposition determined by TGA.

Fig. 11. Effect of heating rate on PVBu decomposition determined by TGA.

The PVBu starting powder exhibits no weight loss when heated in air until 200°C whereas the PVBu solution cast film starts to lose weight at 70°C possibly due to solvent retained in the film

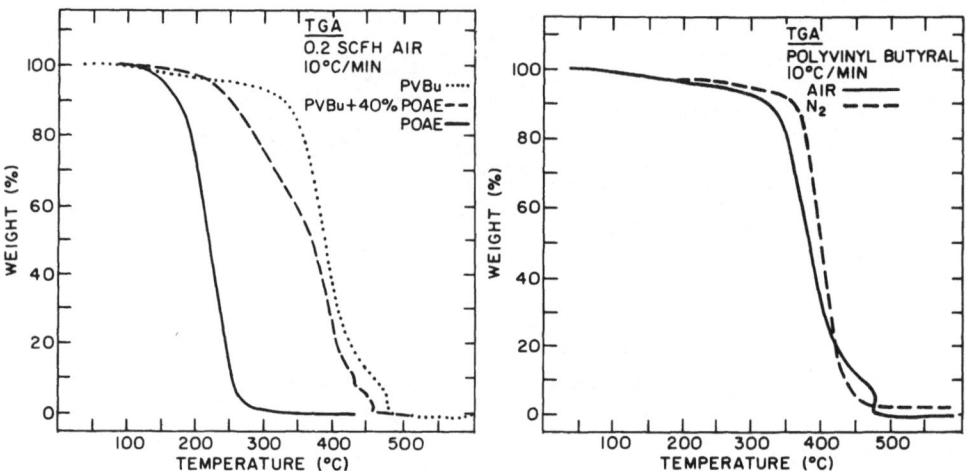

Fig. 12. Effect of plasticizer addition on PVBu decomposition determined by TGA.

Fig. 13. Effect of atmosphere on PVBu decomposition determined by TGA.

after drying or due to moisture pickup. The decomposition however occurs at the same rate and at the same temperatures. The effect of heating rate on the decomposition of 100 mg samples of 0.1cm thick PVBu film is shown in Figure 11. It is seen that decreasing the heating rate from 20°C/min to 1°C/min results in a decrease in the initial and final decomposition temperatures of 50°C. In Figure 12 addition of 40% plasticizer decreases the initial decomposition temperature by 100°C and the final temperature by 25°C. Figure 13 shows the effect of atmosphere upon PVBu decomposition where it is observed that decomposition in nitrogen begins 40°C higher than in air but occurs faster. It also ends 30°C lower than in air and leaves ∿2 wt% of black tar-like residue. When the residue was reheated in air it decomposed completely between 400 and 450°C.

CONCLUSIONS

It has been shown that each of the four thermal analysis techniques discussed gives unique information on polymer properties of organic binders. DSC gives information on binder Tg, Tc, Tm, and the nature of decomposition reactions. TMA can determine Tg as well as show the physical response of a binder to a mechanical load. DMA, besides being the most sensitive technique for determination of Tg, also measures the viscoelastic response of polymers as represented by tensile modulus and mechanical loss. TGA gives information on binder decomposition such as the effect of atmosphere, heating rate, and plasticizer on initial and final decomposition temperatures and amount of residue. A qualification should be noted though when examining properties of polymers for ceramic processing. When the binder is added to a ceramic powder the nature of the binder response to temperature and applied loads changes. Thus, the effect of ceramic powder on binder properties should also be considered before binder selection. Important parameters include ceramic powder particle size, powder surface chemistry, and the amount of binder and plasticizer additions.

ACKNOWLEDGEMENTS

The use of the thermal analysis instrument schematics supplied by Du Pont Co. is acknowledged.

REFERENCES

1. R.A. Dimilia and J.S. Reed, Am. Ceram. Soc. Bull., 62 (4), 484-488 (1983).
2. C.W. Nies and G.L. Messing, J. Amer. Ceram. Soc., 67 (4), 301-304 (1984).
3. G.Y. Onoda, J. Am. Ceram. Soc., 59 (5), 236-239 (1976).

4. S.L. Levine, <u>Am. Ceram. Soc. Bull.</u> , <u>48</u> (2), 230-231 (1969).

5. K. Otsuka and S. Ogihara, <u>Yogo-Kyokai-Shi</u> , <u>92</u> (4), 210-218 (1984).

6. G.Y. Onoda, in <u>Ceramic Processing Before Firing</u> , G.Y. Onoda and L.L. Hench (eds.), New York: John Wiley and Sons, 235-251 (1978).

7. A.G. Pincus and L.E. Shipley, <u>Ceram. Ind. Mag.</u> , April, 106-109 (1969).

8. N. Sarkar and G.K. Greminger, Jr., <u>Am. Ceram. Soc. Bull.</u> , <u>62</u> (11), 1280-1288 (1983).

9. S.L. Rosen, <u>Fundamentals and Principles of Polymeric Materials</u> , New York: John Wiley and Sons, 84-97 (1982).

10. <u>Butvar Polyvinyl Butyral-Properties and Uses</u> , Technical Bulletin No. 6070B, Monsanto Polymers and Petrochemicals Co., St. Louis, MO.

11. Annual Book of ASTM Standards, VO8.03 and 14.02 (1984).

12. D. Turnbull and B.G. Bagley, in <u>Treatise on Solid State Chemistry</u> , N.N. Hannay (ed.), New York: Plenum Press, 555-593 (1975).

13. J.P. Lear and P.S. Gill, <u>Theory and Operation of the Du Pont 982 Dynamic Mechanical Analyzer</u> , Analytical Instruments Div., Du Pont Co.

14. K.F. Baker, Du Pont Instruments Application Brief TA-50.

15. P. Burroughs and M.G. Lofthouse, Du Pont Instruments Application Brief TA-74.

THE THERMAL DIFFUSIVITY OF SIMULATED NUCLEAR WASTE GLASS

BELOW THE GLASS TRANSITION TEMPERATURE

Marco Y. Liem, L. D. Pye and D. Bickford[*]

Alfred University and [*]E. I. DuPont Company

Alfred, New York and Aiken, South Carolina

ABSTRACT

The thermal diffusivities of three simulated nuclear waste glasses were studied from room temperature to the glass transition temperature. A non-steady state method similar to Plummer's was used. The technique involves a planar heat source which causes a temperature rise in a sample. The rise is measured at two positions in the sample and the thermal diffusivity can be calculated from the ratio of these two temperatures. The only quantities to be measured are the temperature as a function of time and the distance separating the thermocouples. Semi-computer automation of the experiment enabled multiple measurements to be made which otherwise would have been impractical. Data were collected by a digital multimeter interfaced to a computer which facilitated numerical analysis.

Our experimental techniques and methods of analysis were confirmed by measuring the thermal diffusivity of vitreous silica and comparing our results to those reported in the literature. Our measured value of 0.0081 cm^2/sec (at 200°C) was within five percent of the average of calculated literature values. The measured thermal diffusivity of the simulated nuclear waste glasses ranged from 0.0018 to 0.0031 cm^2/sec at 150°C. The lower thermal diffusivity of these glasses, compared to vitreous silica, is probably related to their higher iron contents.

INTRODUCTION

Thermal diffusivity, D, is defined as the ratio of thermal conductivity, K, to specific heat per unit volume. The quantity-

$$D = K/s \ d$$

where s is the specific heat, and d the density, was first called thermal diffusivity by Lord Kelvin. Conductivity is important in steady state conditions. However, when transient heat flow is important, it is necessary to know the diffusivity rather than conductivity. Also, in discussing the phonon scattering process, diffusivity is more significant because the mean free path of the scattered phonon is more directly related to diffusivity than to conductivity.

Although D and K are simply related, the experimental techniques for measuring them are very different. Thermal conductivity, measured by steady state methods, requires measuring both a thermal flux and a temperature gradient. Thermal diffusivity, with units of $(length)^2/time$, requires measuring the time it takes for a thermal disturbance to propagate a known distance. In principle, time and distance can be measured more accurately than thermal gradients and heat fluxes.

In the nuclear waste isolation program, a thorough knowledge of the thermal properties of the glasses containing nuclear waste is of great importance. Nuclear waste is generated in both the defense and electrical power industries. The safe disposal of nuclear waste is a critical issue of our times. Accordingly, the need to find a suitable matrix for long-term storage of nuclear waste has led to various alternatives. One of these is borosilicate glass which has numerous favourable characteristics including high resistance to corrosion, low leachability, ability to accept all known radionuclides, high heat capacity, good resistance to radiation damage and relative ease in processing. One method used is to simply mix calcined waste with a special glass frit, melt, and pour the mixture into a stainless steel canister which serves as the first of a series of barriers. These canisters can be quite large (1.0 x 0.2 m) [15] and a knowledge of the thermal diffusivity of the interior glass is important in determining thermal gradients which might arise during annealing or storage.

In this work we measured the thermal diffusivities of three simulated compositions provided by the E.I. duPont de Nemours Savannah River Laboratory at Aiken, South Carolina. The three glasses used had varying amounts of iron and are referred to as HiFe (high iron content), TDS (intermediate iron content) and HiAl (low iron content). Measurements were made from room temperature to the glass transition temperature. The thermal diffusivity of vitreous silica was also measured and compared with previously reported values as a means of confirming our experimental procedures and analysis.

There are three categories of experimental methods for determining thermal-transport properties of materials. These are a) steady-state temperature methods which are used to measure

thermal conductivity and non-steady state methods which include
b) periodic and c) transitory heat flow, used for measuring thermal
diffusivity. In periodic methods the thermal energy supplied to a
sample is modulated with a fixed period. The temperatures at all
points in the sample vary with the same period and D is determined
from measurements of the amplitude and phase relationships in a
sample. The periodic heat flow methods can be subdivided into
longitudinal and radial heat flow. In either case, a sinusoidal
temeperature variation is introduced and the temperature is
monitored at two positions seperated by a known distance. The phase
lag of the heat may be used to measure the thermal diffusivity.

In transitory temperature methods the addition of thermal
energy causes a transitory temperature change as the sample tries to
reach a new equilibrium. D can be determined by measuring the
temperature as a function of time at different points in the sample.
The mathematics used to solve the heat equation depend on the
particular set of boundary conditions chosen. Different transitory
methods include long rods, flat plates, cylinders, comparative
methods, moving line sources and flash methods.

For this work the flat plate method was used. The flat plate
sample geometry is desirable for both experimental and mathematical
reasons. The mathematics for a semi-infinite solid may be applied
and sample preparation is uncomplicated. The only quantities to be
measured are the distance between thermocouples and the temperature
as a function of time at two locations.

A brief discussion of thermal-transport theory is presented
below along with the mathematical basis for measuring thermal
diffusivity in a semi-infinite solid. The apparatus, sample
preparation and experimental techniques are then described.

THEORY

Heat transfer through a dielectric solid takes place either
by conduction or radiation. Radiation involves the transfer of heat
through a medium by transmission, absorption and re-radiation of
electromagnetic energy. At elevated temperatures this method of
heat transfer becomes important and is generally treated in terms of
photon conductivity. Conduction occurs by the transfer of kinetic
energy via thermoelastic waves. The kinetic energy of an atom, ion
or molecule is transferred to an adjacent particle by collision.
The amount of energy transferred depends on the vibrational energy
of the interacting atoms and should be considered in terms of phonon
conductivity. Phonon conduction involves a mean free path concept
which is the distance a phonon will travel before being scattered by
other phonons. By analogy with the corresponding expression in the
kinetic theory of gases, one can define for phonon conduction the
quantity, l, the mean free path, by the equation:

$$K = 1/3 \ cvl$$

where K is the thermal conductivity, c the heat capacity per unit volume, and v the average phonon velocity. In glasses, the random structure limits the mean free path to the dimensions of a single-structure element (i.e. the silica tetrahedron). Values of 1, calculated from experimental values of conductivity, heat capacity, phonon velocity and density are in the range of 5.2A for vitreous silica and 3.3A for a borosilicate glass [7]. The mean free path differences are due to increased phonon scattering in multicomponent glasses as compared with vitreous silica.

A semi-infinite solid, with an initial uniform temperature, is heated by a source of constant flux. Carslaw and Jaegar [11] give an expression for the temperature rise as:

$$T(xn) = 2F/K \ (D \ t)^{1/2} \cdot ierfc(xn) \tag{1}$$

where:
- T = temperature
- F = heat flux
- K = thermal conductivity
- D = thermal diffusivity
- t = time
- x = thickness, the heat source is at x = 0
- n = $1/2(Dt)^{1/2}$
- ierfc = complimentary integrated error function

$$ierfc(y) = \pi^{-1/2} exp(-y^2) - y[1 - 2 \ \pi^{-1/2} \int_o^y e^{-B} \ dB]$$

Experimental conditions which must be met for this equation to hold true are 1) constant flux, 2) unidimensional heat flow, 3) infinite thickness (in practical terms, a slab of sufficient thickness so that the heat flow is essentially the same as it would be in an infinite slab) and 4) no interfacial resistance to the flow of heat. The diffusivity is determined by measuring the temperature rise at the heat source, x = 0, and at the heat sink, a distance x, from the heater. For x = 0, the integrated error function reduces to $\pi^{-1/2}$ so:

$$T(0) = 2F/K \ (Dt/\pi)^{1/2} \tag{2}$$

The ratio, R(xn), of the temperature rises is therfore:

$$R(xn) = T(xn)/T(0) = \pi^{1/2} \ ierfc(xn) \tag{3}$$

Equation 3 shows that the ratio of the temperature rises is a function of the diffusivity, thickness and time. The integrated error function is well tabulated. Therefore, if the thickness and time are known the diffusivity can be determined. Values for the ratio may be plotted as a function of the dimensionless parameter xn

as shown in Figure 1. For a given sample thickness a curve as shown
in Figure 2 is obtained giving the ratio as a function of the diffusivity and time. The diffusivity may be determined either
graphically or by computer.

EXPERIMENTAL APPARATUS

A drawing of the specimen assembly used is shown in Figure 3.
The semi-infinite solid has a sandwich construction with a sample
measuring 12.5 cm long by 7.5 cm wide and 1.0 cm thick. Above and
below the sample are thicker specimen pieces which house the
thermocouples and function as a thermal guard. The planar heat
source is a 5-mil Chromel sheet silver-soldered to Inconel rods
which carry current from the transformer to the heater. Two 500 VA
tranformers in parallel with a step down ratio of 10:1 were used to
supply power to the heater. A V-10 Variac on the primary side of
the transformer was used to adjust the power. A current in the
range of 4 to 6 amps gave the best results. The assembled sandwich
was placed in a stainless steel frame. The top half of the frame
slides up to accomodate different assembly thicknesses and also to
hold the sandwich securely together. The frame was loaded with a
modest weight (5000 g) to insure good contact between the heater and
specimen and to minimize the air interface between the two. It was
removed during temperature changes so as to minimize the chance of
sample fracture. Excess pressure could cause point loading and
fracture of the specimen. The assembly was placed in a Sybron
Thermolyne 10500 Electric Furnace.

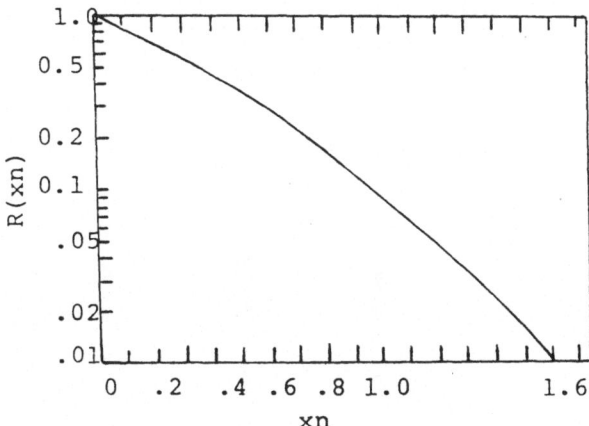

Figure 1. The ratio of the temperature rises as a function of the
dimensionless product xn.

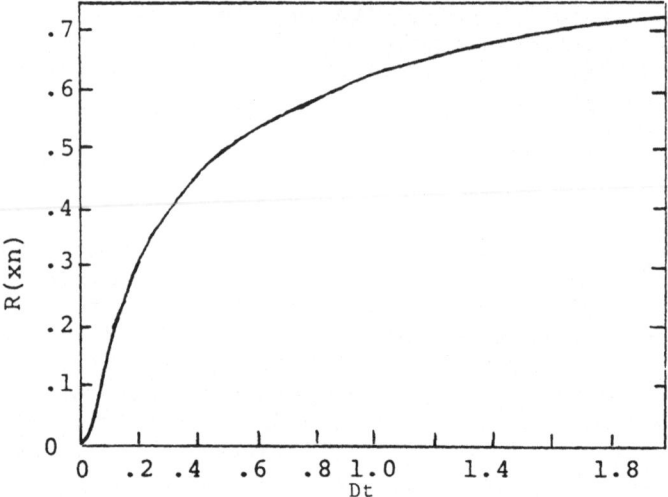

Figure 2. The ratio of the temperature rises as a function of the diffusivity and time.

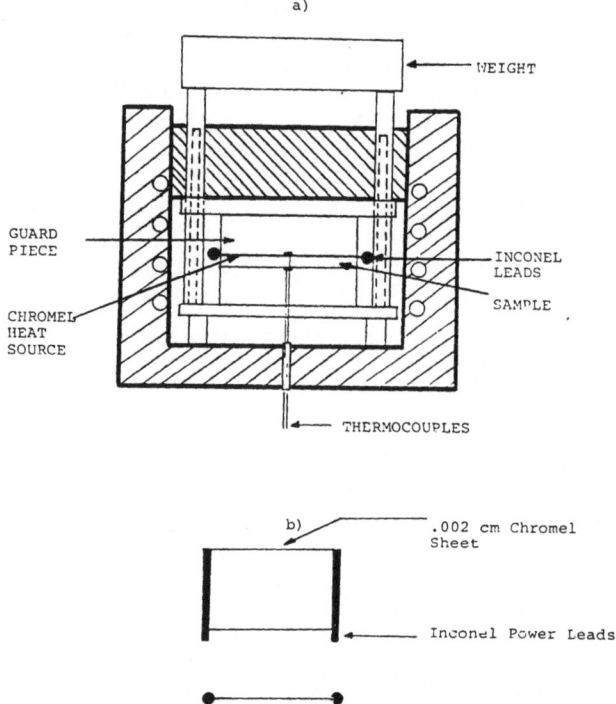

Figure 3. a) Schematic diagram of the assembly used for thermal diffusivity measurements, b) Heater.

The thermocouples were made from 0.006 cm chromelalumel in single bore 0.012 cm tubing. Grooves of sufficient size were cut into the guard pieces of the sample material to accept the tubing. The thermocouple for the heat source lay on top of the heater and that for the heat sink underneath the sample. The guard pieces ranged in thickness from 2.0 to 4.0 cm. The total thickenss is sufficient to satisfy the boundary conditions for an infinite thickness. For diffusivities higher than 0.15 cm^2/sec this condition no longer holds true [1]. Figure 4 shows a photograph of the unit with vitreous silica samples in place.

A schematic of the experimental arrangement is shown in Figure 5.

Figure 4. Photograph of the assembly with vitreous silica sample in place.

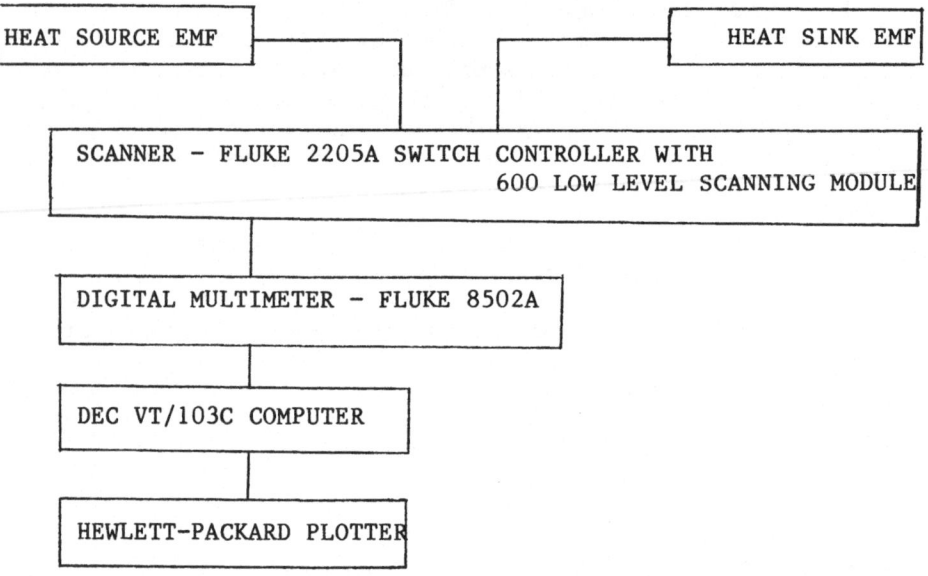

Figure 5.   Schematic for data collection.

The emf outputs were scanned using a Fluke 2205A Switch Controller linked to a Fluke 8502A Digital Multimeter.   The multimeter was interfaced with an RS 232 to a VT/103/C DEC Computer which collected the emf outputs as a function of time.   The computer converted the emf's to temperature and stored the values for subsequent analysis.

GLASSES STUDIED

The glasses received from Savannah River Laboratory had been previously batched, melted and finely ground.   Samples were prepared for this study by remelting the glass at 1100°C for approximately 1/2 hour which was sufficient for fining.   The glass was poured into stainless steel or graphite molds and was subsequently annealed for two hours.   Annealing temperatures were determined through related thermal expansion studies.   The annealed pieces were ground and polished by the Swift Glass Co. of Elmira, New York, to 1.0 cm thickness with a tolerance of 0.003 cm.

Samples of vitreous silica, kindly supplied by Corning Glass, were first measured in order to determine how well the method worked.   Values for thermal diffusivity were calculated from thermal conductivities and specific heats reported in the literature.   The three simulated nuclear waste glasses varied primarily in their iron content.   The nominal compositions of several glasses are shown in Table I.

TABLE I   NOMINAL GLASS COMPOSITIONS (Wt%)

OXIDE	VITREOUS SILICA	CORNING 7900	SYNTHETIC TEKTITE	HIFE	TDS	HIAL
SiO_2	99.0	96.0	75.0	49.91		54.29
B_2O_3		3.0		6.84		7.39
Al_2O_3			11.5	1.77		9.28
Na_2O			1.2	11.54		11.76
K_2O			2.0			
PbO			1.8			
FeO			4.4			
Fe_2O_3			0.5	14.54	11.8	6.35
TiO			0.5			
CuO			0.40			
Li_2O				4.76		5.14
MgO				0.74		0.80
ZrO_2				0.91		1.01
MnO_2				6.02		2.17
CaO				1.13		1.14
NiO				1.83		0.68

MEASUREMENTS

Measurements were taken at intervals of approximately 30°C up to the glass transition temperature. Prior to each measurement the sample was allowed to come to thermal equilibrium, i.e. the temperature difference between the heat source and heat sink was at most 0.5°C (generally much less). This temperature was recorded as the zero-time temperature. Power was then applied to the heater, and the temperatures of the heat source and heat sink, were recorded at two second intervals. Figure 6 shows temperature as a function of time for the heat source and heat sink in a typical experiment.

It is seen that there is an initial rapid temperature rise at the heat source (from 106.5°C to 107.5°C) whereas the temperature of the heat sink rises gradually. The ratio of temperature rises is:

$$R(xn) = T(xn)/T(0)$$
$$T(xn) = T_t - T_{t=0} \text{ for the heat sink}$$
$$T(0) = T_t - T_{t=0} \text{ for the heat source}$$
$$t = time$$

These ratios were calculated for a series of time intervals ranging from 10 seconds to perhaps 220 seconds. The times are chosen such that the values of the ratio are between 0.1 and 0.6. These values correspond to the time where the integrated error function (from which Figure 1 is derived) is most sensitive. As

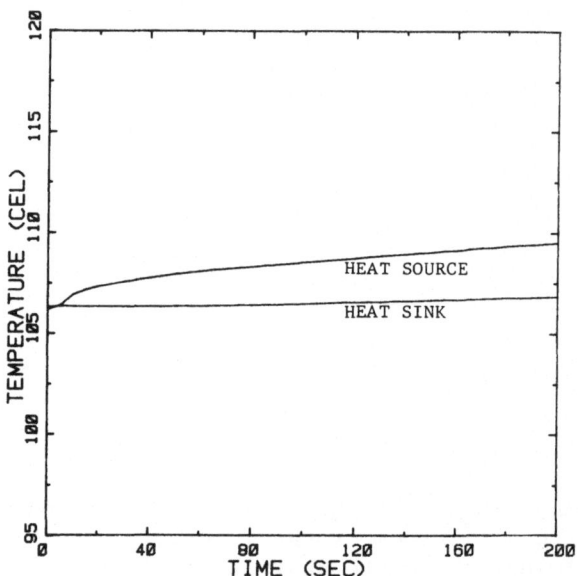

Figure 6. A typical temperature of the heat source and sink as a function of time.

mentioned earlier, a knowledge of R(xn) as a function of time allows calculation of the diffusivity D. It was observed that D varied by a small amount (<10%) at the extremes of the ratio. Therefore, the values of D reported in this work were taken where there was essentially no change of D with time.

RESULTS AND DISCUSSION

There are numerous reported values for the thermal conductivity
of vitreous silica and other glasses. Direct measurements of the
thermal diffusivity of glasses, however, have not been widely
reported. Some work has been carried out with molten glass [12.,
14], but for solid glass, data are rare. Values reported for the
thermal conductivity of vitreous silica vary by as much as two
orders of magnitude [3] at a given temperature, whereas values of
specific heat and density [13] are consistant. From these data,
thermal diffusivities were calculated [Table II] and are plotted in
Figure 7 along with the data obtained in this work. It is seen that

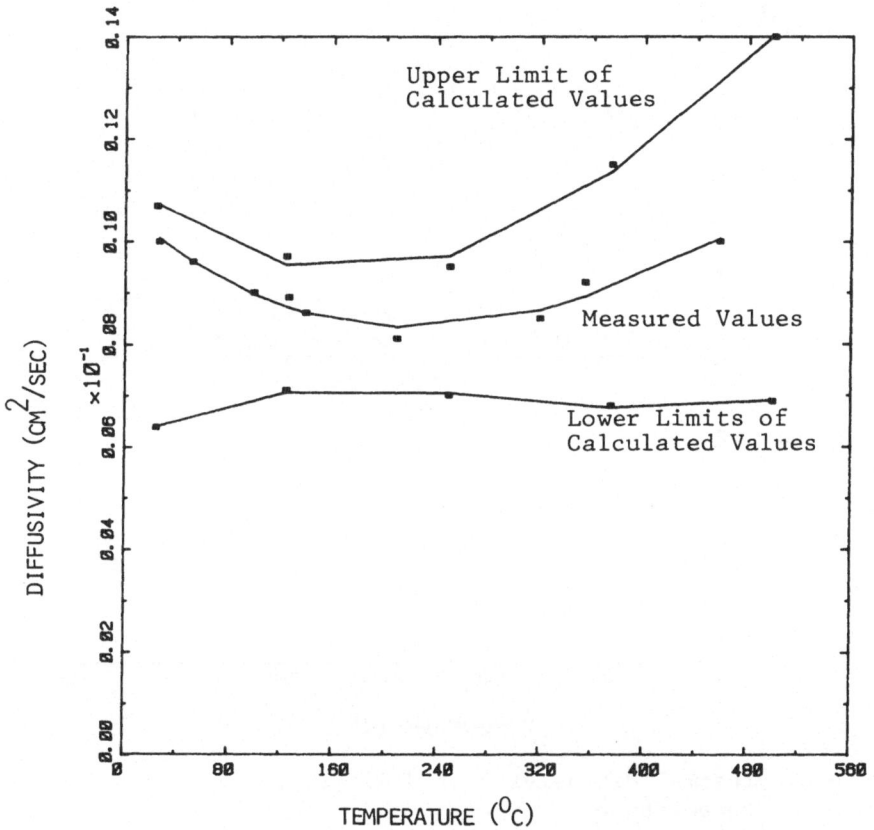

Figure 7. Thermal diffusivity as a function of temperature for
vitreous silica, including the range of values calculated
from the literature.

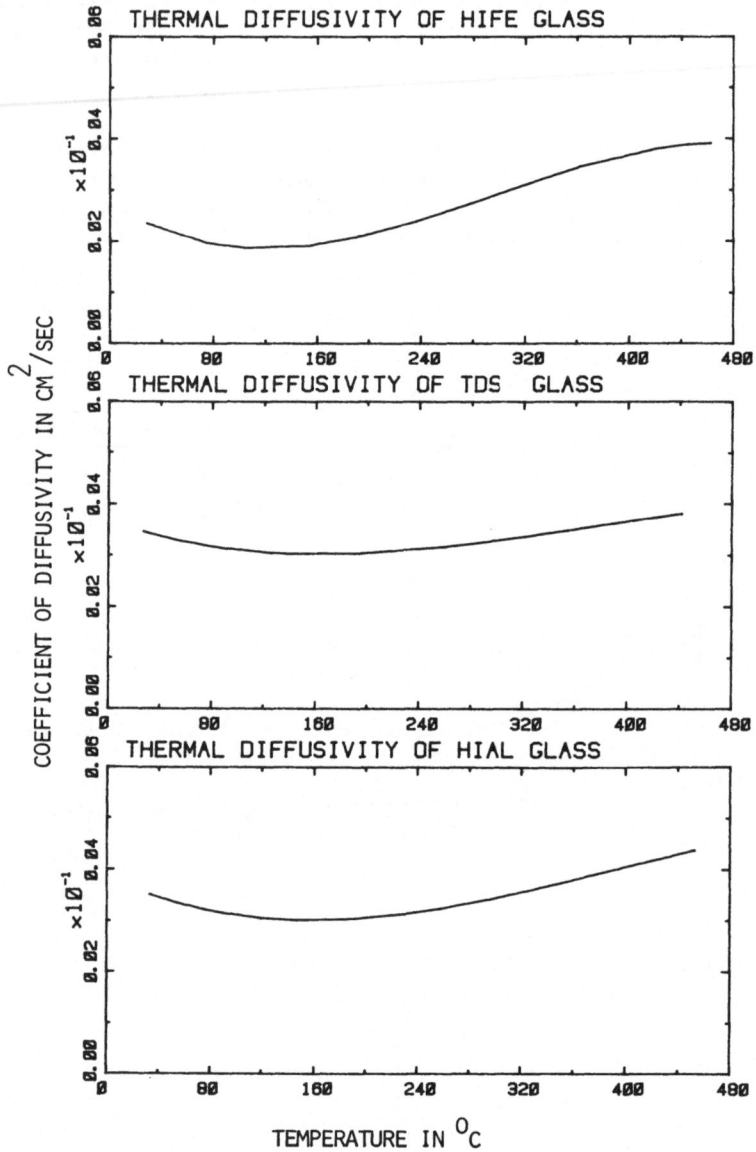

Figure 8.   Thermal diffusivity of HiAl, TDS, HiFe, as a function of
temperature.

the measured values fall in the center of the calculated values, suggesting that our experiments and methods of analysis are acceptable. The results also compare favourably to those reported by Plummer for Corning Code 7900 which is 96% silica.

The results for the three simulated nuclear waste glasses are given in Table III and plotted in Figure 8. Figure 9 shows all the glasses from this study along with a synthetic tektite [1]. Second (HiAl, TDSB) and third (HiFe) degree polynomials were fitted to the data. There is very little deviation from these curves, with

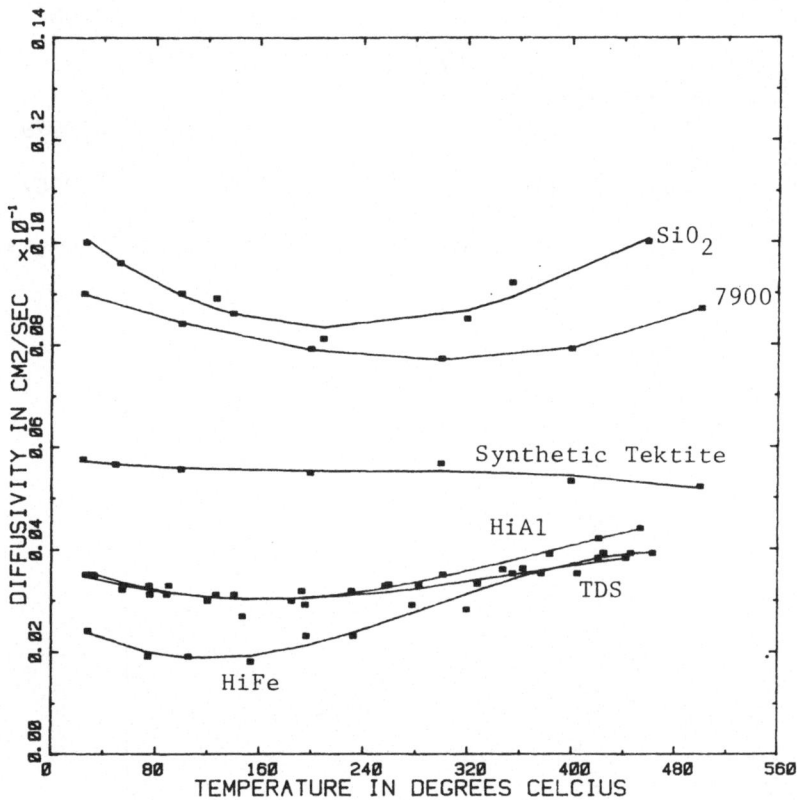

Figure 9. The thermal diffusivity as a function of temperature for all six glasses.

TABLE II. Calculated and measured thermal diffusivity of vitreous
silica and Corning Code 7900. The calculated values
reflect the range of reported conductivities

TEMP °C	VITREOUS SILICA		7900
	$D(cal)^+ (cm^2/sec)$	$D(meas)$	$D^{++} (cm^2/sec)$
25	0.0064–0.0107		0.0090
27		0.0100	
53		0.0096	
96		0.0090	
100			0.0084
125	0.0071–0.0097		
127		0.0089	
140		0.0086	
200			0.0079
210		0.0081	
250	0.0070–0.0095		
300			0.0077
320		0.0085	
355		0.0092	
375	0.0068–0.0115		
400			0.0079
459		0.0100	
500	0.0069–0.0140		0.0087

+ [13] Calculated values reflect range of reported thermal
  conductivities.

++ [1]

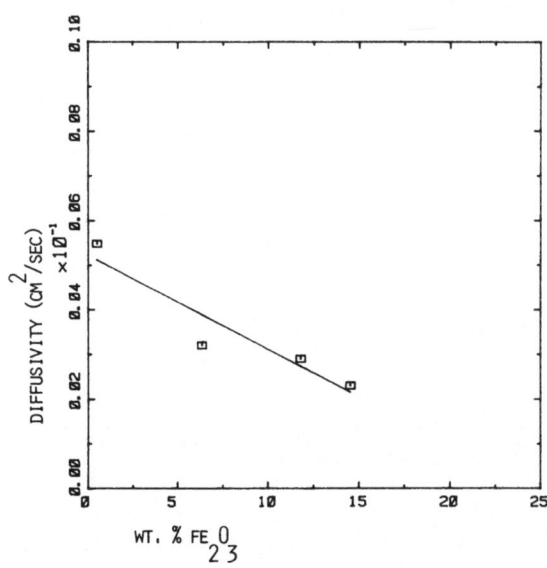

Figure 10.   Thermal diffusivity as a function of iron content.

TABLE III. Measured Thermal Diffusivity Coefficients for Three
Nuclear Waste Glasses and Reported Values for a
Synthetic Tektite [1] [in centimeters squared per
second]

HiFe		HiAl		TDS	
TEMP. (°C)	DIFF.	TEMP. (°C)	DIFF.	TEMP. (°C)	DIFF.
				27	0.0035
29	0.0024	33	0.0035	32	0.0035
		55	0.0033	55	0.0032
75	0.0019	75	0.0033	76	0.0031
106	0.0019	90	0.0033	89	0.0031
		120	0.0030	127	0.0031
154	0.0018	147	0.0027	141	0.0031
197	0.0023	185	0.0030	196	0.0029
		193	0.0032		
233	0.0023	231	0.0032		
		257	0.0033	260	0.0033
278	0.0029			283	0.0033
		301	0.0035		
320	0.0028			328	0.0033
363	0.0028	347	0.0036	355	0.0035
		383	0.0039	377	0.0035
				405	0.0035
421	0.0039	421	0.0042	425	0.0039
446	0.0039	453	0.0044	442	0.0038
463	0.0039				

SYNTHETIC TEKTITE

TEMP. (°C)	DIFF.
25	0.00575
50	0.00565
100	0.00555
200	0.00548
300	0.00565
400	0.00530
500	0.00520

more scatter in the HiFe. This may be due to a decrease in the accuracy of the measurements as the diffusivity decreases. The synthetic tektite has a small amount of iron (0.5% Fe_2O_3). The effect of iron content on thermal diffusivity of the glasses may be seen at 100°C in Figure 10. The diffusivity decreases as the iron content increases, in agreement with that observed by Derby [12].

The diffusivities of the vitreous silica and 7900 rise quicker than the other glasses because of the radiative component having an effect at a lower temperature. The iron glasses are opaque and are not as affected by photon conductivity. At around 400°C the diffusivities tend to converge. At high temperatures (900°C and up) the diffusivities of glasses with similar compositions as these are linear with respect to temperature [12].

At about 160°C the diffusivities approach a minimum value because the conductivity of glass increases slowly with temperature while the specific heat rises rapidly at first and then levels off. In the temperature range examined, phonon conduction is the primary mechanism for heat transport. As the iron content is increased the mean free path apparently decreases thereby lessening the thermal diffusivity.

SUMMARY AND CONCLUSIONS

A fairly straightforward method for measuring thermal diffusivity was used. The primary advantages are that the equations and experimental techniques are quite simple. The boundary conditions are easily satisfied for materials with low diffusivities such as glass. Sample preparation is uncomplicated. The sample size, however, may pose a problem for certain materials. Semi-automation allowed numerous measurements to be made and much data to be collected. This permitted sufficient duplication of measurements to support the results. Comparisons with thermal diffusivities reported in the literature were favourable. However, the information available on thermal diffusivity of glasses is still very limited. It is hoped that our results will encourage further investigations in this field.

REFERENCES

1. W.A. Plummer (Corning Glass Works, Corning, N.Y.) "A Thermal Diffusivity Measurement Technique," (unpublished work).
2. W.A. Plummer, D.E. Campbell and A.A. Comstock, "Method of Measurement of Thermal Diffusivity to 1000°C," J. Am. Ceram. Soc., 45 [7] 310-16 (1962).
3. E.H. Ratcliffe, "A Survey of the Most Probable Values for the Thermal Conductivities of Glasses Between About -150 and 100°C, Including New Data on Twenty-Two Glasses and a Working Formula for the Calculation of Conductivity from Composition" Glass Technol., 4 [4] 113-28 (1963).

4. E.H. Ratcliffe, "Thermal Conductivities of Fused and Crystalline Quartz," Br. J. Appl. Phys., 10 [1] 22-25 (1959).
5. C. Kittel, "Interpretation of the Thermal Conductivity of Glasses," Phys. Rev., 75 [6] 972-74 (1949).
6. K.L. Wray and T.J. Connolly, "Thermal Conductivity of Clear Fused Silica at High Temperatures, J. Appl. Phy., 30 [11] 1702-05 (1959).
7. W.D. Kingery, "Heat-Conductivity Processes in Glass," J. Am. Ceram. Soc., 44 [7] 302-04 (1961).
8. R. Gardon, "A View of Radiant Heat Transfer in Glass," J. Am. Ceram. Soc., 44 [7] 305-12 (1961).
9. W.L. Kennedy, P.H. Siddles and G.C. Danielson, "Thermal Diffusivity Measurements on Finite Samples," Adv. Energy Conv., 2 53-58 (1962).
10. G.C. Danielson and P.H. Sidles, "Thermal Diffusivity and Other Non-steady state Methods"; pp.149-210 in Thermal Conductivity; Vol.2. Edited by R.P. Tye, Academic Press, New York 1969.
11. H.S. Carslaw and J.C. Jaeger, Conduction of Heat in Solids; Ch.II. Oxford University Press, London 1959.
12. J.U. Derby, "The Measurement of Thermal Diffusivity of Simulated Glass Forming Nuclear Waste Melts "; M.S. Thesis, New York State College of Ceramics at Alfred University, Alfred, New York, 1983.
13. A. Goldsmith, T.E. Waterman and H.J. Hirschhorn, Handbook of Thermophysical Properties of Solid Materials, Vol.III pp.891-96. The Macmillan Company, New York, 1961.
14. A.F. Van Zee and C.L. Babcock, "A Method for the Measurement of Thermal Diffusivity of Molten Glass," J. Am. Ceram. Soc., 44 244 (1951).
15. Scientific Basis for Nuclear Waste Management; Vol.1. Edited by G.J. McCarthy. Plenum Press, New York, 1979.

CHARACTERIZATION OF MATRIX-DISPERSOID REACTIONS IN Si$_3$N$_4$-TiC COMPOSITES

Galina Zilberstein and S. Thomas Buljan

GTE Laboratories, Incorporated
40 Sylvan Road
Waltham, Massachusetts 02254

ABSTRACT

Particulate composites offer increased flexibility for unique material property tailoring. Silicon nitride due to its high strength, fracture toughness and thermal shock resistance provides an ideal base for composite material development. Silicon nitride-titanium carbide composites combine the outstanding properties of the silicon nitride matrix with the high hardness of TiC as a dispersoid for exceptional wear resistance in their application as cutting tools and wear parts. Successful composite development, optimization and quality control requires an in-depth understanding of matrix-dispersoid chemical as well as mechanical interactions. Chemical aspects of Si$_3$N$_4$-TiC interphase reaction have been examined. The composition of reaction products formed during processing of Si$_3$N$_4$-TiC composites and effect of impurities on the character and magnitude of reaction is discussed based on results of optical microscopy and electron microprobe analysis.

INTRODUCTION

Silicon nitride-titanium carbide particulate composites form a new class of wear resistant materials. They combine high hardness, fracture toughness, abrasion and corrosion resistance and in recent years have been used in many difficult wear applications. Cutting tools for grey cast iron machining[1-4] provide a prominent example of successful utilization of this new development.

Design and property tailoring of particulate composites, such as Si_3N_4-TiC, require an in-depth understanding of chemical and mechanical interactions between the materials's components. This study concentrates on the chemical aspect of interactions between the Si_3N_4-based matrix and the TiC dispersoid, in the presence of sintering aids and impurities, during densification.

Densification of Si_3N_4+TiC composites proceeds through liquid phase sintering. The liquid, which forms at 1300-1500°C, consists of Y_2O_3 and Al_2O_3 added as sintering aids and SiO_2, which is commonly present in starting Si_3N_4 powders. The liquid may also contain some TiO_2 originating from the TiC powder, and various impurities. Liquid formation is considered an event marking the beginning of densification processes, which also give rise to a number of chemical reactions between the components. Thermodynamic evaluation of reactions proceeding under given hot pressing conditions has indicated high probability of TiN and SiC formation as stable products of Si_3N_4+TiC reaction. However, in a complex system such as Si_3N_4-metal oxide-TiC, in addition to the above mentioned phases, the products of TiC+metal oxide reaction, as well as the products resulting from an interaction between all three Si_3N_4, metal oxides and TiC might be expected to form.

It has been previously shown experimentally that the silicon-yttrium-oxinitrides formed in Si_3N_4-Y_2O_3 hot pressed material contains appreciable amounts of impurities, whereas β-Si_3N_4 crystals are found only in a pure form.[5] Since densification of Si_3N_4-TiC composites occurs in the presence of somewhat similar liquid and results in a complete α-β transformation of Si_3N_4, it was anticipated that the products of Si_3N_4+TiC reaction would either be located in the intergranular glass phase or more likely be concentrated at TiC particle surfaces as reaction zones.

RESULTS AND DISCUSSION

In order to identify reaction products of the composite, processed Si_3N_4, Y_2O_3, Al_2O_3 and TiC powders were hot pressed in argon to full density for an extended period of time. Optical examination of a polished crossection of the composite (Figure 1) reveals the presence of well defined reaction zones of approximately uniform width around every TiC particle. The zones appear to have polycrystalline morphology and to consist of fine grains of at least two phases, one having a characteristic light purple coloration and the other greyish blue. The crystals of both phases constitute a mixture with well defined grain boundaries. The reaction zone has a distinctly different color than the TiC grain interior and is separated from it by clearly defined interface line. The appearance of the smaller TiC particles (1-3 μm in diameter), and occasionally of some larger ones (Figure 1, center of the micrograph), closely resemble those of reaction zones; this suggests that these particles are mainly composed of

the Si_3N_4+TiC reaction products. Since the majority of TiC particles in the studied composite microstructure, under given hot
pressing conditions, develop reaction zones of similar morphology
and of approximately uniform thickness at their grain boundaries,
it may be concluded that such zones represent a general case of
reaction product formation.

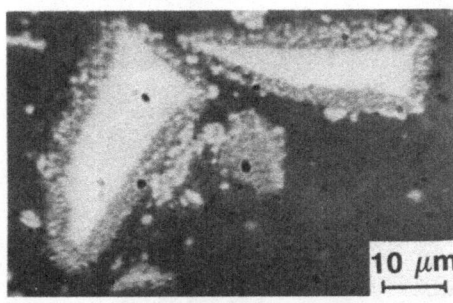

Figure 1: Microstructure
of Si_3N_4IiC com-
posite not pressed
in argon for an
extended period of
time.

Electron probe microanalysis (EPMA)* was employed to examine
the distribution of Ti, Si, Y, Al, Fe, O, N, and C in reaction
products. A back-scattered electron image of a particle illustrating the general case of reaction (A) and one type of completely transformed particle (C) and x-ray maps of the elements
are shown in Figure 2.

As can be seen, locations of Si, Ti and C in the microstructure are well defined. In particle A, titanium and carbon concentrations in the core are high with a definite gradient toward
lower concentrations of both elements in the reaction zone. Although in lower concentrations than in Si_3N_4 matrix, silicon is
also detected in the reaction zone, indicating its definite participation in reaction product formation. Due to the instrumental limitations, nothing definite could be said about the presence of nitrogen and oxygen in reaction products. No yttrium,
aluminum or iron segregations are detected either in the reaction
zone or in the immediate vicinity of this particle. For more detail characterization of the composition of reaction products,
two specific examples of reacted TiC particles were examined by
means of electron probe microanalysis (Particle B in Figure 3 and
Particle C in Figure 2). Both were thought to represent only the
products of reaction.

* Philips Electron Probe 4500

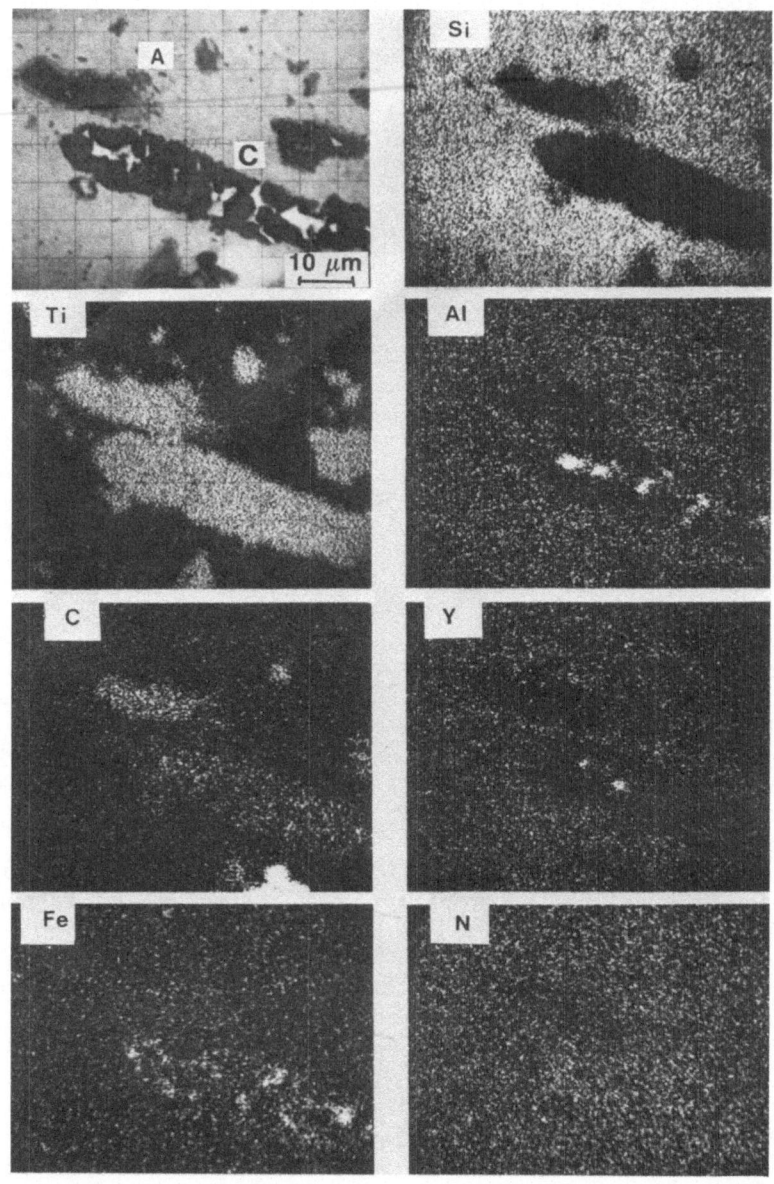

Figure 2: Back scattered electron image (upperLeft) and x-ray dot maps of the elements of TiC particles A and C in uncontaminated Si_3N_4-TiC composite.

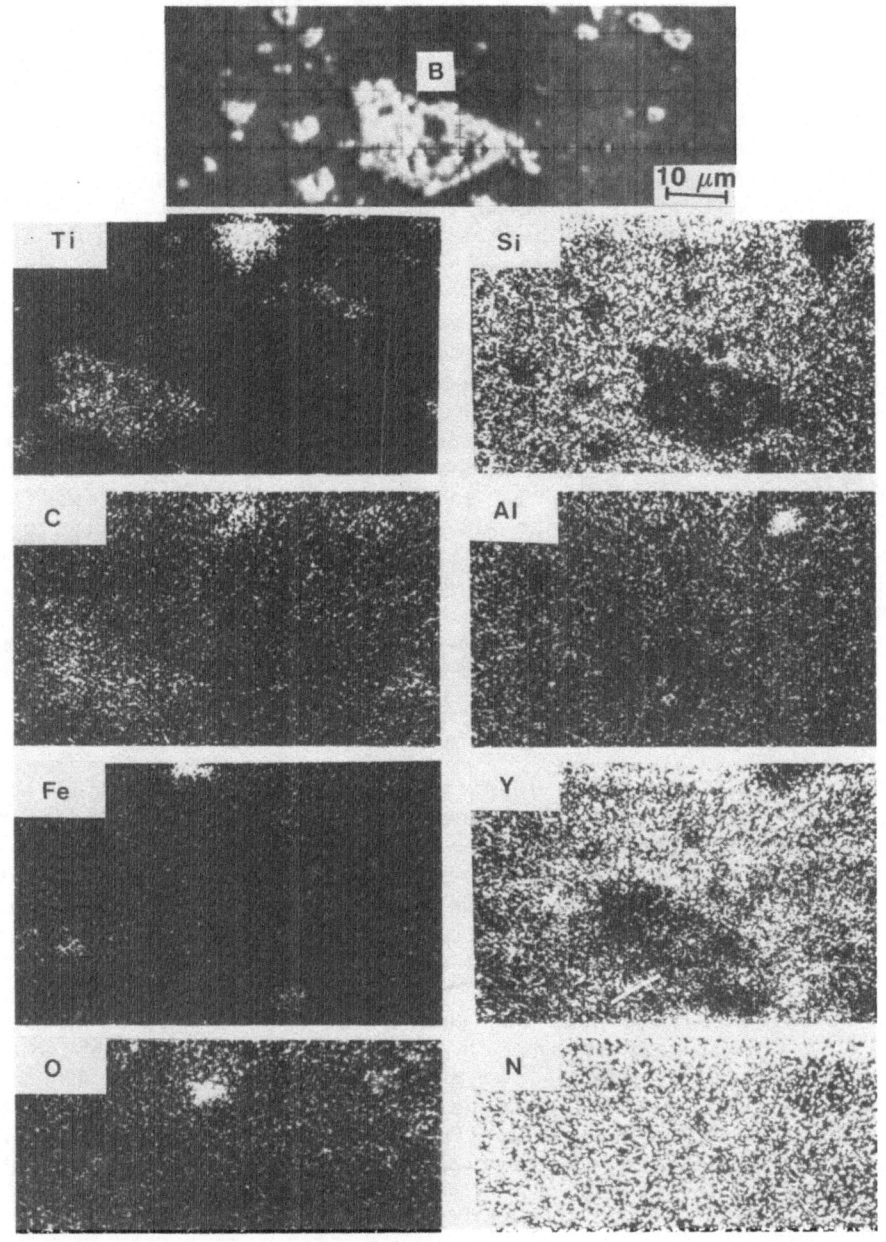

Figure 3: Back scattered slectron image (top) and x-ray dot maps of the elements of TiC particles B is uncontaminated Si_3N_4-TiC composite.

Morphologically, particle B (Figure 3) appears to be composed of three phases. In phase I, titanium, carbon and nitrogen are detected, suggesting titanium carbonitride (TiC_xN_y) formation. Silicon and carbon found in phase II indicate silicon carbide (SiC) formation. Phase III, similar in appearance to the TiC particle core, indeed shows the presence of only titanium and carbon, evidently corresponding to an unreacted portion of the TiC grain. Aluminum and yttrium segregations are also detected within the particle suggesting that formation of TiC_xN_y and SiC could have been enhanced by the altered composition and lowered viscosity of the liquid.

A similar analysis was carried out on TiC particle C (Figure 2, center) which seems to be composed of two phases. The x-ray microanalysis indicates the presence of high Ti, low carbon, nitrogen and possibly oxygen, all four elements having relatively uniform distribution, again suggesting TiC_xN_y or possibly $TiC_xO_yN_z$ formation. The regions between TiC_xN_y particles are found to be rich in aluminum and yttrium. The agglomerate appears Si-free, indicating an absence of SiC and apparently lower SiO_2 content in the $Al_2O_3-Y_2O_3$ rich liquid (glass) phase, which was heavily contaminated with iron. Lowered carbon concentration, absence of silicon and other carbon containing products, and increased porosity in the immediate vicinity of this agglomerate may suggest a formation of some gaseous carbon products in this case. Iron here seems to have acted as a strong catalyst for nitridation.

SiC crystals were also chemically extracted from several samples of hot pressed Si_3N_4-TiC composites and examined by scanning transmission electron microscopy. Electron microscopic examination of SiC residues indicate the presence of Beta and a variety of Alpha polytypes. None of the specific structural types were found to predominate. The particle size of SiC in the examined residues was of the order of 0.1 μm.

X-ray microanlytic data has shown that reaction zones forming at TiC grain boundaries in Si_3N_4-TiC composites hot pressed in argon consist primarily of two phases, TiC_xN_y and SiC. The results also indicate that, depending on liquid phase composition and the presence of impurities (Fe), the composition, amount, and morphological pattern of reaction products can vary quite substantially.

Room and elevated temperature mechanical properties of the composite strongly depend on the composition of the intergranular phase, coherency of the matrix-dispersoid interface, as well as the properties of the dispersoid. Alteration of the composite through enhanced reaction would result in a failure to attain the design properties. Taking into consideration the observed influence of compositional variations on reaction kinetics the effect

394

of most common contaminants of the composite precursor materials, silicon, carbon and iron, was further examined. A small cavity in the middle of a contaminant-free $(Si_3N_4+Y_2O_3+Al_2O_3)$-TiC powder compact was filled with silicon, iron or graphite powders, covered with another contaminant-free pellet and hot pressed at elevated temperature in an inert atmosphere. Resultant microstructures were examined by optical microscopy and the distribution of elements in reaction products was evaluated by electron probe microanalysis using x-ray dot mapping.

A main feature of the silicon-contaminated composite microstructure (Figure 4) is a high concentration of voids which evidently resulted from silicon evaporation during hot pressing. No signs of enhanced reaction could be observed. Nonetheless, the presence of free silicon in Si_3N_4-TiC materials should be considered undesirable because of its potentially deleterious effect on the composite's mechanical properties.

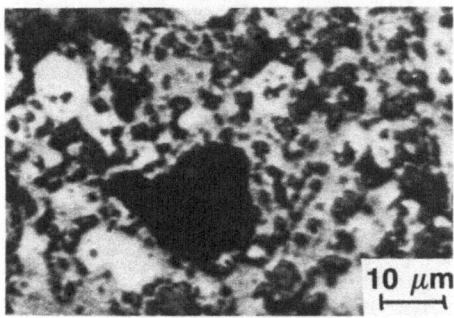

Figure 4: Mircostructure of silicon-contaminated Si_3N_4-TiC Composite.

In the iron-contaminated material, the microstructure developed in the iron filled cavity consists, for the most part, of low hardness, low melting point phases, iron silicides,[6] iron-aluminum-silicides, and titanium carbonitride with high nitrogen content. Silicon carbide is present as a minor phase (Figure 5). Around the cavity, in the areas with somewhat lower level of iron contamination, a pattern indicative of highly catalyzed matrix-dispersoid reaction is observed. As can be seen in Figure 6, the cores of initial TiC grains are now fully occupied by SiC, whereas TiC_xN_y forms only peripheral layers, in contrast to what is observed in contamination-free material (see Figure 1). Optical examination of the areas located at some distance from the cavity reveals an unaltered microstructure similar to the contamination-free microstructure. This observation, together with the absence of iron in these areas, points out that the catalytic effect of iron on the matrix-dispersoid reaction was restricted to a relatively narrow zone around the iron filled cavity.

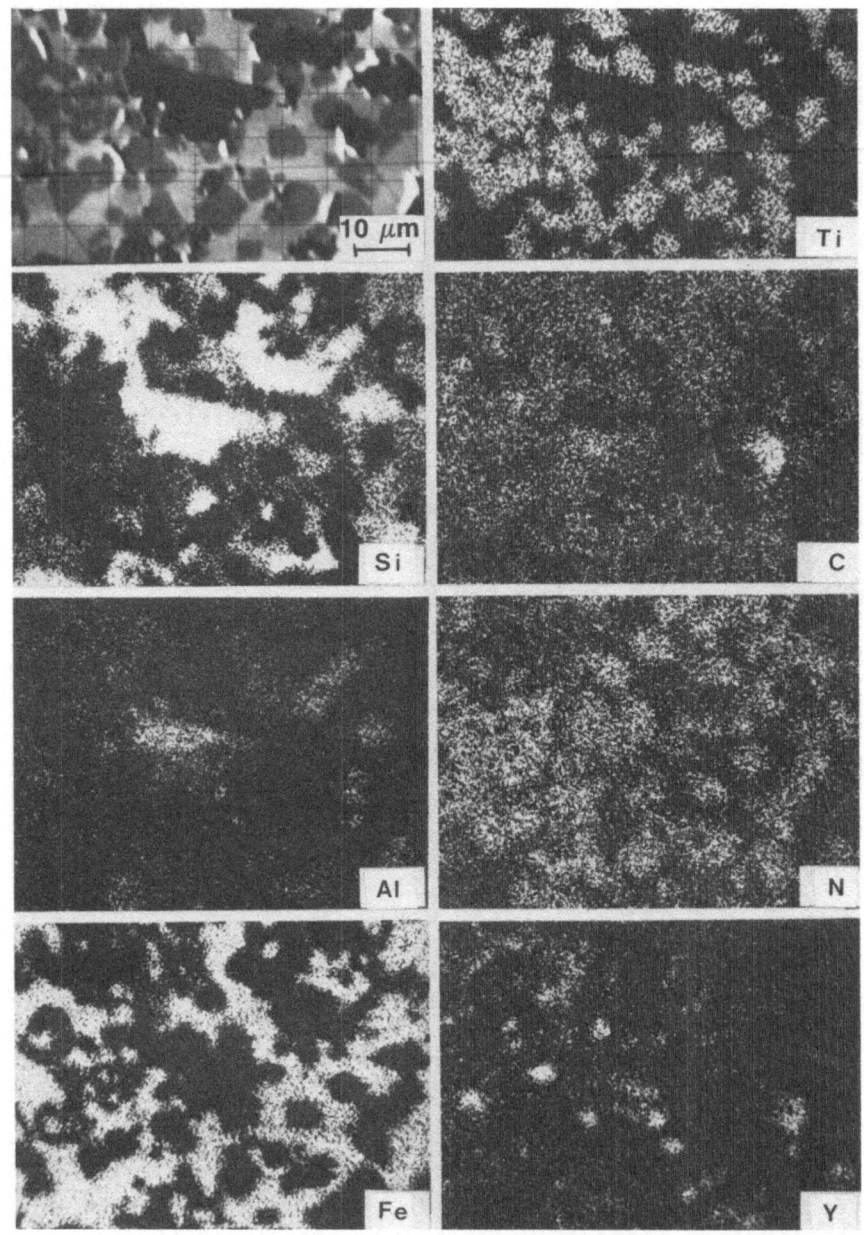

Figure 5:   Back scattered electron image (upper left) and x-ray dot maps of the elements of the microstructure developed in iron filled cavity of iron-contaminated Si_3N_4-TiC composite.

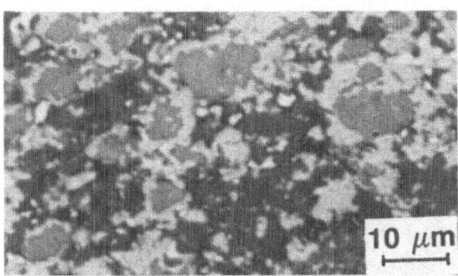

Figure 6: Mircostructure of iron-contaminated Si_3N_4-TiC composite developed around the iron filled cavity.

Optical microscopy examination of carbon-contaminated material reveals four distinct reaction zones which are schematically presented in Figure 7. Zones I and II are located in the carbon filled cavity, zone III along the periphery of the cavity, and zone IV at some distance ($\approx$10 mm) away from the source.

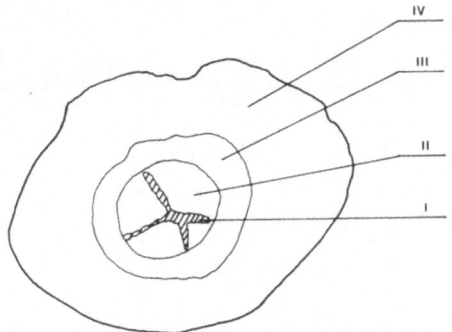

Figure 7: Schematic representation of reaction zones in carbon-contaminated Si_3N_4-TiC composite.

In the carbon-filled cavity, the major phases comprising the microstructure, as indicated by x-ray microanalysis (Figure 8), are SiC, TiO_xN_y and free Si. Unreacted carbon and acicular-shaped crystals of (Si,Al)(O,N) are also present. The microstructure in zone III and to a lesser degree, in zone IV located at $\approx$10 mm distance from the carbon filled cavity exhibits a pattern of enhanced matrix-dispersoid reaction which can be observed as layers of SiC followed by TiC_xN_y on TiC particle cores (Figure 9). Such a substantially altered microstructure in all examined areas may apparently be attributed to the presence of free carbon virtually throughout the specimen. We emphasize that free carbon is a particularly undesirable contaminant in Si_3N_4-TiC composites because of its ability to permeate through a large volume of material. It is interesting to note that in zones III and IV of carbon-contaminated material the reaction products SiC and TiC_xN_y are assembled in a pattern similar to that in the iron-contaminated composite, shown in Figure 6. These observations show that

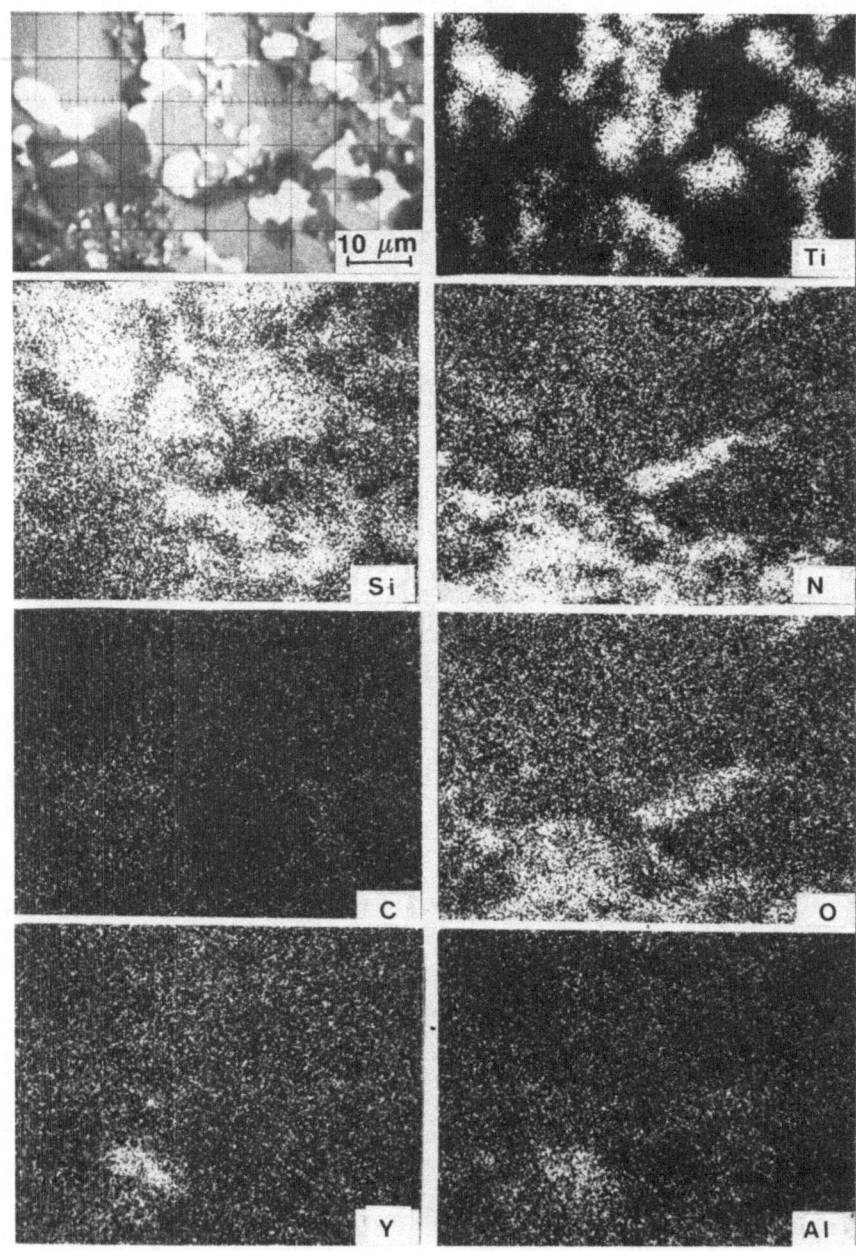

Figure 8: Back scattered electron image (upper left) and x-ray dot maps of the elements of zone I of carbon-contaminated Si_3N_4-TiC composite.

both iron and carbon have a strong catalytic effect on the matrix-dispersoid reaction, and also suggest that a similar mechanism of reaction product formation may operate in both cases.

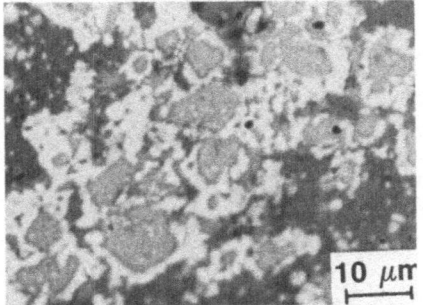

Figure 9: Microstructure of carbon-contaminated Si_3N_4-TiC composite developed around the carbon filled cavity (zone III)

Specifically, in carbon or iron contaminated material, silicon carbide forms in place of the original titanium carbide cores, and titanium carbonitride forms at the periphery. In contrast, in contamination-free material, the matrix-dispersoid interaction is confined to the TiC surface resulting in the two phase SiC-TiC_xN_y reaction zone. This suggests that the rate of reaction product formation as well as disposition of reaction products is mainly controlled by the rate of carbon diffusion.

Careful selection of high purity raw materials and adjustments in processing have resulted in a development of a series of fully dense Si_3N_4-TiC composites in which the formation of both reaction products is minimized and which show substantial improvement in mechanical properties. The microstructure of such a composite, which has found an application as a cutting tool material for grey cast iron machining, is shown in Figure 10. As can be seen from Figure 10(a), the reaction zones are not resolved in optical examination of this material at 1000X magnification but can be observed in Figure 10(b) only as thin (0.1-0.3 µm) diffuse bands along the periphery of TiC particles at higher magnification.

CONCLUSIONS

During sintering of complex Si_3N_4 matrix composites containing a TiC dispersed phase, the composition of the intergranular liquid phase plays an important role. It defines both the kinetics of the reaction and the morphological makeup of the reaction products. In the Si_3N_4-Y_2O_3-Al_2O_3-TiC composite, matrix-dispersoid interaction, invariably results in the formation of SiC and Ti(C,N). Carbon and reactive impurities capable of changing the viscosity of the liquid phase or a carbon potential in the system have a strong influence on the reaction kinetics and resultant microstructure.

Careful selection of raw materials and processing parameters reduces the reaction and results in a composite with exceptional mechanical properties and wear resistance.

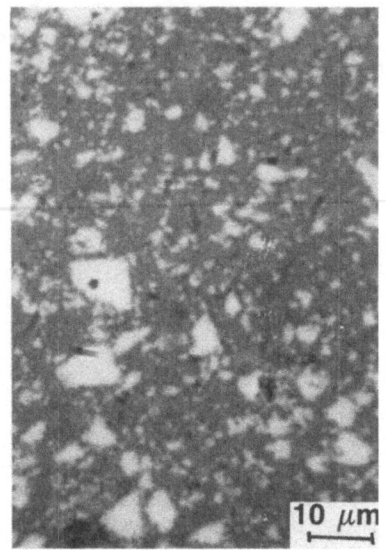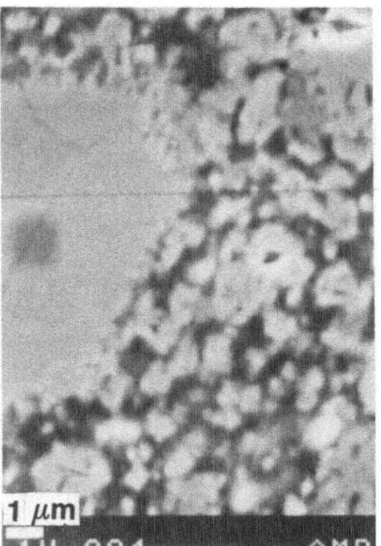

Figure 10: Optical (a) and scanning electron (b) photomicrographs of the microstructure of Si_3N_4-TiC composite material for cutting tools.

ACKNOWLEDGEMENTS

Authors are indebted to Mr. E. Geary, Mr. G. Robinson and Mr. G. Werber for their valuable assistance in the course of this study. We would also like to thank Dr. J. R. McColl for his support and critical reading of the manuscript.

REFERENCES

1. S.T. Buljan and V. K. Sarin, "Improved Productivity Through Application of Silicon Nitride Cutting Tools," The Carbide and Tool Journal, vol. 14 (1982).

2. V. K. Sarin and S. T. Buljan, "Advanced Silicon Nitride Based Ceramic for Cutting Tools," SME Technical Paper No. MR83-183.

3. S. T. Buljan and V. K. Sarin, "Machining Performance of Ceramic Tools," Cutting Tool Materials, Proc. of Intern. Conf. on Cutting Tool Materials, ASM, 1980.

4. S. T. Buljan and V. K. Sarin, "Silicon Nitride Based Composites," Proc. Intern. School on Composite Materials, Bombay, India, 1984.

5. D. R. Clarke and G. Thomas, "Microstructure of Y_2O_3 Fluxed Hot Pressed Silicon Nitride," J. Am. Cer. Soc., vol. 61, No. 3-4, 1978.

6. G. V. Samsonov and I. M. Vinitski, Handbook of Refractory Compounds, IFT/Plenum Data Company, 1980.

KIRLIAN PHOTOGRAPHY

MATERIAL SCIENCE, TESTING AND ICHNOLOGY

Daniel B. Sass and David C. Meissner

Divisions of Physical Science (Geology) and Psychology
Alfred University
Alfred, New York

ABSTRACT

Until recently Kirlian photography was used primarily to record effects generated by the application of electrical energy to living tissue and preserving the attendant phenomena on light sensitive film (see Hyzer, 1974). The resulting patterns and colors were interpreted as symptomatic of the status of the parent organism in terms of its quality of life (Aaronson, 1974). The effects of the interaction between life processes and electricity were considered to be inexplicable in terms of established scientific principles; our research suggests that this pessimistic view may no longer be tenable.

Kirlian photographs taken of "wafers" cut from fossiliferous rocks exhibit patterns and colors similar to those produced from living organisms or near relatives. Investigations into the probable cause of a phenomenon herein called a red "hotspot," which occasionally appears in fossilized tracks, trails and burrows (=ichnofossils) of organisms long-extinct, produced results which suggest that the Kirlian phenomenon may related to the preservation of their chemical residuum, its spacial orientation and carbon content.

PREVIOUS WORK

The chance discovery that a rock wafer cut from a Pennsylvanian fern (200 mybp) would produce an aura similar to that of its modern counterpart, when used for Kirlian photography, has opened some new vistas and directions in Kirlian research. Initial

investigations on randomly selected fossil material led to the
standardization of photographic and experimental techniques and
reproducible results (see Sass and Meissner, 1983). The unexpected
intermittent appearance of an enigmatic red "hotspot" within the
"habitat" of burrowing organisms led the research into the field of
Ichnology - dealing with burrows, tracks, and trails of extinct
organisms. Three ichnofossils were selected for intensive study:

   1.   The burrow of an infaunal marine organism (Skolithus sp.)
from the Upper Devonian Wiscoy Sandstone of Hornell, New York (350
mybp).
   2.   A burrow of (Conostichus broadheadi): affinities unknown.
   3.   The root of a fossilized mangrove (Rhizophora sp.) which
apparently served as a habitat during the Pleistocene Epoch (1
mybp).

   Rock wafers approximately 0.5 mm thick and 3.0 mm in diameter
were cut from fossils in transverse and longitudinal orientation to
each habitat. The specimens were prepared and photographed in a
manner described by the authors (Sass and Meissner, 1983, pg. 1048,
Fig. 1). As noted above, the red "hotspots" occur only randomly in
samples cut from the specimens of some ichnofossils (Figs. 1, 2,
3).

   Gross chemical, X-ray and trace element analyses of the
habitats and their matrices revealed no evidence of possible causes
of the red "hotspots." Separate analyses of burrows (=habitat=ich-
nofossil) and their matrices, disclosed the presence of amino acids
in the burrows and virtual absence in the matrix. Such a selective
distribution diminishes the possibility that the amino acids are
contaminants distributed by ground water or other agencies. The
results cited led to the tentative conclusions that:

   1.   The red "hotspots" related to the organic chemical
remnants left in some ichnofossils.
   2.   The preservation of the concentration of such material
was vicarious.
   3.   The organic material might represent the original
occupant, its prey and/or their exuviae and excretions.
   4.   Each substance might react singularly to electrical
stimulation when photographed by the Kirlian method.

   The direction of further investigation was influenced by the
publication of two articles on subjects only peripherally related
to the project in question. Cheryl Simon (1982) cited the opinion
of Alfred G. Duba of Lawrence Livermore National Laboratory that
high electrical conductivity in rock is enhanced by heating and the
presence of carbon at grain boundaries. Michael Carrick (1983)
reported that in 1978 the Tokyo Metropolitan Police demonstrated
that fumes from Cyanoacrylate Ester adhesives (Super Glue) would

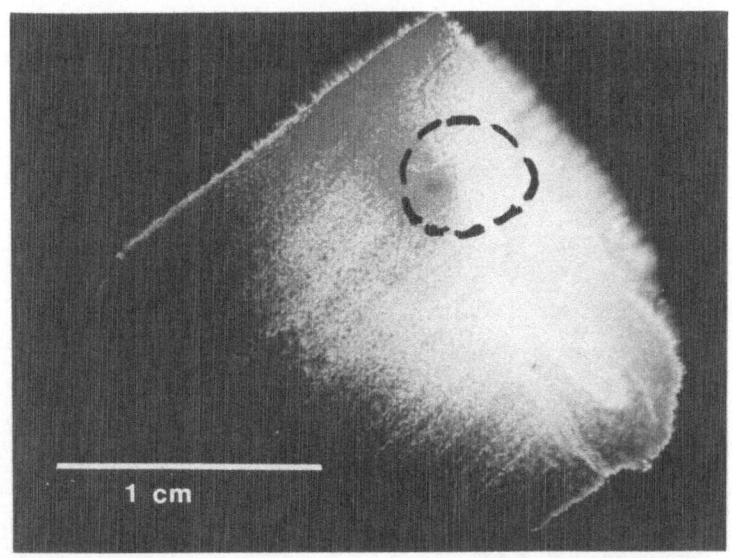

Fig. 1.   Skolithus sp., Black and white print from an Ektachrome
          200 ASA color slide of an oblique longitudinal section.
          Alternating dark and light lines to the left are sedimen-
          tary layers. Area within the dashed circle appears as a
          red "hotspot" on Ektachrome.

develop and enhance latent finger prints [reportedly by reactions
with amino acids].

CURRENT WORK

     The reports cited above and our own research suggested two
suppositions and three questions as guides for our additional
experimentation.

     1.   Could it be supposed that the amino acids in the burrows
might, on occasion, be distributed spacially in death as they were
essentially in life?
     2.   Could it be supposed that fossil carbon, resulting from
the decomposition of its former living host (whatever the source)
influences the location and intensity of electrical discharges
through rock wafers used as subjects for the Kirlian process?
     3.   Might each amino acid have a distinctive "signature" for
standard parameters regulated for the Kirlian process?
     4.   Conversely, might each amino acid respond differently to
light of various wavelengths registered by special film emulsions?

5. Might the fumes from Cyanoacrylate Esters enhance latent amino acids of more venerable vintage than those associated with mankind?

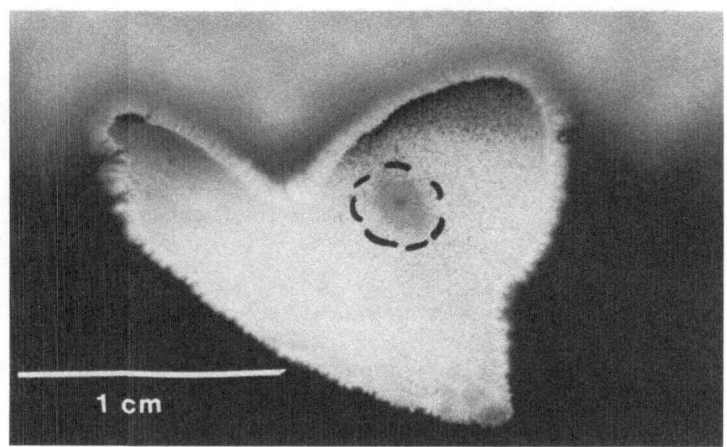

Fig. 2. Conostichus broadheadi, black and white print from an Ektachrome 200 ASA color slide of a longitudinal section of a burrow. The burrow is upside down. Area within the dashed circle appears as a red "hotspot" on Ektachrome. Note the fringe of streamers on the margin of the habitat. (Reproduced with permission of the Paleontological Society.)

SPECULATION

The primacy of amino acids in fossils as related to original organic sources has been well established by Ableson (1956, 1957), Mitterer (1971) and others. Their occasional presence within some ichnofossils has been substantiated by the writers through the intense application of the multiple working hypothesis advocated long ago by T. C. Chamberlain. It is somewhat less tenable, but none-the-less possible, that the distribution of the amino acids and their constituents could be oriented in a two-dimensional death array which reflects their three-dimensional arrangement in life. Unusual circumstances of preservation have been reported by Stuermer (1970) and others. The recovery of DNA from a remnant of an extinct wild ass (Equus quagga), related to the zebra, is another

406

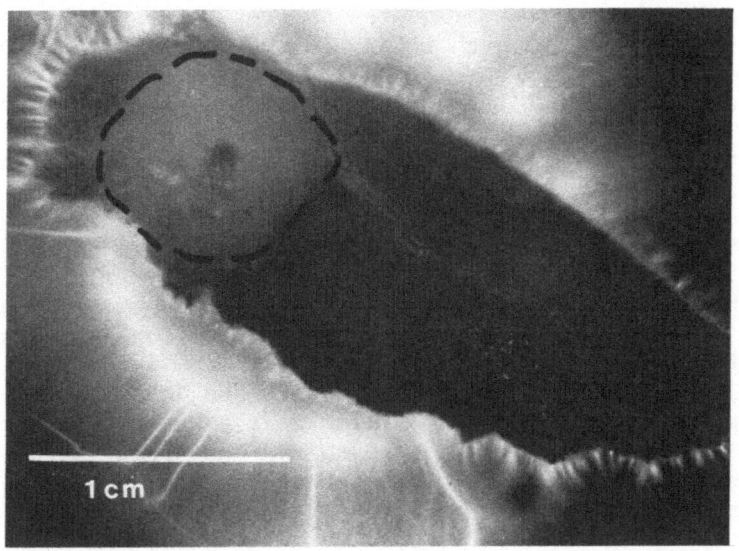

Fig. 3. Rhizophora sp., black and white print from an Ektachrome
         200 ASA color slide of a longitudinal section of a
         fossilized mangrove root. Area within the dashed circle
         appears as a red "hotspot" on Ektachrome. Note the
         development of streamers on the edge of the root and the
         stub on which it is mounted in the background.

case in point. A trained observer with the appropriate technology
can incrase the incidence of such discoveries.

EXPERIMENTATION

     To test the supposition that carbon could concentrate and
channel electrical energy, an experiment was designed which
involved making an artificial rock. Geometric patterns of
dispersed carbon, were encapsulated between two thin layers of clay
slip made from the Alfred Shale. The resulting wafers, with the
carbon patterns concealed, were kiln fired and used as specimens
for Kirlian photography. The images, captured on Ektachrome 200
ASA film, matched the concealed arrays of carbon; indicating
some control of the electrical input.

     To test the potential effect of varying light and voltage
values as signatures for diverse amino acids, the relatively
chemically neutral Nunda (Devonian) Sandstone was selected as a
habitat. Wafers 0.5 mm in thickness x 2 cm in diameter were

cut from Nunda specimens, mounted on brass stubs with electrical conducting paste, and each wafer "innoculated" with a drop of 1% sterile solution of a specific amino acid; thus creating a simulated favored habitat. The resulting Kirlian photographs, regardless of the adjustment of the physical and electrical parameters, displayed only an undiagnostic blue aura. The use of infrared film produced a variegated red wash on both the simulated and fossil habitats, accentuating structural details. The significance of this response is yet to be determined.

The most instructive experiments involved the response of wafers of both simulated and fossil habitats to Super Glue fumes. The wafers were exposed to the fumes for from several weeks to several months in an attempt to enhance the latent amino acids. Upon rephotographing in the Kirlian mode, the blue auras from the Nunda Sandstone appeared as pink to red "hotspots." The images in the fossil habitats (Figs. 1-3) were enchanced and expanded with displays of new auras and color distirbutions which will require additional study.

Finally, a computer enhancement of the fossil red "hotspots," in progress, shows promise of distinguishing shades of red and pink which, upon contouring, may reveal the physical nature of the organic entity presumed to be responsible for the red in-situ "hotspot."

SUMMARY AND CONCLUSIONS

The evidence presented suggests that Kirlian phenomena are derived from, or related to, the chemistry of both extant and extinct organic entities. Experimental data points to amino acids and carbon as possible causitive agents via their control of electrical energy used in securing the photographic record. The intra-habitat red "hotspot" described above, probably indicates a concentration of chemical residuum of the former occupant, its prey and/or their excretions. The exposure of specimens to the fumes from Cyanoacrylate Ester adhesives can enhance the traces of amino acids from both extant and extinct organisms in dramatic fashion. Computer enhancement of the red "hotspots" may, under ideal conditions, produce vestiges of the morphology of the organisms from whose chemistry the phenomenon is apparently derived.

GLOSSARY

Aura: An annular glow or luminous envelope with attendant colors when filmed on appropriate media.

Habitat: The internal physical boundary of an ichnofossil.

Ichnofossil: The track, trail, or burrow which denotes the former presence of an organism.

MYBP: Million years before the present.

Red "Hotspot": A special kind of aura within the physical boundaries of the habitat or ichnofossil. It has a dark center surrounded by an envelope of red color.

Signature: A diagnostic color and/or aura for a particular substance exposed to Kirlian technology.

Streamer: Ray-like extensions of energy from the outer margins of the aura.

Wafer: A rock slice, generally 0.5 mm or less in thickness.

REFERENCES

Aaronson, S., 1974, Pictures of an unknown aura, The Sciences, 14:15-22.

Abelson, P. H., 1956, Paleobiochemistry, Scientific American, 195:83-92.

Abelson, P. H., 1957, Organic consitutents of fossils, in: "Paleoecology," H. S. Ladd, ed., Geological Society of America, Memoir 67, 2:87-92, Waverly Press, Baltimore.

Carrick, M., 1983, Cyanoacrylate glue fuming, In-house Document, Lightning Powder Co., Inc.

Hyzer, W. G., 1974, Kirlian experimentation, Photo Methods for Industry, 17, No. 4:8, 47, 48.

Mitterer, R. M., 1971, Comparative amino acid composition of calcified and non-calicified polychaete worm tubes, Comprehensive Biochemical Physiology, 38B:405-409.

Sass, D. B. and Meissner, D. C., 1983, Electrography of trace fossils, Journal of Paleontology, 57:5, 1047-1049.

Simon, C., 1982, Disintegrating diamonds, Science News, 121:392.

Stuermer, W., 1970, Soft parts of trilobites and cephalopods, Science, 1170:1300-1302.

INDEX